Charles Wesley Purdy

Practical Uranalysis and Urinary Diagnosis

A Manual for the Use of Physicians Surgeons, and Students

Charles Wesley Purdy

Practical Uranalysis and Urinary Diagnosis
A Manual for the Use of Physicians Surgeons, and Students

ISBN/EAN: 9783337140441

Hergestellt in Europa, USA, Kanada, Australien, Japan

Cover: Foto ©berggeist007 / pixelio.de

Weitere Bücher finden Sie auf **www.hansebooks.com**

PRACTICAL URANALYSIS

AND

URINARY DIAGNOSIS

A Manual for the Use of Physicians, Surgeons,
and Students

BY

CHARLES W. PURDY, LL.D., M.D.

QUEEN'S UNIVERSITY

FELLOW OF THE ROYAL COLLEGE OF PHYSICIANS AND SURGEONS, KINGSTON; PROFESSOR OF CLINICAL
MEDICINE AT THE CHICAGO POST-GRADUATE MEDICAL SCHOOL. AUTHOR OF "BRIGHT'S
DISEASE AND ALLIED AFFECTIONS OF THE KIDNEYS"; ALSO OF "DIABETES:
ITS CAUSES, SYMPTOMS, AND TREATMENT"

Sixth Edition, Thoroughly Revised

WITH NUMEROUS ILLUSTRATIONS, INCLUDING PHOTO-
ENGRAVINGS AND COLORED PLATES

PHILADELPHIA
F. A. DAVIS COMPANY, PUBLISHERS
1904

Philadelphia, Pa., U. S. A.
The Medical Bulletin Printing-house,
1914-16 Cherry Street.

PUBLISHERS' NOTICE TO SIXTH EDITION.

In issuing a new edition of Purdy's "Uranalysis" the publishers desire to state that arrangements have been made whereby this standard treatise will be maintained in its position as an authority on the subjects treated.

The demand for a sixth edition of the work so soon after the issue of the fifth edition and the very recent death of the author have seemed, however, to make it advisable that the revision should not, at this time, touch the general scope or plan of the book, but that such changes only should be made as were necessary to set forth clearly the meaning of the author. To this end a few evident omissions have been supplied, several obscure sentences rewritten, and a number of typographical errors and mistakes in chemical statements corrected. For these changes we are indebted to Dr. Walter S. Haines and Dr. William H. German, of Chicago.

PHILADELPHIA, September 1, 1901.

(iv)

PREFACE TO FIFTH EDITION.

THE present edition of this work consists of a careful and thorough revision of the last one, with the addition of much original and new matter, including a new chapter on the microscope and its use in uranalysis. The author has much pleasure in now fulfilling a promise made in the first edition, viz.: to extend the range of centrifugal analysis so that it should include more complete as well as more practical data for urinary work. Although requiring over five years' labor for its perfection, centrifugal analysis has now been elevated to a scientific process, justly entitled to rank among the so-called exact methods of analysis. By means of the improved methods and new tables herewith introduced, the quantities of albumin and chlorine, of phosphoric and sulphuric acids in the urine may be simply and rapidly determined, both relatively and absolutely, with a degree of accuracy equal to that of any other method.

The physician is often led to purchase a microscope for the chief or sole purpose of urinary examinations. In so doing the next step is to learn how to make satisfactory examinations of urine therewith. In purchasing a work purporting to explain the use of the instrument, as a rule little or nothing will be found therein relative to microscopical examinations of the urine. In turning to systematic treatises on the urine, as a rule, an equal lack of information on the subject is met with. The above mentioned are some of the considerations for introducing in the present edition a separate section intended to furnish a general idea of the microscope itself and its special use in urinary work. While this section is by no means claimed to be exhaustive, the author hopes that the beginner will find therein some useful and practical suggestions to aid him in his work.

The chemical department of the work has been carefully revised, a few quantitative methods have been added where previously omitted, and nearly the whole subject of testing for albumin in the urine, both qualitative and quantitative, has been rewritten. In short, an effort has been made to improve the work more especially in its practical bearings on clinical medicine, as well as to bring it thoroughly up to date. The labor has been no inconsiderable task, comprising over a hundred and twenty pages of manuscript, but this task has been greatly lightened by the pleasant recollection of the very generous reception extended to former editions of the work by the profession in general.

The author makes acknowledgment with pleasure of the valuable aid of his assistant, Mr. Carl Irenæus, who has devoted much time and pains in carefully working out the many details of the tables on centrifugal analysis.

57 EAST TWENTIETH STREET,
June, 1900.

PREFACE.

Our present knowledge of the urine and of diseases of the urinary organs may be said to be altogether abreast with other departments of scientific and practical medicine. At present, however, this knowledge is only accessible to the student through somewhat extended search through general works on Medicine, Surgery, Pathology, Physiological Chemistry, Microscopy, etc., in addition to the various works devoted to this special subject. European writers,—especially those in our language,—following the sharp division between medical and surgical diseases, have invariably considered the present subject, both in general and special works, either from an exclusively medical or surgical point of view, and American writers thus far, without exception, have followed this example. But in America the whole profession is taught and qualified to practice both medicine and surgery, and, therefore, the above-mentioned custom compels the student, in order to gain a complete knowledge of this subject, to study several authorities, entailing increased expense and time, if not, indeed, confusion. Believing, therefore, that American authors should, so far as is possible, deal with the whole subject comprised in the titles of special works, it has been the aim of the author in the present work to furnish the student, physician, and surgeon, in one moderate-sized volume, the essential features of our knowledge of the urine and urinary diagnosis, thoroughly up to date, and in the most systematic, practical, and concise form. In carrying out this object an effort has first been made to bring out prominently the relations of the chemistry of the urine to physiological processes and pathological facts. Thus, in dealing with normal urine, each constituent has been considered, so far as at present is known, in the following order: Its chemical nature and composition; its source in the economy; the significance of its increase or decrease in the urine, with the relations of these to metabolic processes, food-supply, physical surroundings, and tendency toward disease; and, finally, the most approved methods of its detection and determination have been described. In dealing with abnormal urine each morbid constituent has been considered, so far as at present is known, in the following order: Its chemical nature and composition; its source in the economy; the clinical significance of its appearance in the urine; and, lastly, the most approved methods of its detection and determination have been described. This method aims at teaching not only how to detect, isolate, and determine the constituents of the urine, normal and abnormal, but also to determine the presence of disturbed physiological processes, to detect the presence of pathological changes, and to measure the degree of both.

The second division of the work—Urinary Diagnosis—aims at a concise description of the special features of the urine that indicate the presence of special pathological processes in progress in the economy, whether they be local or general, medical or surgical, together with a brief enumeration of the leading clinical symptoms of each disease, and, in most cases, an epitome of their nature and etiology.

Having compassed the whole text, it is designed that the investigator will next be in a position to utilize the information thereby furnished to the best practical diagnostic purposes, and give him the mastery over the diseases considered,—*for, as a rule, he who has accurately diagnosticated disease, has already constituted himself its conqueror.*

The author has freely quoted the views of standard authorities, endeavoring in all cases to make due acknowledgment of the same throughout the text. Should, however, any of the latter have been overlooked, he desires here to express his obligations for all knowledge derived from fellow-laborers in the same field of work.

Some repetitions of matter will be noted, both in the text and foot-notes, the object being to save references to other parts of the work and render it more convenient as an open hand-book for the laboratory table.

Over twenty-five years' experience, coupled with a somewhat liberal examination of the literature of this subject, as well as considerable practical observation and experiment, have enabled the author to contribute some original matter and methods, which, he trusts, will prove to be an advance in certain practical departments of the subject.

Finally, an appendix has been added upon the subject of urinary examinations for life-insurance. Believing, from some experience as a medical director, that on the one hand life-insurance associations are often unjustly called upon to pay insurance on uninsurable lives, and on the other hand that applicants are often deprived of the privileges of life-insurance to which they are justly entitled, especial pains have been taken with this department of the work, with the earnest hope of contributing, in some degree, to the amelioration of these two forms of injustice.

57 EAST TWENTIETH STREET,
CHICAGO, September, 1894.

CONTENTS.

PART I.—ANALYSIS OF URINE.

SECTION I.

SECTION II.

SECTION IX.

PART II.—URINARY DIAGNOSIS.

SECTION X

SECTION XI.

SECTION XII.

APPENDIX A.

APPENDIX B.

LIST OF ILLUSTRATIONS.

LIST OF TABLES.

PART I.

ANALYSIS OF URINE.

SECTION I.

GENERAL CONSIDERATIONS.

THE recent advances in our knowledge of physiological chemistry, with the more extended and refined use of the microscope, have lent great precision to the study of the composition of the urine, and thereby furnished us with a keener insight into the relationship of the urine to the organism, both in health and in disease. The variations in nutrition and waste are accurately recorded in the urine hour by hour, and by an intelligent interpretation of modern methods of uranalysis these physiological tides may now be read as accurately as we can number the pulsations of the heart. By the same methods we are now enabled to measure the hourly inroads upon the organism made by disease with a precision often greater than is afforded by the pulse or the clinical thermometer. Wherever in the economy pathogenic processes seriously disturb nutrition or normal metabolism, the results are recorded in the urine, because the urine, more eminently than any other excretion, represents the equation of these changes.

The accurate study of the urine, therefore, has become one of the essential features in advanced clinical medicine. Indeed, through uranalysis alone can an almost daily increasing number of diseases be determined, their intensity be gauged, and their progress toward recovery, or their tendency toward a fatal termination, be predicted. While it is impossible to diagnosticate all diseases from the urine, it is, nevertheless, true that no serious disease can be in progress in the economy without giving rise to more or less marked changes in the character of the urine, and therefore we can no longer afford to exclude urinary analysis from the scientific investigation of any serious form of disease.

In order to fully comprehend the relations of the urine to the organism under the influence of the various pathological conditions, it is first necessary to become acquainted with the

(1)

physiology of secretion and excretion of the urine, as well as the normal composition of the latter, together with those fluctuations which are included within the range of health. While a number of points in the physiology of the secretion and excretion of the urine still remain undetermined, it may be stated that our present knowledge of the subject indicates that the process is partly a physical and partly a vital one. The older theory of Bowman, based on the anatomical construction of the kidneys, taught that the epithelial cells of the urinary tubules constitute the true secretory structure of the kidneys, while the glomeruli act as mere filters for the escape of the watery elements from the blood. According to Bowman, therefore, the filtrate from the glomeruli consists almost solely of water, which aids in extracting the other constituents of the urine from the epithelium of the tubules in its passage along the latter.

Ludwig, on the other hand, explains the process on purely physical grounds, basing his theory on the varying degrees of blood-pressure in the glomerular circulation and the interchange of constituents by diffusion or osmosis in the urinary tubules. Assuming that the relative blood-pressure in the kidney is greatest in the glomerular tufts in consequence of the resistance to the efferent circulation, Ludwig holds that, consequently, a free exudation of water takes place from the tufts, with, perhaps, some dissolved salts. This renders the blood much concentrated—thickened—when it reaches the capillary plexus surrounding the convoluted tubes, while within the latter is now the thin, aqueous filtrate from the tufts. It will be noted that such conditions form all the essential elements for active osmosis,—within the tubules thin, watery fluid, and in the surrounding capillaries thickened blood, while between them is interposed a thin membrane,—the tubular wall. An interchange of elements consequently occurs, by means of which water from the urinary tubules passes into the blood; while, on the other hand, the products of retrograde tissue changes—urea and salts—pass from the blood into the tubules, mingling there with the thin fluid and constituting the urine.

Unfortunately for the theory of Ludwig, he leaves out of

consideration any function on the part of the renal epithelium, which violates analogical reasoning, because the renal epithelium possesses the anatomical peculiarities of glandular epithelium, the function of which is secretory or selective wherever else met with throughout the economy. Moreover, both clinical and pathological experiences teach that the renal epithelium possesses a distinct and important function in the elaborating processes of the kidney; for, in diseases which destroy or remove this epithelium from the urinary tubules, urea and allied products are retained in the system, and the phenomena of uræmia are evoked. Finally, the interesting experiments of Heidenhain have conclusively proved that the renal epithelium possesses a distinct selective power, as follows: If a slightly-concentrated solution of indigo-sulphate of sodium be injected into the blood of an animal, a blue color will soon after be communicated to the urine and the epithelium of the convoluted tubules and ascending limbs of Henle's tubes, while the Malpighian structures do not present the slightest trace of blue. If the spinal cord of an animal be first divided and the indigo injection be subsequently made, the following phenomena may be observed: No urine whatever reaches the bladder, but the blue color passes into the kidney and may be seen in the convoluted tubes and ascending limbs of Henle's tubes as before. Ten minutes after the injection the coloring matter is found solely in the epithelial cells in the locations above noted. An hour after the injection the epithelial cells are found colorless, the blue matter having passed into the lumens of the tubes, where, in the absence of water from the glomeruli, it concentrates into crystals. This establishes a distinct eliminative power on the part of the renal epithelium altogether independent of the glomeruli, because the latter were paralyzed by section of the spinal cord.

It seems altogether likely, therefore, that the chief specific principles of the urine are eliminated by the renal epithelium, precisely as are the coloring matters in Heidenhain's experiments. This seems the more probable now, since it has been established that the chief urinary constituents—urea, uric acid, etc.—exist preformed in the blood. Our present knowledge on this subject warrants the conclusion that the production of the urine is chiefly an elaborating or secreting process, regulated in its fluidity by the glomerular system. In other words, that the water and some of the salts are secreted by the glomeruli, the peculiar anatomical construction of which permits a varying degree of activity corresponding chiefly with the varying degrees of blood-pressure and blood-fluidity; while, in the main, the solid excretory products of the urine are eliminated by the epithelium of the renal tubules, through their vital, selective, or secretory power, as in all other glandular structures of similar anatomical construction.

Composition of the Urine.—The constituents of normal urine are derived from the elements of retrograde tissue metamorphosis in the healthy state of the organism, together with certain waste

AMOUNTS OF URINARY CONSTITUENTS PASSED IN TWENTY-FOUR HOURS.

CONSTITUENTS.	WEIGHT, 66 KILO-GRAMMES.	PER KILOGRAMME OF BODY-WEIGHT.
Water................................	1500.00 grammes.	23.000 grammes.
Total solids..........................	72.00 "	1.100 "
Urea.................................	33.18 "	0.500 "
Uric acid............................	0.55 "	0.008 "
Hippuric acid........................	0.40 "	0.006 "
Creatinin............................	0.91 "	0.014 "
Pigment and other organic matters......	10.00 "	0.151 "
Sulphuric acid.......................	2.01 "	0.030 "
Phosphoric acid......................	3.16 "	0.048 "
Chlorine	7–8.00 "	0.126 "
Ammonia	0.77 "	
Potassium...........................	2.50 "	
Sodium	11.09 "	
Calcium.............................	0.26 "	
Magnesium..........................	0.21 "	

products of substances introduced into the system in the form of food and beverages. The normal constituents of the urine, according to the basis of classification adopted by Hoppe-Seyler, are as follow : 1. Urea and related substances,—uric acid, allantoin, oxaluric acid, xanthin, guanin, creatinin, thio- (sulpho-) cyanic acid. 2. Fatty and other non-nitrogenous substances,— fatty acids of the series $C_n H_{2n} O_2$; oxalic, lactic, glycero-phosphoric acids; very small quantities of certain carbohydrates— sugar (Brucke). 3. Aromatic substances,—the ethereal sulphates of phenol, cresol, pyrocatechin, indoxyl, and skatoxyl; hippuric acid, aromatic oxyacids. 4. Other organic substances,—pigments; ferments, especially pepsin; mucus, and humic substances. 5. Inorganic salts,—chlorides of sodium and potassium, potassium sulphate; sodium, calcium, and magnesium phosphates; silicic acid, ammonium compounds, and calcium carbonate. 6. Gases,—nitrogen and carbon dioxide.

The quantitative composition of the human urine is best expressed in the classical table of Parkes, as on preceding page.

The proportion between the solids of the urine and the water, according to Becquerel, is as follows: Water 967 grains, solids 33 grains in each 1000 grains. Of the solid matters the organic elements amount to about 24.865 grains, while the inorganic constituents are about 8.135 grains in each 1000 grains.

Changes in the Urine upon Standing.—Considering the somewhat complex composition of the urine, holding as it does in solution both organic and inorganic compounds which are subject to organic as well as chemical alterations, as might be expected, the urine is subject to more or less rapid changes after it has been voided. The rapidity of these changes depends chiefly upon the reaction and concentration of the urine, the temperature of the room in which it is kept, and the degree of access to micro-organisms. A normal acid urine usually first precipitates the amorphous urates, then uric acid, and frequently oxalate of calcium. Under ordinary circumstances these changes take place within three days after the urine is voided. At ordinary temperatures (70° to 74° F.—21.1° to 23.3° C.), after twenty-four to forty-eight hours' standing, and much sooner under high temperatures (say, 100° F.—37.8° C.), the urine begins to become dull and opaque from the presence of micro-organisms,—*fission fungi*. These multiply, and at the end of four or five days, as a consequence, ammoniacal decomposition sets in; that is to say, through the activity of the fungi the urea is gradually transformed into carbonate of ammonium. The urine becomes more and more alkaline from the liberation of ammonium carbonate. The amorphous urate deposit now becomes transformed into urate of ammonium, uric-acid crystals are substituted by characteristic prismatic crystals of ammonio-magnesian phosphate (triple phosphate), and amorphous granules of calcium phosphate are deposited in quantity at the bottom and sides of the vessel. Finally, the bacterial activity diminishes as the urine becomes strongly alkaline, and the micro-organisms ultimately perish. Urines of low density and feeble acidity undergo the above-described changes more rapidly, and, moreover, as a rule, do not deposit urates.

A change sometimes occurs in acid urine, consisting of progressive acidity, in which the urine darkens in color and deposits uric acid and urates and sometimes calcium-oxalate crystals, with the frequent presence of yeast-fungus and bacteria. This was formerly termed the *"acid fermentation,"* but, according to Scherer, it is caused by the mucus, which acts as an enzyme, or ferment, producing an acetic-acid or lactic-acid fermentation, with precipitation of uric acid and acid urates.

Collection of Urine for Analysis.—For the purpose of quantitative determination of the urinary constituents it is essential to have a sample of a mixture of the whole product of the kidneys for twenty-four hours. The varying degrees of solid and fluid contents of the urine at different hours of the day render the observance of the above rule strictly essential if we desire a sample of urine that will represent the average product of the kidneys. In order to guard against the early changes in the urine which have just been described, the urine should be collected in a *perfectly-clean* vessel,—preferably a half-gallon bottle, —which should stand in a cool but dry room during the collection, and the bottle should be corked after each addition of urine.

For qualitative determination of the *morbid* products of the urine—as sugar, albumin, etc.—it is preferable to collect a freshly-voided sample of the urine about three hours after a meal, *not the urine voided in the morning on rising.* Of all urines, that voided in the morning on rising is the least likely to contain albumin or sugar. When these substances are only occasionally present in the urine, they are most likely to be found after food and exercise. Finally, in collecting a sample of urine for purely microscopical examination, it should be perfectly fresh and as concentrated as possible. It has been shown that the urine soon undergoes ammoniacal changes upon standing, the result of which is to render it more or less strongly alkaline. Now, morphological

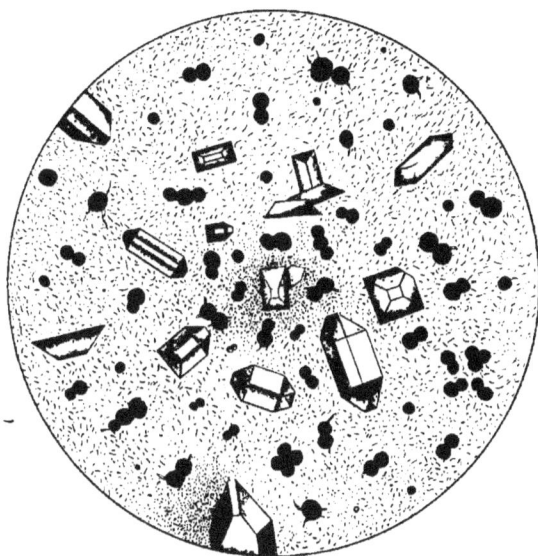

Sᴇᴅɪᴍᴇɴᴛ ᴏғ Aʟᴋᴀʟɪɴᴇ Fᴇʀᴍᴇɴᴛᴀᴛɪᴏɴ. (After
Hoffman and Ultzmann.)

elements, as epithelium and casts, are soluble in alkaline solutions; so that they may, if present when the urine is voided, become unrecognizably altered, or even disappear, if the foregoing precautions be not observed. Concentration of the urine may be obtained by directing the patient to abstain, as much as consistent with comfort, from the use of fluids for twenty-four hours.

PHYSICAL CHARACTERS OF THE URINE.

Color.—The average color of normal urine of specific gravity 1.020 and 1500 cubic centimetres' volume for twenty-four hours is straw or wine yellow,—amber-colored. This, however, is subject to considerable variations at different times of the day, and under varying circumstances included within the range of perfect health. Thus, from an almost colorless (watery) appearance the urine may range through the yellows and reach reddish brown. The pale, watery urine in health contains *relatively* small amounts of coloring matter, as well as urea and salts. It is seldom very acid, often neutral or feebly alkaline, and it is most often brought about by copious drinking. Highly-colored urine, on the contrary, is usually concentrated, containing *relatively* large quantities of solids and coloring matters. Its specific gravity is high, and its reaction is usually sharply acid. It results from diminished excretion of water by the kidneys, while the solids and coloring matters are normal or increased. In health it may occur after hearty meals, vigorous exercise, or when the skin has been unduly active and little fluids imbibed.

Pathologically the urine is subject to much wider variations in color than in health. This may be due either to increase or diminution of the normal coloring matters, on the one hand, or, on the other, to the addition of abnormal pigments. Abnormally light-colored urine is often due to polyuria, as in diabetes, hysteria, and convulsions; or it is often observed in diseases of the kidney which not only increase the normal amount of water in the urine, but which also reduce the solids and coloring matters,—notably, interstitial nephritis and amyloid degeneration of the kidneys. Highly-colored urine, approaching red, is most often induced by acute pyrexia and inflammations. This is due in part to concentration of the urine, though largely also to the

presence of uro-erythrine. The distinctly red tints of the urine
are always due to the presence of foreign coloring matters, most
often blood. The dark-brown tints may be due to the presence
of methæmoglobin in diseases of the kidney attended by hæmorrhage. The urine in cases of melanotic cancer sometimes becomes almost black, especially after standing for some time.
Green urine, of dull hue, is common in jaundice, the color being
due to the presence of biliverdin. The urine is frequently of a
greenish hue in diabetes, notably when the urine contains a high
percentage of sugar. Blue urine, of dull tint, is not uncommon
in cholera and typhus, owing to the presence of indigo. Finally,
certain drugs, when swallowed, affect the color of the urine to
a more or less marked degree: thus, rhubarb and senna cause
brown or reddish tints; carbolic acid sometimes causes a black
color in the urine, notably after the urine has stood some time;
the same results follow the ingestion of naphthalin, hydrochinon,
resorcin, and pyrocatechin. Lastly, santonin, when swallowed,
always causes a yellow color in the urine, of decided hue.

J. Vogel has, at considerable labor, constructed a scale of
colors of the urine from nature, which has, in a manner, become
standard for comparative purposes. These colors he expresses

I	II	III	IV	V	VI	VII	VIII	IX	
1	2	4	8	16	32	64	128	256	Pale yellow.......= I
	1	2	4	8	16	32	64	128	Light yellow......= II
		1	2	4	8	16	32	64	Yellow ....= III
			1	2	4	8	16	32	Reddish yellow ...= IV
				1	2	4	8	16	Yellowish red.....= V
					1	2	4	8	Red ....= VI
						1	2	4	Brownish red .....= VII
							1	2	Reddish brown....= VIII
								1	Brownish black...= IX

as (1) pale yellow, (2) light yellow, (3) yellow, (4) reddish yellow, (5) yellowish red, (6) red, (7) brownish red, (8) reddish
brown, and (9) brownish black. (See Frontispiece.) He divides
these into three groups, the first three being yellow urines, the
second three reddish urines, and the last three brown or dark
urines. In comparing the color of the urine with the scale the
urine should first be filtered if not perfectly clear, and it should

be examined by transmitted light in a glass vessel at least three or four inches in diameter. It is claimed that the shades of color in Vogel's scale correspond to certain relative amounts of coloring matter in the urine, and the test-table on preceding page has been constructed for the purpose of color analysis.

Application.—The table indicates how much coloring matter equal parts of urine of different colors contain relatively. Thus, if a certain volume of pale-yellow urine contain 1 part of coloring matter, the same volume of yellowish red contains 16 parts; of red, 32 parts; of brownish black, 256 parts. It further indicates that 1 volume of yellow urine contains as much coloring matter as 4 volumes of pale yellow; 1 of red = 32 of pale yellow, etc. If, therefore, one person pass 1000 cubic centimetres of yellow urine in twenty-four hours, and another 4000 cubic centimetres of pale-yellow urine in the same time, both secrete an equal amount of coloring matter. In order to make an approximate comparison by figures, Vogel places the quantity of coloring matter which 1000 cubic centimetres of pale-yellow urine contain = 1.

Example.—1800 cubic centimetres of urine of yellow color are passed. 1000 cubic centimetres of pale-yellow urine equal one part of coloring matter.

But yellow, according to the table, contains four times as much; therefore the following proportions: $1000 : 4 = 1800 : x = 7.2$ as the amount of coloring matter in 1800 cubic centimetres of yellow urine, the coloring matter in 1000 cubic centimetres of pale-yellow urine being considered as the unit.

Halliburton gives the following concise table of color variations of the urine, with their causes. (See next page.)

Odor.—The odor of normal freshly-voided urine is peculiar, —of slightly aromatic nature,—due, it is believed, to phenylic, taurylic, damoluric, and damolic acids in minute quantities. There is considerable difference in the intensity of the uric odor in health, always being most pronounced in concentrated urine. If the urine become alkaline from standing, it acquires a peculiar, repulsive, putrescent odor, in which ammonia is plainly distinguishable. The former is due to the decomposition of mucus and other organic matters, while the ammonia is in the

Color.	Cause of Coloration.	Pathological Condition.
Nearly colorless.	Dilution, or diminution of normal pigments.	Nervous conditions : hydruria, diabetes insipidus, granular kidney.
Dark yellow to brown-red.	Increase of normal, or occurrence of pathological, pigments.	Acute febrile diseases.
Milky.	Fat-globules.	Chyluria.
	Pus-corpuscles.	Purulent diseases of the urinary tract.
Orange.	Excreted drugs.	Santouin, chrysophanic acid.
Red or reddish.	Unchanged hæmoglobin.	Hæmorrhages, or hæmoglobinuria.
	Pigments in food (logwood, madda, bilberries, fuchsin).	
Brown to brown-black.	Hæmatin.	Small hæmorrhages.
	Methæmoglobin.	Methæmoglobinuria.
	Melanin.	Melanotic sarcoma.
	Hydrochinon and catechol.	Carbolic-acid poisoning.
Greenish yellow, greenish brown, approaching black.	Bile-pigments.	Jaundice.
Dirty green or blue.	A dark-blue scum on surface, with a blue deposit, due to an excess of indigo-forming substances.	Cholera, typhus ; seen especially when the urine is putrefying.
Brown-yellow to red-brown, becoming blood-red upon adding alkalies.	Substances which are introduced into the system with senna, rhubarb, and chelidonium.	

form of carbonate, resulting from bacterial decomposition of a part of the contained urea,—$CON_2H_4 + 2H_2O = (NH_4)_2CO_3$.

Certain substances, when ingested, impart to the urine peculiar and unnatural odors. Thus, a characteristic odor is acquired after eating asparagus, and an odor not unlike violets is produced by the administration of turpentine-oil. The odor of cubebs, copaiba, sandalwood-oil, garlic, tolu, etc., are more or less communicated to the urine when taken internally. These odors may be serviceable by indicating that the patient has taken certain medicines or foods. Beauvis and others have claimed that the peculiar odors after asparagus and turpentine-oil do not appear in the urine in organic diseases of the kidney. This, if true, might be valuable for diagnostic purposes, but more extended observation has not confirmed the assertion.

Pathologically the odor of the urine renders some information. Thus, if the urine be ammoniacal when voided, it is strong evidence of the existence of cystitis. In suppurating conditions of the upper urinary tract the urine is often peculiarly offensive (putrid), in consequence of its contained decomposing pus, blood, and organic elements. In diabetes the urine often has the odor of acetone. Urine containing cystin possesses an odor, at first, like sweet briar, but subsequently becomes very offensive.

Transparency.—The normal freshly-voided urine may be said to be always macroscopically clear; after standing, a mucous cloud, more or less pronounced, usually appears, which is unchanged by alkalies, heat, or mineral acids. Pathologically the urine may become cloudy from various causes, as the precipitation of urates, carbonates, phosphates, or organic products, as blood, chyle, excess of mucus, pus, bacteria, etc. If the cloudiness of the urine disappear upon the application of gentle heat, it may be concluded that the turbidity was due to the presence of precipitated urates. If, on the contrary, the turbidity increase upon the application of heat, it is due either to precipitation of the earthy phosphates or to albuminous cell-elements, as pus or blood. If the phosphates be the cause of the cloudiness, the latter rapidly clears up by the addition of an acid. If, on the other hand, it become more turbid upon the addition of the acid, it may be concluded that pus, blood, or albuminous cell-elements are the cause of the opacity. If the urine remain unchanged by the acetic acid, or if the turbidity be very slightly increased thereby, it may be concluded that mucus or micro-organisms are the cause of the turbidity.

Consistence.—Normal urine is always of aqueous consistence; that is to say, it drops and flows as does water. Pathologically the urine may become thick and viscid, so that it flows from the vessel slowly and with difficulty, not separating into drops. Such is usually the case when the urine contains a large amount of mucus or pus, and especially if, in addition, the urine be alkaline. Diabetic urine, if heavily laden with sugar, is of diminished consistence, as is evidenced by its tendency to froth when agitated ; and the same may be said of highly-albuminous urine. In chylous urine, owing to the contained molecular fat, the fluid often becomes much thickened. In fibrinuria the urine, after standing, sometimes becomes thickened into a jelly-like consistence ; so that it may stick to the vessel when the latter is inverted.

Specific Gravity.—The average specific gravity of normal urine of 1500 cubic centimetres (50 ounces) in volume for twenty-four hours is about 1.020. By this is meant taking distilled water at 15° C. (60° F.) as 1. the normal standard for urine is about 1.020. Slight variations from this standard are consistent with perfect health, and depend chiefly upon the character of the food taken, as well as the quantity thereof, and the rapidity of tissue metamorphosis. If the diet consist largely of nitrogenous foods, they furnish a higher relative amount of solids than fluids to be excreted by the kidneys, and, consequently, the specific gravity of the urine will be somewhat increased. Active muscular exertion also tends to raise the specific gravity of the urine. Copious diaphoresis may bring about a concentrated condition of the urine, with an accompanying increase of specific gravity. The specific gravity of the urine may be lowered by fasting or by imbibing large quantities of fluids. Very marked departures from the normal specific gravity of the urine often constitute pathological factors of great importance. In nearly all forms of organic albuminuria (Bright's disease) the tendency is toward a lowered specific gravity of the urine. An important fact to be noted in this connection is that, in most cases of so-called *functional albumi-*

nuria, the specific gravity of the urine is above the normal standard. Prognostically, in cases of nephritis, a marked reduction of the specific gravity of the urine should always be regarded as of serious import. If the specific gravity of the urine be markedly increased, it is strongly indicative of melituria. Should it reach as high as 1.030, or above, search should always be made for sugar.

The specific gravity of the urine may be taken with the urinometer, but more accurately by the Westphal Mohr or Sartorius balance. Only approximately correct results are possible with the urinometer; but considerations of convenience have induced most (formerly including the author) to sanction its use. The best of urinometers are often far from correct, and therefore the Westphal or Mohr balance is strongly advised where practicable.

The urinometers made by Squibb are among the best of such instruments. They are standardized at 25° C. (77° F.) and a thermometer is furnished with each instrument for temperature corrections (see Fig. 1). In taking the specific gravity of the urine with the urinometer the jar is filled about three-fourths full of urine, and any froth appearing at the top is removed by

FIG. 1.—SQUIBB'S URINOMETER.

filtering-paper or a pipette. The urinometer is next introduced into the urine and touched gently with the finger-tip, so that it sinks and rises a few seconds until it finds the correct level. When it comes to a rest the scale is read off on a level with the eye, and the figure on a level with the surface is marked.

The Westphal balance (Fig. 2) is extremely accurate, carrying out the specific gravity to the fifth figure (fourth decimal). To take the specific gravity of the urine by this instrument proceed as follows:—

After the instrument is mounted, and the beam rests in equilibrium, the glass plummet which contains the thermometer is suspended from the hook on the right-hand end of the beam together with one of the large riders (*A*). This balances with distilled water at 15° C. and represents 1. Next pour the urine into the jar until the twist in the platinum wire is below the surface, then begin the weighing. Place the second rider (*B*) in the first notch on the left of the scale on the beam, and, if the plummet rises, remove the rider to the second notch. If now the beam balances, and the temperature of the urine is 15° C. the specific gravity of the urine is exactly 1.020. If the plummet still rises, however, take the third-size rider (*C*) and find the notch on the beam where it rests in equilibrium, or very nearly so. If the beam balances with the third rider (*C*),—say, in the fourth notch,—the specific gravity is exactly 1.024. Should, however, the plummet still rise slightly, take the fourth rider (*D*, smallest one) and find the exact balance and, if in the sixth notch, the specific gravity will be 1.0246. In other words, the second rider (*B*) gives the third figure (second decimal) of the specific gravity; the third rider (*C*) finds the fourth figure (third decimal) of the specific gravity; and the fourth rider (*D*) gives the fifth figure, or fourth decimal. With a little practice, determinations of the specific gravity of the urine by means of the Westphal balance will be found rapid, simple, and absolutely correct.[1] Care should be exercised that no air-bubbles become attached to the plummet, and the proper temperature corrections should be made as directed below.

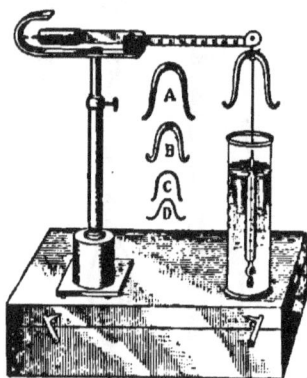

FIG. 2.—WESTPHAL BALANCE.

The temperature of the urine immediately after being voided ranges from 85° to 95° F. (29.5° to 35° C.); therefore, in taking the specific gravity of *freshly-voided* urine, before cooling, its temperature should be observed, and for every 7 degrees of temperature the thermometer indicates above that upon which the instrument is standardized 1 degree should be added to the specific gravity of the urine in addition to that indicated by the instrument.

Chemical Reaction.—Normal mixed urine—that is to say, the whole twenty-four hours' product—is always acid. The acidity is due to acid sodium phosphate, and not, as formerly supposed, to free acid. This acid sodium phosphate is derived from the alkaline sodium phosphate of the blood; the uric, hippuric, sulphuric, and carbonic acids of the urine take up part of the sodium, leaving an acid salt.

The degree of acidity of the urine varies at different times of the day, especially with regard to food. Soon after a meal the

---

[1] All determinations of specific gravity of the urine in the author's laboratory are made with the Westphal balance.

acidity begins to diminish, and in from three to four hours the alkaline tide usually reaches its height; occasionally, though rarely, the acidity may become so diminished at such times that the urine gives an alkaline reaction with test-paper. Freshly-voided urine may be alkaline either from fixed alkali (alkaline salts of potassium or sodium) or from volatile alkali (alkaline salts of ammonium). It is important to distinguish between these two conditions, as in the first case it merely reflects a condition of the blood, while in the second case it is nearly always associated with chronic inflammatory conditions of the lower urinary tract, notably the bladder. If red or violet litmus-paper turn blue in contact with urine just voided, and remain blue upon drying, the reaction is due to fixed alkali. If, on the other hand, the paper return to the original color upon drying, the reaction is due to volatile alkali (ammonia).

The urine is rendered alkaline by the administration of alkaline carbonates or the salts of vegetable acids. The urine may be rendered alkaline, usually to a less extent, by the following circumstances: Soon after a full meal; after the discharge of gastric juice in abnormal ways,—through fistula or by copious vomiting; after hot baths and free perspiration; upon a vegetable diet. With vegetarians, as with herbivorous animals, the food contains an excess of alkaline salts or vegetable acids. These acids are converted into carbonates in the blood, which, passing into the urine, cause an alkaline reaction.

The acidity of urine is increased by the ingestion of acids, saccharin, a purely meat diet, and prolonged muscular exercise. It may be developed, as already shown, by acid fermentation, and in certain pathological conditions free fatty acids may appear and render the urine sharply acid (lipaciduria).

Occasionally it happens that the urine is amphoteric,—i.e., the same urine turns red litmus-paper blue and blue litmus red. This seemingly paradoxical reaction, according to Huppert, depends upon the presence of acid and neutral phosphates in variable proportions.

Quantity.—The average quantity of urine of a healthy individual who eats and drinks in moderation, and lives in a temperate atmosphere, is about 1500 cubic centimetres (50 ounces) in twenty-four hours. The relative quantity varies considerably

2

with the time of day, most being passed in the afternoon, less in the morning, and least at night. The volume of urine for twenty-four hours varies much in conditions of health, according to certain circumstances. It is decreased by unusual activity of the skin and bowels, as well as by rest and abstaining from fluids. It is decidedly increased by imbibing large quantities of fluids, the use of diuretic drugs, to a less extent by cold, atmospheric moisture, exercise, and liberal eating.

Pathologically the urine is increased in diabetes, cirrhosis of the kidney, amyloid or waxy kidney, pure cardiac hypertrophy, pyelitis, hysteria, and convulsions. The quantity of urine is decreased in acute nephritis, cyanotic induration from cardiac defect, acute fevers, and inflammations. The urine may be more or less completely suppressed in the acute forms of nephritis, in the algid stage of cholera and yellow fever, by violent fevers and inflammations, by shock or collapse from internal injuries,—as rupture of the liver, spleen, bowels, or other viscera,—by the reflex shock or the congestion following catheterization (urinary fever), and by obstructive diseases of the urinary passages, notably the ureters. Finally, it is important to observe that in nephritis more or less complete suppression of the urine, often followed by uræmia and death, may result from the administration of anæsthetics such as chloroform and ether, notably the latter.

In estimating the quantity of urine it should be carefully collected for twenty-four hours, in an accurately-covered vessel, in order to exclude dust and prevent evaporation, the patient being directed to void and collect the urine previous to each movement of the bowels.

Solids.—In determining the solids of the urine observations should be made upon a sample of the mixed product of the kidneys for the whole twenty-four hours. The most accurate results are obtained by taking a given quantity of urine,—say, 20 cubic centimetres,—in a previously-weighed porcelain dish, and evaporating it over a water-bath. It should then be dried in a warm chamber for an hour or so, and then allowed to cool, when it should be weighed. This should be repeated a number of times until there is no further loss of weight from drying; then the difference in weight between the empty dish and that containing the dried solids constitutes the weight of the solids in 20 cubic centimetres of the urine. From this the solids of the whole volume of urine may be readily reckoned.

The foregoing method being somewhat tedious, and, besides, consuming too much time for practical work, approximate results may be more readily obtained by multiplying the last two figures of the specific gravity of the urine by the co-efficient of Häser, which is 2.33. This gives, approximately, the number of grammes of solids in each 1000 cubic centimetres of the urine.

Example.—If the twenty-four hours' urine be 1500 cubic centimetres, and the specific gravity be 1.020 with Häser's co-efficient, then we have as follows :—

$20 \times 2.33 = 46.60$ grammes in 1000 cubic centimetres of urine;

therefore, $\dfrac{46.60 \times 1500}{1000} = 69.9$ grammes of solids.

TABLE FOR ESTIMATING TOTAL SOLIDS FROM SPECIFIC GRAVITY.

SPECIFIC GRAVITY.	SOLIDS FOUND BY WEIGHING.	SOLIDS FOUND BY MULTIPLYING BY 0.233.
	Per Thousand:	Per Thousand:
1.0160	37.4	37.28
1.0260	62.0	60.58
1.0154	35.1	35.88
1.0261	60.2	60.81
1.0213	48.6	49.63
1.0230	56.4	53.59
1.0225	49.3	52.42
1.0240	54.1	55.92
1.0257	60.4	59.68
1.0275	63.9	64.07
1.0217	48.5	50.56
1.0223	52.15	51.96
1.0140	31.08	32.62
1.0236	56.64	54.98
1.0133	30.87	30.99
1.0134	31.06	31.22
1.0238	57.09	55.45
1.0250	60.47	58.25
1.0164	37.26	38.21
1.0135	33.35	31.45
1.0210	48.54	48.93
1.0137	32.55	31.92
1.0085	19.16	19.80
1.0110	24.96	25.63
	Average:	
1.0200	46.59	46.52

From these determinations it will be found that, by dividing the mean quantity of solid constituents found in 1000 grammes of urine by the last three decimals of the mean specific gravity, we obtain the quotient 0.23295, for which we may conveniently substitute the number 0.233, as Häser suggests. By multiplying with this quotient the three last decimals of the specific gravity carried out to four places of decimals, we obtain the figures in the third column of the table. The variations from the results obtained by actual weighing may be seen by a glance at the table. If, however, as is usual, the specific gravity of the urine be determined only to three decimals, the second and third figures multiplied by 2.33, as suggested by Häser, give approximately the amount of solid matters in 1000 parts of urine.

A material reduction of the solids of the urine in cases of renal disease indicates a tendency to uræmia, and therefore puts the physician on his guard against this dangerous complication. The diagnosis between diabetes insipidus and hydruria may be determined by the amount of solids in the urine. The so-called *renal inadequacy*—an obscure term introduced by Sir Andrew Clark, which usually means unrecognized interstitial nephritis—and the *anazoturia* of Willis are both indicated by a reduction of the solids of the urine.

During the course of acute fevers and inflammations the quantity of solids in the urine furnishes very valuable information as a guide both for prognosis and treatment. The amount of tissue metamorphosis, as evidenced by the quantity of solids in the urine, is a good indication of the severity of the disease. If these changes be too active, measures are indicated for restraining them. If exudations are to be removed, a copious excretion may indicate to us that the chemico-vital changes ending in elimination are progressing sufficiently without artificial aid. Again, by insufficiency of the urinary solids defective elimination may be detected when the thermometer indicates a high ratio of tissue metamorphosis in progress, and we are thereby admonished to employ measures to re-establish elimination.

No definite deductions are to be drawn, from the quantity of solids present, as to the relative amount of any special product, especially that of urea. Since the amount of urea normally constitutes about one-half of the solids excreted by the kidneys, it has been suggested that an approximate knowledge of the amount of urea is to be gained by merely dividing the whole quantity of solids by 2. Such a rule should never be suggested as a guide in pathological conditions, because, under such circumstances, the various solids of the urine are often present in widely different *proportions*, as well as quantities. In addition to this, the special solid constituents of the urine possess widely different specific gravities,—notably that of urea from sodium chloride, which is as 2 to 3. For these reasons, even when the total solids are determined with accuracy, the amount of urea, nitrogen, or other constituent, if sought, can only be determined by special quantitative methods, which will be described in the succeeding section of this work.

Having ascertained the actual quantity of solids in the urine, in order to make deductions therefrom of any definite value in health or disease, careful regard must be had to certain conditions and features connected with each individual case before we can determine the degree in which the quantity of solids is excreted above or below the average or normal standard for such individual case. Those that chiefly influence the normal balance of excretion can be reduced to a ratio that will afford approximate results of definite and practical value. The most prominent of these conditions are: the weight of the individual, the age, the diet, and the amount of exercise taken. We may place the average weight at 66 kilogrammes (145 pounds) avoirdupois, the age being from 20 to 40, the diet that of ordinary mixed foods, and the exercise being that of the usual healthy man about ordinary labor. To the above standard the following general rules may be applied :—

1. The average excretion of solids per weight of 145 pounds being 61.14 grammes (945 grains), a proportional reduction or addition should be allowed, according to the weight of the subject examined.

2. Deduct 10 per cent. from the average solids in persons between 40 and 50 years of age, 20 per cent. if between 50 and 60, 30 per cent. if between 60 and 70, and 50 per cent. above 70.

3. *For Diet.*—In persons who have fasted for two or more

days, as in some fevers and other diseases, deduct one-third from the average solids. If the diet be spare, deduct one-eighth to one-sixth; if rather plentiful, but still below that of health, deduct one-tenth.

4. *For Exercise.*—If there be total rest, deduct one-tenth from the average solids; but if merely quietude, deduct one-twentieth.

The average standard of excretion of solids by the kidneys (61.14 grammes—945 grains—for weight of 66 kilogrammes—145 pounds avoirdupois) represents the mean of the combined observations by Becquerel, Parkes, Bocker, J. Vogel, Lehman, Gorup Besanez, Ranke, and Rummel, obtained by evaporating the urine and determining the solids by actual weighing. The rules for correction are based on the proportions laid down by Parkes.

Estimation of Acidity of the Urine.—Since the acidity of the urine is due to acid sodium phosphate, it cannot be quite correctly estimated by the ordinary acidimetric method, owing to the action of the alkali employed upon the acid sodium phosphate, a mixture of neutral and acid sodium phosphate resulting at first, producing the so-called amphoteric reaction and rendering the recognition of the exact end-reaction impossible. A slight excess of NaOH must therefore be added and the reading taken when the reaction has become faintly alkaline, the degree of acidity found being a trifle too high.

Method.—Take 100 cubic centimetres of urine in a flask from the twenty-four hour specimen and titrate with a one-tenth normal solution of sodium hydrate, using a sensitive litmus-paper as an indicator until a faintly-alkaline reaction is produced. The number of cubic centimetres of the one-tenth normal solution employed multiplied by 0.0063 will give the percentage of acidity in terms of oxalic acid. The total acidity thus found corresponds to from 2 to 4 grammes of oxalic acid per day.

SECTION II.

COMPOSITION OF NORMAL URINE.

ORGANIC CONSTITUENTS.

UREA—$CO(NH_2)_2$.

UREA, or carbamide, was first prepared synthetically from ammonium cyanate—$(NH_4)CNO$—by Wohler (1828). From the urine it was first prepared in an impure state by Rouelle, and subsequently by Fourcroy and Vanquelin. Urea crystallizes in colorless, quadrilateral, or six-sided, silk-like prisms with oblique ends, or, when rapidly crystallized, in delicate white

FIG. 3.—CRYSTALS OF UREA.

needles, which melt at 120° C. (248° F.). They contain no water of crystallization, and are permanent in the air, soluble in cold water, the solution being neutral in reaction. With nitric acid urea unites to form nitrate of urea ($CON_2H_4HNO_3$), which crystallizes out in octahedral, lozenge-shaped, or hexagonal plates, which are less soluble in water than are urea crystals.

Urea owes its origin in the economy partly to retrograde tissue metamorphosis, including the blood, and partly to splitting

(21)

up of unassimilated nitrogenous principles of the food. Thus, the greater portion of nitrogen taken into the system in the way of food is excreted by the kidneys in the form of urea. It is therefore the most bulky single constituent of the urine, ranging in quantity, according to circumstances, from 20 to 40 grammes (300 to 600 grains) in twenty-four hours in the healthy adult.

That the liver constitutes the chief seat of urea formation is now pretty generally accepted as fact. This was originally claimed to be the case by Meissner, and more recently confirmed by Brouardel, Schroeder, and Minkowski. It is not improbable, however, that the spleen and perhaps the lymphatic and secreting glands to some extent participate in the urea formation. Formerly it was erroneously supposed that urea was formed in the kidneys; but it is now known that after complete extirpation of the kidneys the formation of urea continues as before, and accumulates in the blood. Likewise, in diseases of the kidney entailing suppression of the urine, urea continues to be formed and accumulates in the organism. The evidence derived from pathology strongly points to the liver as the chief seat of urea formation. Thus, in diabetes, we know that metabolism of the hepatic cells is greatly increased, causing an abundant formation of sugar as well as urea, which pass into the blood and are excreted by the kidneys. On the other hand, in degenerative changes in the liver the urea formation is markedly diminished. Thus, in acute yellow atrophy of the liver the urea in the urine is greatly diminished—sometimes absent. The relations of degenerative changes in the liver to urea formation have recently been much elucidated by Noel Paton,[1] who points out that two functions of the liver exist—bile formation and urea formation—and, moreover, that they bear a direct relationship to each other. It has already been stated that urea owes its origin in the economy to retrograde tissue metamorphosis and nitrogenous principles of the food; in other words, the proteid constituents in the organism. Of the intermediate steps in this transformation but little is definitely known, although much has been written on this subject, most of which is conjectural. The excretion of urea reaches its maximum

---

[1] British Medical Journal, vol. ii, 1886, p. 207.

quantity upon an exclusive meat diet; much less is excreted upon a mixed diet, and least of all upon a strictly vegetable diet.

Variations in the quantity of urea excreted, in a measure, constitute an expression of the changes in nitrogenous tissue metabolism, and as such possess definite clinical value. Thus, in acute fevers and inflammations, until the crisis of the disease is reached there is greatly increased elimination of urea. On the other hand, in chronic diseases (cachexias), when tissue metamorphosis is retarded through malnutrition, the excretion of urea is diminished. Similar results follow in diseases involving the integrity of the liver,—the elaborating source of urea. In Bright's disease urea excretion is diminished in consequence of impairment of the structure of the kidneys. Preceding, usually for some time and during uræmic attacks, the excretion of urea is markedly diminished, forming a valuable indication of the approach of this dangerous complication. Mental and muscular activity hasten urea excretion by accelerating tissue waste ; and hence urea excretion is more active during waking than during sleeping hours. The variations in the amount of exercise, the quantity and quality of food taken, atmospheric vicissitudes, the degree of activity of the other excretory organs, etc., render the relative and absolute amount of urea excreted almost as variable as the amount of water. If the mode of life be very regular and equable, the amount remains pretty uniform ; but great care is necessary to maintain all the physiological conditions even. Under the latter circumstances the mean amount of urea excreted in twenty-four hours, by healthy adult males between the ages of 20 and 40 years, is 33.18 grammes (512.4 grains). Close calculations give the average excretion of urea as .015 to .035 gramme per hour for each kilogramme of body-weight. The absolute quantity of urea excreted by women is below the average of men. The same is true with children ; but, on the other hand, the relative quantity excreted by children is much higher than by either men or women.

Urea may be separated from the urine as follows: First evaporate, then add strong, pure nitric acid in excess, keeping the mixture cool during acidulation. Pour off the excess of fluid from the crystals of urea nitrate formed ; strain through muslin and press between heavy filter-paper. Add to the dry product barium carbonate in excess, and add sufficient alcohol to form a pasty consistence. Dry on a water-bath, and extract with alcohol; filter; evaporate the filtrate on a water-bath, and set aside to crystallize. The result is nearly pure urea, plus the coloring matters of the urine. The simplest method of decolorizing the product is to filter through animal charcoal, and afterward purify by recrystallization.

Detection.—1. A very simple method of detecting urea is to place a drop or two of the suspected fluid upon a glass slide, and after adding a drop of nitric acid gently warm over a spirit-lamp. If urea be present, upon evaporation the microscope will show the characteristic crystals of nitrate of urea, of rhombic or hexagonal forms, singly or in masses, their acute angles being eighty-two degrees.

2. Add to the suspected fluid three parts of an aqueous solution of furfur aldehyde, and subsequently a few drops of strong hydrochloric acid, and warm. A series of colors—yellow, green, violet, purple, and red—is produced, settling at length into a brown, sticky mass if urea be present.

3. To a few drops of the suspected fluid in a test-tube add an equal quantity of solution of hypobromite of sodium, and a rapid evolution of bubbles takes place if urea be present.

4. Warm a few crystals of urea in a test-tube; add a trace of sodium or potassium hydrate and a drop of a dilute solution of cupric sulphate. A violet or rose-red color is produced,—the biuret reaction.

Determination of Urea.—The methods devised for the purpose of determining the quantity of urea in the urine are very numerous; most of which, however, are founded upon one of three principles: 1. The mercuric nitrate, or Liebig's method. 2. The hypobromite, or Knop-Hufner method. 3. The differential density method, obtained by measuring the amount of urea by the specific gravity of the urine lost through decomposition of the contained urea.

Liebig's Method.—The combination between urea and mercuric oxide—$(CON_2H_4)_2$ Hg $(NO_3)_2 + 3$ HgO—forms a white precipitate, insoluble in water and weak alkaline solutions. It is, therefore, necessary to prepare a standard solution of mercury, and to have an indicator by which to detect the point when all the urea has entered into combination with the mercury, the latter slightly predominating. This indicator is sodium carbonate, which gives a yellow color with the excess of mercury, in consequence of the formation of hydrated mercuric oxide.

Theoretically, 100 parts of urea should require 720 parts of mercuric oxide; but practically 772 parts of the latter are necessary to remove all the urea, and at the same time to show the yellow color with alkali; therefore, the solution of mercuric nitrate must be of empirical strength in order to give accurate results. The following solutions are required for testing :—

1. Standard mercuric nitrate solution. Dissolve 77.2 grammes (1158 grains) of red oxide of mercury (weighed after drying over a water-bath), or 71.5 grammes (1072 grains) of the metal itself in dilute nitric acid. Expel the excess of acid by evaporation to a syrupy consistence. Make up to 1000 cubic centimetres with distilled water, adding the water gradually. This solution is of such strength that 19 cubic centimetres will precipitate 10 cubic centimetres of a 2-per-cent. urea solution. Add 52.6 cubic centimetres of water to the litre of the mercuric nitrate solution, and shake well; then 20 cubic centimetres (instead of 19) equal 10 cubic centimetres 2-per-cent. urea solution,—i.e., 1 cubic centimetre equals 0.01 urea.

2. Baryta mixture. Prepare by mixing two volumes of solution of barium hydrate with one volume of solution of barium nitrate, both saturated in the cold.

Analysis.—Take 40 cubic centimetres of urine, add to this 20 cubic centimetres baryta mixture and filter off the precipitate of barium salts(phosphates and sulphates). Take 15 cubic centimetres of the filtrate (corresponding to 10 cubic centimetres of urine) in a beaker. Discharge into it the mercuric nitrate solution from a burette, until, on mixing a drop of the mixture with a drop of saturated solution of sodium carbonate on a white tile, a pale-lemon color appears. Then read the amount used from the burette, and calculate thence the percentage of urea.

Corrections.—This method only approaches accuracy when the quantity of urea present is about 2 per cent.,—the normal

percentage of urea in the urine. The chlorine in the urine
must be estimated, and the quantity of urea indicated reduced
by subtraction of 1 gramme (15 grains) of urea for every 1.3
grammes (19 grains) of sodium chloride found. If the urine
contain less than 2 per cent. of urea, 0.1 cubic centimetre of mercuric nitrate solution must be deducted for every 4 cubic centimetres used; if more than 2 per cent. of urea, a second titration
must be performed, with the urine diluted with half as much
water as has been needed of the mercurial solution above 20 cubic
centimetres. Suppose, then, 28 cubic centimetres have been used
in the first titration, the excess is 8 cubic centimetres; therefore,
4 cubic centimetres of water must be added to the urine before
the second titration is made. When ammonium carbonate is
present, first estimate the urea in one portion of urine, and the
ammonia by titration with normal sulphuric acid in another;
0.017 gramme of ammonia equals 0.030 gramme of urea. The
equivalent of ammonia in terms of urea must be added to the
urea found in the first portion of the urine.

Modifications.—Rautenberg[1] and Pflüger[2] have devised the
following modifications of Liebig's original method, just described :—

Rautenberg's method consists in maintaining the urea solution neutral throughout by successive additions of calcium
carbonate.

Pflüger's modification is the one most generally employed,
which is as follows : A 2-per-cent. solution of urea is prepared;
10 cubic centimetres of this are placed in a beaker, and 20 cubic
centimetres of the mercuric nitrate solution are run into it in a
continuous stream; the mixture is then placed under a burette
containing normal sodium carbonate, and this is added, with
constant agitation, until a permanent yellow color appears. The
volume so used is noted as that necessary to neutralize the acidity
produced by 20 cubic centimetres of the mercurial solution in
the presence of urea. A plate of glass is then laid on black
cloth, and some drops of a thick mixture of sodium bicarbonate

¹ Ann. Chem. Pharm., vol. cxxxiii, p. 55.
² Zeit. anal. Chem., vol. xix, p. 375.   Pfeiffer (Zeit. Biol., vol. xx, p. 540)
has carefully compared the different methods.

(free from carbonate) and water placed upon it at convenient distances. The mercurial solution is added to the urine in such volume as is judged appropriate, and from time to time a drop of the white mixture is placed beside the bicarbonate, so as to touch but not mix completely. A point is at last reached when the white gives place to yellow; both drops are then rubbed quickly together with a glass rod, and the color disappears; further addition of the mercurial solution is then made to the urine till a drop rubbed with the bicarbonate remains permanently yellow. Now neutralize by the addition of the normal sodium carbonate to near the volume found necessary in the preliminary experiment. If this be quickly done, a few tenths of a cubic centimetre of mercuric nitrate will be found sufficient to complete the reaction. If much time has been lost, however, it may occur that, notwithstanding the mixture is distinctly acid, it gives, even after the addition of sodium carbonate, a permanent yellow, although no more mercuric nitrate be added. The analysis, under such circumstances, must be repeated, taking the first titration as a guide to the quantities which are necessary. Pflüger's correction for concentration of urea is different from Liebig's, and is as follows:—

$V^1 =$ volume of urea solution $+$ volume of sodium-carbonate solution $+$ volume of any other fluid free from urea which may be added.

$V^2 =$ volume of mercuric nitrate solution used.

$C =$ correction $= - (V^1 - V^2) \times 0.08$.

This formula holds good for cases in which the total mixture is less than three times the volume of mercuric nitrate solution used; with more concentrated solutions the formula gives results too high.

Hypobromite Method.—This is a far more simple and ready method to manipulate. The principle of the method introduced by Knop depends upon the fact that, when urea in solution comes in contact with sodium-hypobromite solution, nitrogen is set free as a result of the total decomposition of the urea. Thus:—

$$CON_2H_4 + 3NaBrO = CO_2 + 2H_2O + 3NaBr + 2N.$$

The CO_2 is absorbed by the excess of sodium hydrate in the hypobromite solution used.   One gramme of urea furnishes 370 cubic centimetres of nitrogen at 0° C. and under 760 millimetres pressure.   Knop's hypobromite-of-sodium fluid is made as follows: In 250 cubic centimetres of distilled water 100 grammes of sodium hydrate are dissolved, and after cooling 25 cubic centi-metres of bromide are added.

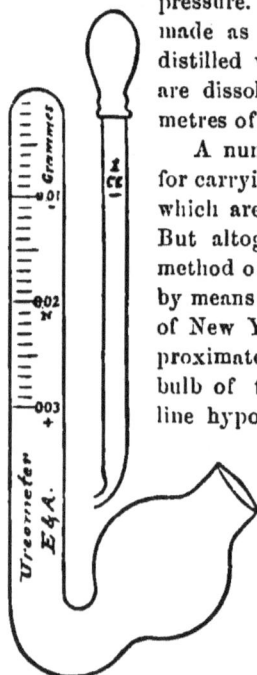

A number of apparatuses have been devised for carrying out this method, the best known of which are those of Hufner, Dupré, and Gerrard. But altogether the simplest and most ready method of carrying out the hypobromite test is by means of the instrument devised by Doremus, of New York, which gives very satisfactory ap-proximate results for rapid clinical work.[1]   The bulb of the instrument is filled with the alka-line hypobromite solution, and by inclining the tube the long arm is filled to the bend at the bulb.[2]   By means of the nipple-pipette, 1 cubic centimetre of the urine to be tested is slowly dis-charged up the long arm into the hypobromite solution.   A rapid de-composition of urea takes place; the bubbles of nitrogen rise in the long arm of the instrument, while the displaced liquid flows into the bulb, which serves as a reservoir for the overflow.   In fifteen minutes the decomposition of urea is

FIG. 4.—DR. DOREMUS'S
UREOMETER.

---

[1] Messrs. Elmer & Amend, 205 and 211 Third Avenue, New York, supply these instruments at very moderate cost.

[2] As the hypobromite solution does not keep well, it is best to keep the bromine and the sodium-hydrate solution separate, as follows: Have on hand a solution of sodium hydroxid, 100 grammes to 250 cubic centimetres of water (6 ounces to the pint).   In another bottle should be kept the bromine.   To pre-pare the solution freshly for use, take 10 cubic centimetres of the sodium-hydroxid solution, and add thereto 1 cubic centimetre of bromine (1 to 10). After the bromine has been thoroughly mixed with the alkali, dilute with an equal volume of water, and the solution is ready for testing.

complete, and the graduation on the long arm will indicate the quantity of urea in the volume of urine tested. Two forms of the instrument are furnished,—one graduated to read fractions of a gramme to the cubic centimetre of urine, and the range is from 0.01 to 0.03 gramme. If it be desired to read the percentage of urea instead of the grammes per cubic centimetre, simply remove the decimal-point two figures to the right,—thus, 0.02 gramme to the cubic centimetre is 2.0 per cent. of urea. The other form of the instrument is similar, save that it is graduated to show the number of grains of urea per fluidounce of urine.

The normal quantity of urea in the urine is about 2 per cent., or 0.02 gramme per cubic centimetre, or 10 grains per ounce.

The *differential density method*, as devised by Dr. George B. Fowler, of New York, is very simple in application. This method is based upon the fact that there is a difference in the specific gravity of urine before and after the decomposition of its urea by the hypochlorites. Every degree of density lost corresponds to 0.77 per cent., or about 3½ grains per ounce. The hypochlorite solution employed is Squibb's solution of chlorinated soda (Labarraque's solution), of which seven parts will destroy the urea in one part of urine, unless the amount is very large, in which case the urine should be diluted with an equal bulk of water and the result multiplied by 2. The process consists in adding to one volume of urine—say, one ounce in a large hydrometer jar—seven volumes of Labarraque's solution, the specific gravity of which has been taken. Decomposition immediately ensues, and at the expiration of a few hours all the nitrogen of the contained urea has escaped. The specific gravity of the quiescent mixture is now noted, and also that of the pure urine. We now have the specific gravity of the mixture of the urine and Labarraque's solution after decomposition. In order to ascertain what it was before decomposition, we resort to the law of proportions. Multiply, therefore, the specific gravity of the Labarraque solution by 7, add the specific gravity of the urine, and divide the sum by 8. Now, from this—the specific gravity of the mixture before decomposition—subtract that obtained after decomposition. Multiply the difference in degrees by $3\frac{1}{2}$, and the result will be the number of grains of urea per ounce of urine; or, better, by 0.77, which gives the percentage of urea. The presence of sugar or albumin in the urine does not interfere with this test.

Uric Acid $(C_5H_4N_4O_3)$.

Uric acid, like urea, is a nitrogenous product, although it exists in the urine in comparatively small amount,—about 0.4 to 0.8 gramme being excreted by the kidneys in the healthy man in twenty-four hours. Upon decomposition uric acid yields about 33.3 per cent. of its weight in nitrogen. It is feebly soluble, requiring 15,000 parts of cold or 1900 parts of boiling water to dissolve it. It is, for the above reason, rarely found in the urine in the free state; more often in the form of crystalline deposit,—reddish sand; most often, however, it exists in combination as urates.

Uric acid is dibasic; that is to say, it contains two atoms of H which may be substituted by two atoms of a monad metal. Urates containing but one atom of potassium, sodium, or ammonium are acid urates; those containing two atoms are neutral salts. Uric-acid crystals usually occur in urines of strongly-acid reaction, although exceptionally they may be found in alka line urine at the beginning of, or in the early stage of, alkaline fermentation. It is common to find uric-acid crystals deposited from highly-concentrated urines (dark-colored urine), such as is often noticed after a diet largely of proteid foods or after free perspiration; and in such cases it is not of special significance. It is otherwise, however, when deposit occurs from increased formation and excretion, resulting after febrile crises, rheumatic arthritis, renal and vesical lithiasis, leukæmia, pernicious anæmia, diabetes, uric-acid diathesis, and conditions of respiratory insufficiency. Crystals of uric acid occurring in physiological and pathological urines present many variations, most of which are well seen in Plate III.

Uric acid crystallizes in the urine in rhombic, rectangular prisms, wedge- and whetstone- shape, of yellowish-red color, constituting, with its salts, the only sediments of the urine thus colored.

[Uric-Acid Crystals. Normal Color. × 450.
(After Peyer.)]

Uric acid corresponds with urea in its protein origin in the organism, but the seat of its formation has given rise to much discussion. Two different views are held upon this subject: 1. That, like urea, it is formed in the tissues, notably in the spleen and liver, and merely excreted by the kidneys. This view is supported by the following facts: (*a*) In the normal condition but a small amount of uric acid is found in the blood. (*b*) In gout, where excretion of the uric acid is diminished, it accumulates in the blood and tissues.

(*c*) After extirpation of the kidneys it continues to be formed. (*d*) The secretion of uric acid is most abundant at the period of digestion, when the liver and spleen are the most active. 2. This view supposes that the kidneys not only constitute the seat of excretion, but also that of formation of uric acid. Garrod has ably supported this view,[1] basing his conclusions, first, on the fact of the small amount of uric acid in the blood of birds and reptiles, and also that he was unable to find more uric acid in the liver and spleen of birds than in those organs in mammals. But Schroeder has recently shown (1) that the liver of birds contains a high percentage of uric acid; (2) that after removal of the kidneys uric acid continues to be formed, and accumulates in the blood and liver; (3) that by passing blood through the liver, immediately after removal of that organ from the body, it is found that the uric acid is much increased; and (4) he regards ammonia as the most important precursor of uric acid, just as in mammals it is the most important precursor of urea.[2] In addition to this, the experiments of Minkowski are still more conclusive. He succeeded in keeping geese alive for from six to twenty hours after extirpation

FIG. 5.—URIC-ACID CRYSTALS.
(After Kuhn.)

---

[1] Lumleian Lectures, Lancet, vol. i, 1883.
[2] Ludwig's Festschrift, 1887, p. 89.

of the liver; after the operation their urine contained but 2 or 3 per cent. of uric acid, instead of the normal 60 or 70 per cent.; the ammonia was correspondingly increased to 50 or 60 per cent., instead of the normal 9 to 18 per cent.

All the facts at present available render the following conclusions probable: 1. That uric acid is formed chiefly in the liver. 2. It is formed by the synthesis of ammonia and lactic acid, which, after the removal of the liver, appear in the urine in equivalent quantities. 3. That the remnant of uric acid in the urine after extirpation of the liver originates from xanthin and similar products.

The quantity of uric acid in the urine should never be presumed to be excessive from the mere fact of deposit of uric-acid crystals in the urine upon cooling, as in fact such may, and very often does, occur when the uric acid is both relatively and absolutely deficient. The conditions of the urine which tend to precipitation of uric acid are as follow: 1. High grade of acidity of the urine. 2. Poverty in mineral salts. 3. Low percentage of pigmentation. 4. High percentage of uric acid. 5. Long standing. Any urine upon standing sufficiently long will deposit uric-acid crystals in consequence of the changes culminating in ammoniacal decomposition.

As already stated, the quantity of uric acid excreted in twenty-four hours by the average healthy man ranges from 0.4 to 0.8 gramme (6 to 12 grains). The proportion of urea to uric acid is stated by Parkes to be about 45 to 1.

Uric-acid excretion is increased by a rich animal diet, especially if combined with limited exercise in the open air; by acute febrile conditions; by lung and heart diseases, attended by dyspnœa; by diseases which impede respiration, as large abdominal tumors, ascites, etc.; by leukæmia; and by the so-called uric-acid cachexy. Uric acid, on the other hand, is diminished in advanced Bright's disease, urina spastica, hydruria, arthritis (especially the gouty form), gout, and, in general, wherever urea is diminished in quantity uric acid is usually also diminished.

Detection of Uric Acid.—1. Uric acid, as such, is easily recognized by the microscope, owing to the characteristic rhombic, yellowish-red crystals. The microscope also most readily distinguishes uric acid from its compounds.

2. The presence of both uric acid and its compounds may be readily detected by means of the *murexide test.* The sediment, or residue, after evaporation, is first treated with a few drops of nitric acid in a porcelain capsule, and, after warming over a spirit-lamp until evaporated almost to dryness, a drop or two of ammonia is added, and, if present, a beautiful purple-red color immediately appears, gradually diffusing over the bottom of the capsule.

The murexide test depends upon the facts that, by the addition of nitric acid and heat to uric acid or urates, first alloxan and subsequently alloxantin are formed, which, on the addition of ammonia, form murexide,—acid purpurate of ammonium.

3. If one or two drops of nitrate-of-silver solution be dropped upon white filter-paper, and then touched with an alkaline solution of uric acid (sol. with sodium carbonate), a black color appears if 0.001 per cent. of uric acid be present; even 0.0005 per cent. of uric acid, if present, will cause a brownish-yellow or grayish stain.

4. A solution of uric acid or urate, warmed with copper sulphate and caustic potash, produces a reddish precipitate of cuprous oxide.

Determination of Uric Acid.—An approximate and sufficiently accurate process for most clinical purposes is Heintz's method, which is also very simple : Take 200 cubic centimetres of urine, and add to it 10 cubic centimetres of hydrochloric acid. Let stand for twenty-four hours in a cool room. Collect the precipitated uric-acid crystals on a previously-weighed filter, and wash with cold distilled water. Dry the filter and uric-acid crystals in a desiccator, and weigh. By subtracting the weight of the filter, the result will be the weight of the uric acid in 200 cubic centimetres of urine. If albumin be present, it should first be removed and the urine should *always be filtered before applying the test,* otherwise subsequent filtration is very difficult.

Occasionally it happens that urine containing uric acid gives no precipitate as above, and a number of other methods have been suggested, but as yet we lack a thoroughly trustworthy and ready method for clinical work. Haycraft's method has heretofore been most employed. This is based upon the fact that uric acid combines with silver as silver urate; the silver urate is collected, dissolved in nitric acid, and the silver is estimated volumetrically by Vollard's method. From the amount of silver found the amount of uric acid is calculated. Czapek found a large error in this method, and Salkowski also regards the process as misleading, because the composition of the silver urate is not uniform, and Gossage confirms this statement. The methods of Fokker-Salkowski and Camerer have been strongly recommended, of late. The following method of Hopkins, which is a modification of Fokker's method, has recently been recommended as the best to date [1]:—

Hopkins's Method.—In this process the uric acid and all the urates are precipitated by saturation with ammonium chloride, which converts them all into urates of ammonium. They are then filtered out and the uric acid separated by the action of hydrochloric acid. The final estimation is then made by titration with a standard solution of potassium permanganate, which is found more accurate than weighing.

The process, as applied to all urines, normal and pathological, is as follows:—

1. *In Normal Urine without Deposit.*—(a) To 100 cubic centimetres of the urine ammonium chloride is added until practically saturated; about 35 grammes are necessary. When a small quantity of the chloride remains undissolved, even after brisk stirring at intervals of a few minutes, saturation is sufficient. As the temperature of the urine again rises, from the depression due to the process of solution, any residual crystals will, for the most part, dissolve; but there is no necessity for adding more.

(b) After having stood for two hours,—better with occasional agitation, to promote subsidence,—the precipitate produced is filtered through a thin filter-paper and washed three or four times with saturated solution of ammonium chloride. The filtrate should remain perfectly clear and bright.

(c) With a jet of hot distilled water the urate, which will be somewhat pigmented, is now washed off the filter into a small beaker, and heated just to boiling with an excess of hydrochloric acid. It is then allowed to stand for the uric acid to separate out completely. Two hours is sufficient, if the liquid be cooled. The acid is then filtered off and washed with cold distilled water. The filtrate should be measured *before* the washing has begun, and 1 milligramme added to the final result for each 15 cubic centimetres of liquid present. This need never be more than from 20 to 30 cubic centimetres.

The acid is now again washed off the filter with hot water, warmed, with the addition of sodium carbonate, till dissolved, and the solution then made up to 100 cubic centimetres. Being transferred to a flask of sufficient capacity, it is mixed with 20 cubic centimetres of strong, pure sulphuric acid, and immediately titrated with one-twentieth normal potassium-permanganate solution. The latter should be added slowly toward the end of the reaction, the close of which is marked by the first approach of pink color, which is permanent for an appreciable interval. Previously, the disappearance of the color will have been instantaneous. The flask should be agitated throughout the operation.

The standard solution is made by dissolving 1.578 grammes of pure potassium permanganate in 1 litre of distilled water. Of this, 1 cubic centimetre is equal to 0.00375 gramme of uric acid. The addition of 20 per cent. of sulphuric acid to the solution produces a temperature suitable to the reaction, and no thermometer need be used.

2. *In acid urine containing cystin* the author recommends the addition of a few drops of strong ammonia, which will obviate all difficulty.

3. *In alkaline urine with abundance of phosphates*, they should be filtered off after complete precipitation by heat. The ammoniacal urate comes down more rapidly in alkaline than acid urine. The only objection to adding ammonia in all cases is its tendency to precipitate the phosphates.

4. *Albuminous urine* does not interfere with accurate determination of uric acid by this method, but requires a little longer digestion with hydrochloric acid.

5. *With highly-pigmented urine* the original urate precipitate

is thoroughly digested with alcohol, and after acidification the filtrate is gradually heated to boiling, and then digested for some time on a water-bath. The separated crystals are then thoroughly washed.

In all cases exceedingly accurate results are claimed, and the process can be carried out with ease and comparative rapidity.

XANTHIN ($C_5H_4N_4O_2$).

This substance contains one atom of oxygen less than uric acid, hence its close alliance chemically with the latter. Xanthin is present in normal urine, although in very small quantity. Neubauer found only 1 gramme in 300 litres of normal human urine. When deposited in the urine spontaneously, it forms brittle scales of yellowish-white color and of somewhat waxy consistence. It is insoluble in alcohol and ether, feebly soluble in water, and readily soluble in alkalies, as well as in dilute nitric and hydrochloric acids. Xanthin is a constituent of a very rare form of calculus, having thus been met with not to exceed half a dozen recorded times, and these always in cases of youth. As a urinary sediment it has been encountered more frequently.

Although xanthin is widely distributed throughout the economy, having been found in most of the viscera as well as in the blood, the conditions which give rise to its increased and decreased excretion by the kidneys, as well as its occurrence as a urinary sediment, are as yet but little known. Durr and Strohmyer state that its amount in the urine is increased by sulphur-baths.

Detection.—When dissolved in dilute hydrochloric acid, upon evaporation xanthin separates into hexagonal crystals. When evaporated to dryness with nitric acid, a yellow residue remains, which turns red with potassium hydrate and reddish violet on being heated.

ALLANTOIN ($C_4H_6N_4O_3$).

This substance is obtainable from uric acid by oxidation with potassium permanganate (care being taken that the temperature does not rise), the potassium permanganate taking up water and oxygen, forming allantoin and carbonic acid :—

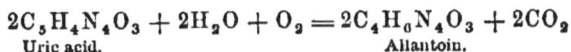

$$2C_5H_4N_4O_3 + 2H_2O + O_2 = 2C_4H_6N_4O_3 + 2CO_2$$

Uric acid.   Allantoin.

Allantoin crystallizes in colorless prisms, which are soluble in hot water, but slightly soluble in cold water, and insoluble in alcohol and ether. It is precipitated from its solutions by mercuric salts. Allantoin occurs in mere traces in normal human urine, except directly after birth, but it is increased by a meat diet, and by administration of tannic acid. It was found by Wohler in the urine of newborn calves, and since then by numerous observers in the urine of newborn infants. Its close chemical alliance with uric acid is shown by the fact that Salkowski found allantoin with urea and oxalic acid much increased in the urine of animals (dogs) by the administration of uric acid.

Detection and Determination.—Allantoin may be separated from the urine by precipitation with lead acetate, filtering, passing sulphuretted hydrogen through the filtrate, filtering again, evaporating the final filtrate to a syrupy consistence, and letting it stand for several days. Allantoin then crystallizes out.

CREATININ $(C_4H_7N_3O)$.

Creatinin and creatin are both constituents of normal urine, and are subject to interchange one into the other, according to certain conditions. In alkaline urine creatin appears in greater quantity, while the reverse is the case with strongly acid urine. Physiologically they may be considered as one body, although chemically they differ in the fact that creatin contains H_2O more than creatinin, as may be seen by their formulæ,—creatinin, $C_4H_7N_3O$; creatin, $C_4H_9N_3O_2$. Since the urine is usually acid, creatinin is generally considered the normal constituent.

Creatinin is a decided base—probably the strongest in the economy—giving a distinctly alkaline reaction with test paper. It crystallizes in large colorless prisms, soluble in water and alcohol, but almost insoluble in ether.

Creatinin is a constant constituent of normal human urine, being excreted in nearly the same amount as uric acid,—0.5 to

0.9 gramme in twenty-four hours (7 to 13 grains). It has been generally considered that creatinin of the urine arises from the creatin of the muscles, because when animals are fed on creatin the creatinin of the urine is increased. But Meissner has shown that when creatin is injected into the blood it appears in the urine as such unchanged; and it would therefore appear that the kidneys have not the power of converting creatin into creatinin; the change probably takes place normally in muscles, the creatinin entering the blood and being excreted by the kidneys. Bunge, however, points out the fact that the relatively small excretion of creatinin cannot account for the large amount of creatin of the muscles,—90 grammes. He therefore considers it more probable that creatinin is ultimately converted into urea, the creatinin of the urine—or creatin, if the urine be alkaline—being derived from the food.

Creatinin is excreted in increased amount upon a meat diet, and in diminished quantity by fasting. Clinically it is excreted in increased quantity in acute diseases,—as pneumonia, the efflorescent stages of typhoid and intermittent fever, and in some cases of

FIG. 6.—CREATININ CRYSTALS.
(After Kuhn.)

diabetes, though not in all. It is diminished in convalescence from acute diseases, in advanced degeneration of the kidneys, and tetanus. In diseases characterized by muscular wasting in general, creatinin is usually diminished.

Detection and Determination.—1. When a solution of sodium nitro-prussiate is added to a dilute solution of creatinin, and subsequently sodium hydroxid is added, a red color appears, which changes to yellow upon standing (Weyl's test). 2. A solution of creatinin, as in the urine, acidulated with nitric acid, gives, with phospho-molybdic acid, a yellow crystalline precipitate, soluble in hot nitric acid. 3. With zinc chloride it gives a characteristic crystalline precipitate (groups of fine needles) consisting of a combination of zinc chloride with creatinin. This test is used for quantitative estimation of creatinin.

The five bodies just considered in detail—viz., urea, uric acid, xanthin, allantoin, and creatinin—constitute the chief constituents of normal urine of the nitrogenous type, or protein derivation. In addition to these, however, thio- (sulpho-) cyanic acid is found in the urine, in very minute quantities, in the form of thiocyanates.

AROMATIC SUBSTANCES OF THE URINE.

These comprise four classes : (a) Hippuric acid and similar aromatic compounds of glycocin. (b) Combinations of glycuronic acid with aromatic substances. (c) Aromatic oxy-acids. (d) Ethereal sulphates.

HIPPURIC ACID $(C_9H_9NO_3)$.

Hippuric acid is monobasic and crystallizes either in the form of fine needles or four-sided prisms and pillars terminated by two- or four- sided beveled surfaces. The typical form is that of a vertical rhomboid prism. It is colorless, odorless, and of slightly bitter taste. It requires 600 parts of cold water to dissolve it, but it is much more soluble in hot water and readily so in hot alcohol and ether. It is soluble in ammonia and alcohol, but insoluble in hydrochloric acid. Hippuric acid is a constant element of normal urine, although present in small amount,—0.5 to 1 gramme being excreted by the kidneys in twenty-four hours.

Hippuric acid is an interesting product in a comparative physiological sense, since it forms a connecting link between the urine of herbivora, omnivora, and the carnivora; being present in comparatively large amount in the former, much less in the second, and absent in the last-named order. It is also interesting in itself, as forming one of the best illustrations of synthesis occurring in the organism.

In man the hippuric acid excreted by the kidneys depends chiefly upon the character and quantity of food eaten, being increased by a vegetable diet,—especially by certain fruits, as cranberries, prunes, green gages, etc. The administration of benzoic acid, oil of bitter almonds, toluol, cinnamic acid, benzylamin, phenylpropionic and kinic acids also cause an increased excretion of hippuric acid. On the other hand, an animal diet greatly decreases its excretion, although it does not disappear from the urine upon an exclusive meat diet. Clinically we find an increased excretion of hippuric acid in diabetes and chorea, as well as in acute febrile processes. Numerous observations go to show that hippuric acid is formed in the kidneys; synthesis, as a rule, failed after removal of these organs.

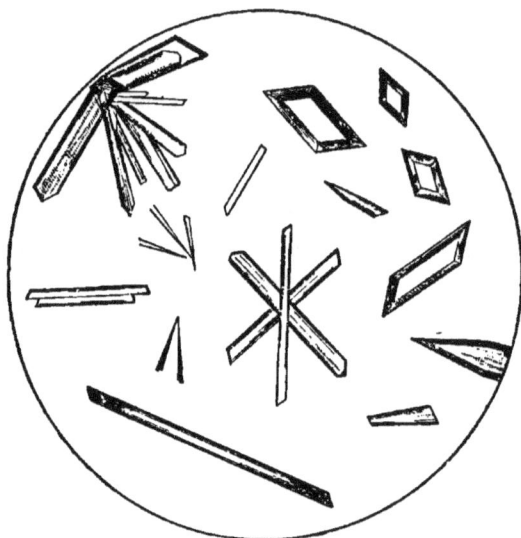

FIG. 7.—HIPPURIC-ACID CRYSTALS. (After Peyer.)

Detection and Determination.—1. Evaporate the urine with nitric acid, and heat the residue in a dry test-tube. If hippuric acid be present an odor like that of oil of bitter almonds is plainly observable, due to the formation of nitrobenzol. 2. If the urine contain an excess of hippuric acid, it may be determined by slightly evaporating and feebly acidulating with hydrochloric acid. Upon standing for a few hours the hippuric acid crystallizes out and may be recognized by the microscope. The hippuric-acid crystals may be readily separated from uric-acid crystals by hot water, if the uric acid be present in sufficient quantity to interfere with the test. 3. Meissner's method consists in precipitating carefully 1000 to 1200 cubic centimetres of urine with strong baryta-water; the excess of the latter is removed by a few drops of sulphuric acid (of which an excess is to be avoided) and then filtered. The filtrate, accurately neutralized with hydrochloric acid, is then evaporated on a water-bath to the consistency of a thick syrup, and the neutral residue, while still hot, is added to 150 or 200 cubic centimetres of absolute alcohol in a closed glass vessel. Any succinic-acid salts are precipitated together with the chloride of sodium, etc., while the hippuric-acid salts remain in solution. After repeated agitation, as soon as the precipitate has settled well the alcoholic solution is decanted and the alcohol completely driven off on the water-bath, the syrupy residue, which on cooling solidifies to a crystalline mass, is put in a closed vessel while it is still hot, acidulated with hydrochloric acid, and the hippuric acid extracted by shaking with not too small amounts of ether (100 or 150 cubic centimetres). After the ether is distilled off, the residue is diluted with water, and heated to boiling with a little milk of lime. The hippuric acid separates from the concentrated filtrate, after the addition of hydrochloric acid, in beautiful crystalline rosettes; they can be obtained entirely free from color by treating with pure animal charcoal.

Combinations of Glycuronic Acid.—This body is closely allied to the carbohydrates, and is apt to be mistaken for sugar in the urine. Normally it occurs in mere traces in urine. It occurs partly in combination with aromatic substances.

Aromatic Oxy-acids.—Of these hydroparacumaric acid or oxyphenylpropionic acid and oxyphenylacetic acid are found in minute quantities in the urine, in combination with potassium.

ETHEREAL SULPHATES.

In 1851 Stadeler discovered, on distilling the urine with dilute sulphuric acid, small quantities of phenol or carbolic acid in the distillate. Hoppe-Seyler showed that phenol is not present in the urine in a free state, but as a compound, from which it is liberated by sulphuric acid in distillation. In 1876 Baumann discovered that this compound consists of an ethereal compound of phenol and sulphuric acid. Baumann also discovered the presence of other like ethereal sulphates in the urine, consisting of compounds of the radicle HSO_3, sometimes incorrectly termed sulphonates. The most important of this series are the ethereal potassium sulphates of phenol, creosol, catechol or pyrocatechin, indole, and skatole. These compounds are present in the urine of herbivora more abundantly than in carnivora or omnivora; but they are present in the urine of all animals in smaller or larger quantities. They originate in two ways: (a) from aromatic substances in food, and (b) in the intestine as the result of putrefactive changes. In man, whose food contains but little aromatics, the origin of these ethereal sulphates is probably chiefly, if not entirely, in the intestines, as above indicated; since, if putrefaction in the intestines be arrested, these substances disappear from the urine. The proportion of the ethereal sulphates to the total sulphates of the urine is about 1 to 10.

Hoppe-Seyler found the ethereal sulphates in the urine in excess under the following circumstances : 1. In deficient absorption of the normal products of digestion, such as occurs in peritonitis and tuberculosis of the intestine, because the products of digestion undergo putrefactive changes, and the putrefactive products are absorbed. 2. Diseases of the stomach, in which the food long remains in the stomach and undergoes fermentative changes, always result in excess of ethereal sulphates in the urine. 3. Putrefactive processes outside the alimentary canal, putrid cystitis, putrid abscesses, putrid peritonitis, etc., have the same effect. The amount of ethereal sulphates is, moreover, in all cases proportional to the severity of the putrefaction, and is increased by the retention and diminished by the discharge of putrid matter ; as, for instance, in opening an abscess. By these and other observations it has been conclusively established that the best guide to the occurrence and amount of putrefaction in progress in the economy is the relation of the ethereal sulphates to the total sulphates.

PHENOL-POTASSIUM SULPHATE ($C_6H_5OSO_3K$).

The above is the form in which phenol or carbolic acid exists in the urine. Phenol (C_6H_6O) was found to be one of the intestinal products of putrefaction by Baumann, as already stated.

This is absorbed and excreted in the above form; some of the sulphate comes from tyrosine, which passes through the stages of paracreosol and paraoxybenzoic acid before conversion into the phenol salt (Baumann).

After the use of carbolic acid, externally or internally, the amount of phenol sulphate in the urine is increased; two substances are formed by splitting up of carbolic acid, called pyrocatechin and hydroquinon. These become, in alkaline urine, dark brown upon exposure to the atmosphere; producing the well-known color of the urine in "carboluria."

Detection.—1. Distil the urine acidulated with sulphuric acid. Phenol appears in the distillate; add bromine-water, a precipitate of tribromophenol appears,—deep yellow. 2. Warm the distillate with Millon's reagent and a cherry-red color results. 3. Add ferric chloride to the distillate and a deep-violet color results.

Determination.—Approximate results may be obtained as follows: 100 cubic centimetres of urine are concentrated over a water-bath to 20 cubic centimetres volume. Sulphuric acid is added in such quantity as to represent 5 per cent. of the mixture. Distil until the distillate is no longer rendered cloudy by addition of bromine-water. The distillate is filtered, if necessary, and colored a permanent light yellow with bromine-water. The mixture is allowed to remain two or three days at a moderate temperature. A precipitate of tribromophenol ($C_6H_2Br_3OH$) forms and is collected on a weighed filter, washed with water, and dried in an exsiccator over sulphuric acid to constant weight; 100 parts of tribromophenol correspond to 28.4 parts phenol.

INDOXYL-POTASSIUM SULPHATE ($C_8H_6NO.SO_3K$).

(INDICAN?)

Indole (C_8H_7N), the basis of the above substance, is formed in the intestines; indoxyl, as a radicle thereof, has the formula C_8H_6NO, which, united with SO_3K, forms the indoxyl-sulphate of potassium of the urine. It occurs in white tablets and plates, soluble in water, but sparingly soluble in alcohol. By oxidation indigo blue is formed therefrom. Indoxyl-potassium sulphate has received the erroneous name of indican, under the mistaken belief that it is identical with vegetable indican. But the latter substance is a glucoside, and only resembles the former in the fact that one of its decompositional products is indigo blue. The blue, green, and some red urines met with in disease probably owe their colors to this salt in different stages of oxidation. It is excreted in excess on an exclusive meat diet, or the ingestion of indole. Clinically an increased excretion of this substance by the kidneys is observed in Addison's disease, cholera, carcinoma of the liver, chronic phthisis, central and peripheral diseases of the nervous system, typhoid fever, dysentery, and the stage of reaction in cholera. In obstructive diseases of the small intestine its excretion is enormously increased. In general, the appearance of large quantities of this substance in the urine implies that abundant albuminous putrefaction is progressing in some part of the system. Sometimes, as urine is undergoing decomposition changes, a bluish-red pellicle, consisting of microscopic crystals of indigo blue and red, is seen,—due to decomposition of the indoxyl-sulphate.

Detection.—1. *McMunn's method:* Equal parts of urine and hydrochloric acid with a few drops of nitric acid are boiled together, cooled, and agitated with chloroform. The chloroform is colored violet, and shows an absorption band, before D, due to indigo blue, and another after D, due to indigo red.

2. *Jaffé's method:* Mix 10 cubic centimetres of strong hydrochloric acid with an equal volume of urine in the test-tube, and while shaking add drop by drop a perfectly-fresh saturated solution of chloride of lime, or chlorine-water, until the deepest attainable blue color is reached. The mixture should next be agitated with chloroform, which readily takes up the indigo and holds it in solution, and the quantity present may be approximately estimated according to the depth of the color.

If the urine contain albumin, it should be removed before applying these tests, otherwise the blue color often arising from the mixture of hydrochloric acid and albumin after standing may prove misleading.

3. Pour 4 cubic centimetres of hydrochloric acid into a small flask, and while stirring add from 10 to 20 drops of urine. If the proportion of indigo be about normal, the resulting color

will be rather light yellow; if in excess, the acid will turn violet or blue,—the more intense will be the color in proportion to the quantity present. If no coloration appear after waiting a minute or two, there is no excess, however deep a color may subsequently appear.

In addition to those considered, the urine contains three other ethereal sulphates,—viz., creosol-potassium sulphate, catechol-potassium sulphate, and skatoxyl-potassium sulphate. These for the most part are present in minute amounts, and possess the same significance as those considered.

URINARY PIGMENTS.

NORMAL UROBILIN.

Urobilin is the chief coloring agent of normal urine, although it is not the exclusive one, as shown by spectroscopic analysis. Normal urobilin is a dark-brown, amorphous, rather resinous substance, sparingly soluble in water, but readily dissolved by alcohol, ether, chloroform, acids, and acidulated water, as well as by solutions of sodium, potassium, or ammonium hydroxid.

The origin of urobilin is still the subject of some difference of opinion. Formerly, it was believed that bilirubin, entering the intestines with the bile, was acted upon by nascent hydrogen resulting from putrefaction, and as a result it constituted a reduction product, which Maly considered identical with hydrobilirubin. It was also supposed that the pigment of the fæces was partly absorbed, carried to the kidneys, and there excreted. Hydrobilirubin, stercobilin, and urobilin were considered identical, but spectroscopic examinations prove them to be different.

McMunn regards the formation of urobilin as the result of oxidation processes by means of nascent oxygen, either in the intestines or elsewhere, rather than due to reduction processes. He bases this view principally on the fact that, by the action of hydrogen peroxide on acid hæmatin, he is able to prepare an artificial product, which shows the same spectroscopic appearances as normal urobilin. The question, then, whether stercobilin and urobilin are products of reduction or oxidation is still unsettled. It is important to remember, however, that urobilin may originate either from bile-pigment or from blood-pigment.

Certain facts point to the inference that stercobilin and uro-bilin originate—at least, to a limited extent—independently. Thus: 1. In animals with biliary fistula, while no bile enters the intestine, yet the urine still contains urobilin. 2. In Copeman and Winston's case of biliary fistula, in which no bile had entered the intestine, the fæces were uncolored by stercobilin, yet the urine still contained urobilin. 3. Some cases recorded by Mott would seem to indicate that the formation of urobilin is seated in the liver. When destruction of the red corpuscles is excessive in the portal circulation, the liver contains an excess of iron, and the ferric residue of hæmoglobin occurs as urobilin in the urine abnormally abundant.

Urobilin exists in normal urine in small amount, the quantity being much increased in acute fevers,—four to five or more times. Typhoid and septic fevers, which cause rapid destruction of the blood-corpuscles, markedly increase the excretion of urobilin. On the other hand, there is diminished excretion of urobilin in-conditions associated with diminished metamorphosis of red blood-corpuscles, as chlorosis and anæmia, in convalescence from acute diseases, as well as in hysteria and nervous diseases. It is an interesting and highly-important clinical fact that increased excretion of urobilin has been observed in intra-cranial hæmor-rhages, hæmorrhagic infarctions, retro-uterine hæmatocele, and extra-uterine pregnancy.

According to the observations of Lawson, the excretion of urinary pigment is much greater in tropical than in temperate climates. Thus, assuming the normal unit to be 4.8 in the aver-age adult in temperate climates, he found in the tropics that it rose to 12 or 14. In pneumonia it has been observed to rise to 16 and 20; in acute rheumatism, from 30 to 32 at the height of the disease; in typhoid fever, from 80 to 100; and in a man who had inhaled arsenated hydrogen, from 600 to 800.

Detection.—Upon rendering the urine alkaline by the addition of ammonia, and, after filtering, adding some chloride-of-zinc solution to the filtrate, a beautiful green fluorescence may be observed by reflected light if urobilin be present. The above solution furthermore exhibits, with the spectroscope, a dark absorption-band between Fraunhofer's lines b and F.

Uroerythryn.

By some this is considered identical with skatole-pigment, but McMunn claims for it certain characteristic reactions. It is chiefly this pigment which colors the urates a reddish tint, and it may be extracted therefrom by boiling alcohol. This solution gives two ill-defined bands before F in Fraunhofer's scale. In the solid state it becomes green with sodium or potassium hydroxid. The origin of uroerythryn in the economy, as well as its relationship to urobilin, have not been determined.

Urochrom.

This name was first applied by Thudicum to the body which he considered the chief coloring agent in the urine. Urochrom is thought by some to consist of impure urobilin. Certain it is, at least, that urochrom contains much urobilin. Urochrom occurs in yellow scales, which partly dissolve in water, less soluble still in alcohol, but soluble in ether and dilute mineral acids and alkalies. The aqueous solution becomes dark on standing, finally changing to a red color, becoming turbid, and depositing resinous flocculi. Urochrom is precipitated from its aqueous solution by nitrate of silver as a gelatinous mass soluble in nitric acid ; acetate of lead gives a white flocculent precipitate. Mercuric nitrate gives a white precipitate, which, on boiling, becomes a pale flesh-color, the supernatant fluid appearing rose-red. By oxidation there is first formed from urochrom a red substance, which corresponds to uroerythrin, and to which the red urine of disease often owes its color.

Other Organic Constituents of Urine.

Oxalic Acid $(C_2H_2O_4)$.—This substance never occurs in the urine in a free state, but always in combination with calcium (oxalate of lime), which is held in solution by the acid sodium phosphate of the urine. It is sometimes absent from the normal urine, though usually present in quantity of about 0.1 gramme for twenty-four hours. When present in excessive quantities in the urine it is precipitated in the form of oxalate-of-calcium crystals. These are distinguished by their form,—quadratic octahedra, with a short principal axis,—often termed "envelope crystals." Occasionally dumb-bell forms are seen. The origin of oxalic acid in the organism is undetermined ; but a close relationship evidently exists between it and uric acid.

4

Succinic Acid $(C_4H_6O_4)$ is the third of the series of acids of which oxalic acid is the primary. Succinic acid has occasionally been met with in the urine, especially after the ingestion of asparagus and asparagin.

Lactic Acid $(C_3H_6O_3)$.—It is doubtful if lactic acid be present in normal urine, but it has been found combined with bases after physiological disturbances and severe muscular labor. In such cases it occurs as sarcolactic acid. It has been found in trichinosis, acute yellow atrophy of the liver, cirrhosis of the liver, diabetes, phosphorus poisoning, rickets, leucocythæmia, osteomalacia, and in animals after extirpation of the liver.

Fatty Acids.—These are present in normal urine, though in mere traces,—0.008 gramme per day. They consist of formic, acetic, butyric, and propionic acids. The excretion of fatty acids by the kidneys is increased during the process of ammoniacal fermentation. Certain febrile diseases cause an increase of these bodies in the urine to 0.06 gramme, and in certain diseases of the liver the increase reaches from 0.6 to 1 gramme per day. These acids apparently exist in the urine in a free state.

Glycero-phosphoric Acid $(C_3H_9PO_6)$ occurs in faint traces in normal urine, about 15 milligrammes per litre. It is said to be increased in nervous disorders and after chloroform narcosis.

CARBOHYDRATES.—Much discussion has taken place over the question as to whether grape-sugar is a constituent of normal urine. One difficulty in determining this point is the fact that the urine contains a number of substances which reduce alkaline solutions of cupric oxide the chief of these being uric acid, hippuric acid, pyrocatechin, glycuronic acid, and creatinin. Abeles showed, however, that none of these substances undergo alcoholic fermentation with yeast, while this does occur with the reducing substance of normal urine. Wedenski, by shaking a large quantity of normal urine with benzoic chloride, obtained insoluble benzoyl compounds of carbohydrates, which give all the reactions of grape-sugar. It may therefore be considered as conclusive, as Brucke first affirmed, that minute quantities of sugar exist in normal urine.

Animal Gum.—This constitutes a product of mucin, and was first found in the urine by Landwher ; also in saliva, the synovia, in colloid cysts, in chondrin, and in connective tissue. It forms an opalescent solution with water ; gives a sticky precipitate with copper sulphate, and also with ferric chloride. It is precipitated by alcohol, and does not reduce alkaline solutions of cupric salts. It yields oxalic acid upon treatment with nitric acid, and lævulic acid after treatment with hydrochloric acid, as does vegetable gum.

Milk-sugar occurs in variable quantities in the urine of nursing mothers and lesions of the mammary glands. Hofmeister precipitated urine with lead acetate and ammonia, filtered, shook the filtrate with silver oxide, filtered, decomposed the filtrate with sulphuretted hydrogen to get rid of the silver, filtered ; to the final filtrate barium carbonate was added, and the mixture evaporated to dryness. Alcohol removed milk-sugar from the residue, and characteristic crystals thereof were obtained by evaporating off the alcohol. Kaltenbach proved that this was milk-sugar by further obtaining therefrom galactose and mucic acid.

Inosite—This substance has been found in normal urine in small quantities by numerous observers. An increased excretion has been noted in Bright's disease and in diabetes. It may be detected as follows : Several litres of urine, feebly acidified, are completely precipitated with acetate of lead and filtered. The filtrate is warmed and precipitated with basic lead acetate. After standing forty-eight hours the precipitate is collected, washed, suspended in water, and treated with a stream of sulphuretted hydrogen ; the lead sulphide is filtered off; uric acid sepa-rates from the filtrate after a few hours ; this is also filtered off. The solution is then evaporated to a syrupy consistence on a water-bath, and absolute alcohol added. The precipitate is dissolved in hot water, and three or four times the volume of 90-per-cent. alcohol added. Ether is cautiously added till a permanent cloud appears ; the inosite crystallizes out, and may be collected. It will then give its characteristic tests, as follows: 1. With a few drops of nitric acid on a platinum dish, treated with ammonia and calcium chloride, after evaporating to dryness a bright red or violet color appears. 2. Add a little mercuric nitrate to a solution of inosite, on a porcelain capsule, a yellow precipitate results. On gently heating, this becomes red ; on cooling, the color vanishes. Proteids, tyrosin, and sugar must be absent.

FERMENTS.—*Pepsin* has been isolated from normal urine by Brucke, Sahli, and others. The morning urine is richest in pepsin. The pepsin of urine forms peptone and all intermediate proteoses from fibrin, the same as does pepsin. *Detection :* Small pieces of fibrin are first soaked in urine until pepsin is absorbed therefrom. If then removed to 0 1 per cent. hydrochloric-acid solution they are rapidly digested. Control ex-periments, with fibrin not soaked in urine, give negative results.

Diastase.—Holvotschiner has obtained minute quantities of ptyalin or a similar diastatic ferment from urine.

Rennet.—Helwes and Holvotschiner have both obtained from urine traces of a ferment which curdles milk.

Trypsin. — This ferment is doubtless absent from normal urine. Sahli alone, of all investigators, claims to have found it; but his results are considered as due to non-prevention of putrefaction in his experiments.

MUCIN.—This is the chief constituent of the mucus of normal urine, derived from the muciparous glands of the urinary passages, but not from the kidneys. In normal urine, mucin occurs in small amount ; but in catarrhal diseases of the urinary tract it is increased, often in abundance. In appearance it is viscid, slimy, and tenacious.

Detection : 1. Mucin is precipitated from its solutions by vegetable acids, the precipitate being insoluble in excess of the acid. 2. Add to one volume of urine three volumes of strong alcohol and let stand for several hours. Filter and wash the precipitate with alcohol, and treat with warm water. The filtrate containing the mucin is then acidified with acetic acid, and if turbidity results it is due to mucin.

INORGANIC CONSTITUENTS.

The inorganic constituents of the urine comprise chiefly the chlorides, carbonates, sulphates, and phosphates. These occur in combination with potassium, ammonium, calcium, and magnesium. Small quantities of fluorine, silicic acid, and iron are also present; and free gases, including carbonic acid, nitrogen, and traces of oxygen. The total amount of these salts in the urine varies from 9 to 25 grammes per day. They are derived from (*a*) the food ingested and (*b*) from the metabolic processes or tissue-changes; the latter more especially with regard to the sulphates and phosphates. The salts of the blood and those of the urine are very similar, save that the blood contains but traces of sulphates, while the urine is rich in these salts.

The sulphates are derived chiefly from the changes occurring in the proteids of the body; the nitrogenous elements of the proteids being excreted as uric acid and urea, while the sulphur becomes oxidized, forming sulphuric acid, which appears in the urine mostly in combination with metallic bases, but also, to some extent, in ethereal combinations with organic radicles, making up the ethereal sulphates just considered.

CHLORIDES.

The chlorides of the urine consist chiefly of chloride of sodium with a small amount of chloride of potassium and ammonium. The amount of chlorides excreted by the kidneys in the healthy adult averages from about 10 to 16 grammes in twenty-four hours,—6 to 10 grammes of chlorine. The chlorides, therefore, next to urea, constitute the principal solid constituent of the urine. Chlorine is very widely distributed throughout the organism, nearly always in combination with sodium, potassium, ammonium, and magnesium. As chloride of sodium it is excreted with the perspiration, saliva, bile, fæces, and urine.

In health the excretion of the chlorides by the kidneys varies in amount according to the quality and quantity of food ingested. Thus, upon a liberal salt diet they are largely increased; the output representing very closely the amount of chloride of sodium taken in. With active exercise there is also increased excretion of chlorides by the kidneys, while, on the other hand, during repose of the body they are diminished.

Clinically the excretion of chlorides with the urine is diminished in all acute febrile conditions, and especially when attended by serous exudations. As a general rule, in such cases there is a steady decrease until the crisis of the disease is reached, after which they gradually increase. A continued increase of chlorides in the urine in febrile states may, therefore, be accepted as an evidence of improvement. In pneumonia the chlorides may disappear from the urine, and their absence, under such circumstances, always indicates a serious condition of the patient. In chronic conditions associated with impaired digestion and dropsy the chlorides are diminished. In the latter case much of the chlorides becomes stored up in the dropsical fluid in the serous cavities and cellular tissues. The chlorides are excreted by the kidneys in excess in diabetes insipidus and in the declining stages of dropsy after the establishment of diuresis.

Detection.—1. The urine is treated with nitric acid and a solution of nitrate of silver added. A caseous precipitate soluble in ammonia, insoluble in nitric acid, shows the presence of chlorides.

2. The above test may be made available for approximative estimation as follows : A standard solution of nitrate of silver, 1 to 8 (1 drachm to the ounce), is first prepared. Take a glass half full of urine, and add a few drops of nitric acid ; then add one or two drops of the standard solution of silver nitrate, and note the changes. If a white, flaky precipitate occur, quickly sinking to the bottom of the glass without diffusing through the urine, the chlorides are undiminished. If a simple cloudiness appear, readily diffusing throughout the urine without the appearance of curdy flakes, the chlorides are diminished to 0.1 per cent., the normal being 0.5 to 1 per cent. Should no precipitate whatever occur, the chlorides are absent. This method,

proposed by Ultzmann, is rather obsolete, in view of the rapid and accurate results by advanced centrifugal methods.

Determination.—*Mohr's Method:* Prepare the following solutions: 1. Standard nitrate-of-silver solution: Dissolve 29.075 grammes of fused nitrate of silver in 1000 cubic centimetres (1 litre) of distilled water; 1 cubic centimetre of this solution is equal to 0.01 gramme of sodium chloride. 2. A saturated aqueous solution of neutral potassium chromate.

Process.—Take 10 cubic centimetres of urine; dilute with 100 cubic centimetres of distilled water; add to this a few drops of the potassium-chromate solution. Drop into this mixture from a burette the standard nitrate-of-silver solution; the chlorine combines with the silver to form silver chloride in the form of white precipitate. When all the chlorides are thus precipitated, silver chromate (red) appears, though not while any chloride remains in solution. The silver nitrate must, therefore, be added until a pink tinge appears. Read off the quantity of standard solution of silver used, and calculate therefrom the quantity of chlorides in the 10 cubic centimetres of urine tested, and from this the percentage.

Corrections.—1. A high-colored urine may cause difficulty in determining the appearance of the pink tinge. This may be obviated by diluting the urine to a normal color. 2. One cubic centimetre should be subtracted from the total number of cubic centimetres of the silver-nitrate solution employed, as the urine contains small quantities of certain compounds more easily precipitable than the chromate. 3. To obviate such errors the following modification of the test as advised by Sutton is employed: Ten cubic centimetres of urine are measured into a thin porcelain dish and 1 gramme of pure ammonium nitrate added; the whole is then evaporated to dryness, and gradually heated over a small spirit-lamp to low redness till all vapors are dissipated and the residue becomes white. It is then dissolved in a small quantity of water, and the carbonates produced by combustion of the organic matter neutralized by dilute acetic acid; a few grains of pure carbonate of lime are added to remove all free acid, and one or two drops of potassium chromate. The mixture is then titrated with deci-normal silver solution (16.966 grammes silver nitrate per litre) until the pink color appears. Each cubic centimetre of silver solution represents 0.005837 gramme of salt; consequently, if 12.5 cubic centimetres have been used, the weight of salt in 10 cubic centimetres of urine is 0.07296 gramme, or 0.7296 per cent. If 5.9 cubic centimetres of urine are taken for titration, the number of cubic centimetres of silver solution used will represent the number of parts of salt per 1000 parts of urine.

Liebig's method consists in estimating the chlorine with mercuric nitrate, but it sometimes fails unless very accurately manipulated, and, moreover, it often gives erroneous results, and therefore it is only referred to here.

Volhard and Falch Method.—This method depends upon the action of soluble sulphocyanides with solutions of silver and ferric salts. Soluble sulphocyanides produce in silver solutions a white precipitate similar to silver chloride, which is insoluble in dilute nitric acid. A like precipitate of sulphocyanide of silver with a solution of nitrate of silver is given by the blood-red solution of sulphocyanide of iron, and the color of the latter at last completely disappears. If, therefore, a solution of sulphocyanide of potassium be added to an acid solution of nitrate of silver, to which a little ferric sulphate has been added, every drop of the sulphocyanide solution at first produces a blood-red cloud, which, however, quickly disappears again on stirring, while the fluid becomes milk-white. It is not until all the silver is precipitated that the red color of the sulphocyanide of iron remains permanent, and the end of the process is reached. The reaction is one of great delicacy; so that it is equally sensitive as Mohr's reaction, while it has the advantage over the latter that titration can be conducted in acid urine.

Solutions Required.—1. *Standard silver solution* is made by dissolving 29.075 grammes of pure silver nitrate in water and diluting to 1 litre. Each cubic centimetre corresponds to 10 milligrammes of chloride of sodium or 6.065 milligrammes of chlorine.

2. *Solution of Ferric Oxide.*—A cold saturated solution of crystallized ferric alum, free from chlorine, or a solution of ferric sulphate, which contains 50 grammes of iron oxide to the litre.

3. *Standard Sulphocyanide-of-Potassium Solution.*—Since sulphocyanide of potassium cannot be readily weighed with accuracy, 10 grammes are dissolved in a litre of water, and this solution is standardized by silver solution. Thus 10 cubic centimetres of the silver solution are measured off, 5 cubic centimetres of the iron solution are added, and then pure nitric acid is added drop by drop until the mixture is colorless. If the sulphocyanide-of-potassium solution be then allowed to flow into it from a burette, each drop at first gives a blood-red color, which at once disappears on stirring. When all of the silver is precipitated as sulphocyanide of silver, the next drop of the sulphocyanide-of-potassium solution gives a permanent red color to the fluid, indicating the end of the reaction. If, for example, to 10 cubic centimetres of the silver solution 9.6 cubic centimetres of the sulphocyanide-of-potassium solution have been used before the red color is permanent, 960 cubic centimetres are measured off, and this is diluted with 40 cubic centimetres of water to make a litre. Both solutions must now be equivalent, which is to be determined by titration.

Analysis.—Five or 10 cubic centimetres of the urine, after the addition of 1 or 2 grammes of nitrate of potassium free from chlorine, are evaporated to dryness on a water-bath. The residue is then heated over a free flame, at first gently, afterward strongly, and a white, fused residue remains. Since the nitrous acid formed in this process prevents the end reaction, the fused saline mass is dissolved in water, acidulated with nitric acid, and then the chlorine is precipitated with an excess of the standard silver solution. After this mixture has been warmed on a water-bath for a time to completely remove the nitrous acid, it is allowed to cool; 5 cubic centimetres of the iron solution are added, and then the sulphocyanide-of-potassium solution, equivalent in strength to the silver solution, is added while constantly stirring until the excess of the silver added is precipitated, which is known by the permanent red color of the mixture. The difference, then, between the number of cubic centimetres of the silver and sulphocyanide solutions corresponds to the chlorine contained in the urine. If, for instance, at first 12 cubic centimetres of the silver solution were added to 10 cubic centimetres of urine, and 4 cubic centimetres of the sulphocyanide solution were required to titrate back the excess, the amount of chlorine in the urine corresponded to $12 - 4 = 8$ cubic centimetres of the silver solution $= 8.0$ grammes of sodium chloride or 4.852 grammes of chlorine in the litre of urine.

PHOSPHATES.

The quantity of phosphoric acid excreted by the kidneys in the healthy adult ranges from 2.3 to 3.5 grammes in twenty-four hours, the average being about 2.8 grammes. Ordinary phosphoric acid (H_3PO_4) is tribasic, containing 3 atoms of hydrogen, which may be replaced by a metal. In the urine phosphoric acid occurs in combination, in part with the alkaline earths—*earthy phosphates*, and in part with the alkalies—*alkaline phosphates.* *The earthy phosphates* are insoluble in water, but soluble in acids, and consist of phosphates of calcium and magnesium. The calcium phosphates are most abundant ; those of potassium scanty. They exist in the urine in quantity of about 1 to 1.5 grammes in twenty-four hours' excretion,—the relative proportion being calcium phosphate about 33, and magnesium phosphate about 67. In acid urine the earthy phosphates are in solution, while in alkaline urine they are precipitated and form a sediment more marked if heat be applied. If, therefore, an alkaline urine is heated in a test-tube, a precipitate forms, which may be mistaken for albumin. It may, however, be distinguished from the albuminous reaction by slight acidification, which readily dissolves the earthy phosphates. This is a frequent source of error in testing for albumin in the urine by means of heat. If the acid magnesium phosphate be acted upon by ammonia the ammonio-magnesium phosphate is formed,—*triple phosphate.* This appears in the urine as fern-leaf or snow-flake crystals, or, after long standing, in the form of prismatic or "coffin-lid" shaped crystals. If urine decompose in the bladder through retention and consequent fermentative changes, ammonia is liberated from the urea, and the free ammonia unites with the acid magnesium phosphates to form the triple-phosphatic crystals so characteristic of chronic cystitis.

The alkaline phosphates of the urine consist of the acid

phosphate of sodium and the phosphate of potassium, that of sodium being most abundant and the potassium phosphate scanty. These (unlike the earthy phosphates) are readily soluble in water and alkaline fluids. The alkaline phosphates form the chief bulk of the phosphates of the urine, ranging from 2.0 to 4.0 grammes in twenty-four hours. As already noted, the acidity of normal urine depends upon its contained acid sodium phosphate, and not upon the presence of a free acid.

In the blood the alkaline phosphates exist as neutral sodium and potassium phosphates ; but, as Ralfe has shown,[1] these are changed into acid salts through a decomposition effected by the act of secretion, in which the bicarbonates and neutral phosphates in the blood are changed into carbonates and acid phosphates, respectively. The acid salts, in obedience to the law of diffusion, pass out in the urine, whilst the carbonates remain in the circulation. The excretion of the alkaline phosphates by the kidneys varies in health according to the quality of food ingested, being in excess upon an animal diet.

The phosphoric acid of the urine is derived in part from the food and in part from the decomposition of lecithin and nuclein. The excretion of phosphoric acid by the kidneys varies with the amount of food taken ; after the midday meal, especially if much meat be eaten, it rises, and reaches its maximum in the evening ; it falls during the night, reaching its minimum about midday.

Clinically, the excretion of phosphoric acid by the kidneys is diminished in gout, in most acute diseases, in kidney lesions, in the intervals of intermittent fever, and during pregnancy. The author regards the diminution of phosphoric-acid excretion by the kidneys almost as constant a feature of the urine in Bright's disease and allied lesions of the kidneys as the presence in the urine of albumin.

An increased excretion of earthy phosphates accompanies diffuse diseases of the bones,—osteomalacia, rickets, etc. ; also in diffuse periostitis and diseases of the nerve-centres.

A condition of so-called *phosphatic diabetes* has been described by Tessier and confirmed by Ralfe and others, in which there is a continued excessive excretion of phosphates by the

---

[1] Lancet, July, 1874.

kidneys, attended by symptoms somewhat like diabetes,—loss of flesh; aching pains in the lower back and pelvic region; dry, harsh skin; tendency to boils, with morbid appetite, etc.

Detection.—1. The *earthy phosphates* may be detected by rendering the urine strongly alkaline with sodium or potassium hydrate, or ammonia, and gently heating, which cause their precipitation in the form of a whitish cloud, that shortly after settles to the bottom of the test-glass. The precipitate is dissolved upon the addition of acetic acid.

2. Ultzmann suggests an approximative method with this test, as follows: A test-tube 2 centimetres (0.787 inch) wide is filled with the urine to the depth of 5.33 centimetres (2.997 inches), to which are added a few drops of strong ammonia or potassium-hydrate solution, and warmed over a spirit-lamp until the earthy phosphates separate. After standing fifteen minutes the depth of the sediment is measured. If the layer be 1 centimetre (0.3937 inch) high, the earthy phosphates are present in normal amount; if 2 or 3 centimetres in depth they are increased; but if the sediment be only a line or so in depth, they are diminished.

The Alkaline Phosphates.—1. First remove the earthy phosphates by precipitation with potassium hydrate or ammonia and filter them off. Next add to the urine one-third of its volume of magnesium fluid.[1] The alkaline phosphates are all precipitated in the form of a snowy deposit.

2. The above test may be made useful for approximative estimation, as Ultzmann suggests, in the following manner: To 10 cubic centimetres of the urine add a third part (say, 3 cubic centimetres) of magnesium mixture. There is formed a precipitate of crystalline ammonium-magnesium phosphate, with which comes down an amorphous mass of calcium phosphates. If there ensue through the entire fluid a milky turbidity, the alkaline phosphates are in normal amount; if we have a copious precipitate, which gives the fluid the appearance of cream, then there is great increase; if the fluid remain transparent, or only slight turbidity ensue, we have a decrease of the alkaline phosphates.

---

[1] *Magnesium fluid* contains, of magnesium sulphate and ammonium chloride, each 1 part; distilled water, 8 parts; and pure ammonia-water, 1 part.

58 ANALYSIS OF URINE.

Determination of Total Phosphoric Acid.—The following solutions are necessary :—

1. A standard solution of uranium nitrate is prepared, consisting of 20.3 grammes pure uranic oxide in 1000 cubic centimetres of distilled water; 1 cubic centimetre corresponds to 5 milligrammes of phosphoric acid.

2. Sodium-acetate solution : 100 grammes of sodium acetate are dissolved in 900 cubic centimetres of distilled water, and to this 100 cubic centimetres of acetic acid are added.

3. Saturated solution of potassium ferrocyanide.

Analysis.—Fifty cubic centimetres of the urine are poured into a beaker, and 5 cubic centimetres of sodium-acetate solution are added. The mixture is warmed over a water-bath and the uranium solution added, drop by drop, as long as a precipitate falls. If this be not easily recognized, the mixture should be stirred, a drop placed upon a porcelain plate, and a drop of potassium ferrocyanide added. If a reddish-brown color do not appear at the line of contact, the addition of uranium solution should be continued, the beaker-glass being again warmed. The limit of reaction occurs when all the phosphoric acid has been precipitated by the uranium solution. After this is reached, the next drop of uranium solution, finding no phosphoric acid, forms a reddish-brown precipitate with potassium-ferrocyanide solution. The quantity of uranium solution used is next read off, each cubic centimetre being equal to 5 milligrammes phosphoric acid, from which can be readily calculated the percentage amount in the urine.

Estimation of Phosphoric Acid Combined with Calcium and Magnesium.—To determine the phosphoric acid in combination with the alkaline earths, 200 cubic centimetres of urine are precipitated with ammonia, collected after twelve hours on a filter, and washed with ammonia-water (1 to 3). The filter is then pierced at the point, and the precipitate washed down with a stream of water into a beaker, and dissolved while warm with as little acetic acid as possible. Add 5 cubic centimetres of sodium-acetate solution, dilute to 50 cubic centimetres, and proceed as in preceding analysis. The difference between the total amount of phosphoric acid and that in combination with the alkaline earths represents the quantity combined with the alkalies,—alkaline phosphates.

SULPHATES.

The sulphates in the urine are of two kinds (1) ordinary neutral sulphates of sodium and potassium, and (2) the ethereal sulphates. Since the sodium salts predominate in the economy, the sulphate of sodium occurs in the urine in greater quantity than the potassium sulphate. The quantity of sulphates excreted by the kidneys in the healthy adult varies from 1.5 to 3 grammes per day. The sulphates, being extremely soluble compounds, are, therefore, never met with in the urine in the form of deposits.

An increased excretion of sulphates by the kidneys occurs after the ingestion of sulphuric acid or its salts, upon active exercise, upon an exclusive meat diet, and inhalations of oxygen-gas. Clinically an increase occurs in acute fevers, with increased urea secretion. The most marked increase is noted in meningitis, encephalitis, and rheumatism. In general, it may be accepted that the excretion of urinary sulphates runs parallel in quantity to that of urea.

Detection.—1. To 10 cubic centimetres of urine add a few drops of hydrochloric acid, then add about 3 cubic centimetres of barium-chloride solution ; a white, milky precipitate is formed of sulphate of barium.

2. More simply still, to 10 cubic centimetres of urine in a test-tube add one-third the volume of acidulated barium-chloride solution,[1] when a white, milky precipitate at once appears in the presence of sulphates.

3. Approximate estimation may be made by the above as follows : If a simple, opaque, milky turbidity result, the sulphates are present in about the normal amount; if more opaque, possessing the appearance of cream, the sulphates are excessive; if but a light translucent cloudiness result, the sulphates are diminished.

Determination.—(a) The volumetric method is conducted by adding to a given volume of urine a standard solution of barium chloride as long as precipitation occurs.

[1] Barium chloride, 4 parts ; acid hydrochloric, 1 part ; distilled water, 16 parts.

Solutions Required.—1. A standard chloride-of-barium solution : 30.5 grammes of crystallized barium chloride to 1000 cubic centimetres of distilled water ; 1 cubic centimetre of this solution corresponds to 0.01 gramme of sulphuric acid.

2. Solution of potassium sulphate, 20 per cent.

3. Pure hydrochloric acid.

Analysis.—One hundred cubic centimetres of urine are rendered acid by 5 cubic centimetres of hydrochloric acid, and brought to boiling in a flask. The combined sulphates are thus converted into ordinary sulphates, and give a precipitate similarly with barium chloride. The chloride-of-barium solution is allowed to drop into the mixture as long as any precipitate occurs, the mixture being heated before each addition of barium chloride thereto. After adding 5 to 8 cubic centimetres of the standard solution, allow the precipitate to settle ; pipette off a few drops of the clear, supernatant fluid into a watch-glass ; add to it a few drops of the standard barium solution. If any precipitate occur, return the whole to the flask and add more barium chloride ; again allow the precipitate to settle and test as before ; continue thus until no more precipitate is formed on the addition of barium-chloride solution.

Excess of barium chloride should be avoided ; when only a trace of excess is present, a drop of the clear fluid removed from the flask gives a cloudiness with a drop of potassium-sulphate solution placed on a glass plate over a black ground. If more than a cloudiness appear, too large a quantity of barium chloride has been added, and the analysis must be repeated.

From the quantity of barium chloride used, the percentage of sulphuric acid in the urine is calculated,—1 cubic centimetre of barium-chloride solution corresponding to 0.01 gramme of SO_3.

(*b*) *Gravimetric determination* is best conducted as suggested by Salkowski. This consists in washing the precipitate of barium sulphate obtained by adding barium chloride to a known volume of urine ; 100 parts of sulphate of barium correspond with 34.33 parts of sulphuric acid (SO_3).

Analysis.—One hundred cubic centimetres of urine are taken in a beaker; this is acidified with 5 cubic centimetres of hydro-chloric acid. Chloride of barium is added till no more precipitation occurs. The precipitate is collected on a small filter of known ash, and washed with hot distilled water till no more barium chloride occurs in the filtrate; *i.e.*, until the filtrate remains clear after the addition of a few drops of sulphuric acid. Then wash with hot alcohol, and afterward with ether. Remove the filter and incinerate with its contents in a platinum crucible. Cool over sulphuric acid in an exsiccator; weigh, and deduct the weight of the crucible and filter ash; the remainder is the weight of barium sulphate formed, from which the SO_3 is readily calculated,—100 parts of barium sulphate corresponding with 34.33 parts of SO_3.

Correction.—As carried out above, a slight error occurs in the analysis from the formation of a small quantity of sulphide of barium. Correct as follows: After the platinum crucible has cooled add a few drops of pure sulphuric acid (H_2SO_4). The sulphide is converted into sulphate. Heat again to redness and drive off the excess of sulphuric acid.

(c) Salkowski's method of estimating the *combined sulphuric acid*—*i.e.*, the amount of SO_3 in ethereal sulphates—is as follows: 100 cubic centimetres of urine are mixed with 100 cubic centimetres of an alkaline barium-chloride solution, which is a mixture of two volumes of solution of barium hydrate with one of barium chloride, both saturated in the cold. The mixture is stirred, and after a few minutes filtered; 100 cubic centimetres of the filtrate ($=$ 50 cubic centimetres of the urine) are acidified with 10 cubic centimetres of hydrochloric acid, boiled, kept at 100° C. on a water-bath for an hour, and then allowed to stand till the precipitate has completely settled; if possible, it should be thus left for twenty-four hours. The further treatment of this precipitate ($=$ combined sulphates) is then carried out as in the last case [see (b)].

Calculations.—Two hundred and thirty-three parts of barium sulphate correspond to 98 parts of H_2SO_4, 80 parts of SO_3, or 32 parts of S. To calculate the H_2SO_4, multiply the weight of barium sulphate by $\frac{98}{233} = 0.4206$; to calculate the SO_3, multiply by $\frac{80}{233} = 0.34335$; to calculate the S, multiply by $\frac{32}{233} = 0.13734$.

This method of calculation applies to the gravimetric estimation both of total sulphates and of combined sulphates.

(d) To obtain the amount of preformed sulphuric acid, subtract the amount of combined SO_3 from the total amount of SO_3. The difference is the preformed SO_3.

Example.—One hundred cubic centimetres of urine gave 0.5 gramme of total barium sulphate; this multiplied by $\frac{80}{233} = 0.171$ gramme = total SO_3. Another 100 cubic centimetres of the same urine gave 0.05 gramme of barium sulphate from ethereal sulphates; this multiplied by $\frac{80}{233} = 0.017$ gramme of combined SO_3. Total SO_3 — combined $SO_3 = 0.171 - 0.071 = 0.157$ gramme of preformed SO_3.

CARBONATES.

Carbonate and bicarbonate of sodium, ammonium, calcium, and magnesium are present in minute quantities in fresh urine of alkaline reaction. The ammonium carbonate may be found in large quantity as a result of alkaline decomposition of the urine. The carbonates of the urine are derived from the food, from lactic, malic, tartaric, succinic, and other vegetable acids in the food. If the urine contain much carbonates when voided it is turbid, or soon becomes so upon standing, and upon sedimentation the precipitate is that of calcium carbonate usually associated with phosphates. Carbonate of calcium constitutes the basis of urinary calculus of great rarity in the human subject, very frequent, however, in herbivora.

Detection.—The presence of carbonates in the urine may be revealed by the evolution of colorless gas, upon the addition of acid, and this gas renders baryta-water turbid.

Determination.—The following method of Marchand may be employed for estimating the free carbonic acid of the urine: One hundred cubic centimetres of urine are placed in a glass flask, closely fitted with a doubly-perforated cork. Through one opening a tube is passed, which dips into the urine, and at the other end is connected with a tube containing some quicklime. Through the other opening in the cork one arm of a doubly-bent tube is passed; this arm does not dip into the urine. The other arm is introduced into an empty flask through a tightly-fitting cork. This flask is connected by a similar tube with a second flask filled with clear baryta-water, and this with a third and fourth filled with baryta-water.

The urine is heated to 100° C. over a water-bath ; any portions boiling over go into the empty flask. The carbonic acid comes off and forms a white precipitate of barium carbonate in the flasks filled with baryta-water. Air is then drawn through the apparatus, any carbonic acid in the atmosphere being removed by the quicklime. The carbonate of barium formed is collected on a filter, washed with distilled water, dissolved in hydrochloric acid, precipitated again by sulphuric acid, and weighed as barium sulphate. From the quantity thus obtained the amount of carbonic acid in the urine can be calculated : 196.65 parts of barium carbonate correspond to 232.62 parts of barium sulphate and 44 parts of carbonic acid.

(b) The *total carbonic acid* may be similarly estimated after strongly acidifying the urine with hydrochloric acid.

The *combined carbonic acid* is the difference between the total and the free carbonic acid.

OTHER INORGANIC CONSTITUENTS.

Iron occurs in the urine in small quantities, but its combination is yet unknown. Free *ammonia* occurs in traces, greatly increased in putrefactive changes of the urine. *Hydrogen dioxide* was first shown in the urine by Schonben. It exists in small amount, and, so far as known, is without special signification. It is detected by tetra-paper, which, if immersed in its solution, will show the presence of ozone by taking a blue color. 2. Dilute indigo solution is bleached by dioxide of hydrogen in the presence of iron-sulphate solution.

GASES.—The urine contains small quantities of gases. *Carbon dioxide,* 4 to 9 volumes free gas ; 2 to 5 combined. *Oxygen,* 0.2 to 0.6 volume ; and *nitrogen,* 0.7 to 0.8 volume. The gases of the urine may be withdrawn by the air-pump.

CENTRIFUGAL ANALYSIS.

With the first edition of this work the author introduced his method of centrifugal analysis for the ready approximate determination of bulk percentages of chlorides, phosphates, sulphates, and albumin in the urine. Nothing was claimed for this method at that time further than rapid approximate bulk measurement

TABLE

For Chlorides in the Urine,

showing the bulk percentages of silver chloride (AgCl) and the corresponding gravimetric percentages and grains per fluidounce of sodium chloride (NaCl) and chlorine (Cl).

Bulk Percentage of AgCl.	Percentage, NaCl.	Gr. per Oz., NaCl.	Percentage, Cl.	Gr. per Oz., Cl.	Bulk Percentage of AgCl.	Percentage, NaCl.	Gr. per Oz., NaCl.	Percentage, Cl.	Gr. per Oz., Cl.
¼	0.03	0.15	0.02	0.1	8	1.04	4.98	0.63	3.02
½	0.07	0.31	0.04	0.19	8½	1.1	5.29	0.67	3.22
¾	0.1	0.47	0.06	0.28	9	1.17	5.6	0.71	3.4
1	0.13	0.62	0.08	0.38	9½	1.23	5.91	0.75	3.6
1¼	0.16	0.78	0.1	0.48	10	1.3	6.22	0.79	3.79
1½	0.19	0.93	0.12	0.57	10½	1.36	6.53	0.83	3.97
1¾	0.23	1.09	0.14	0.66	11	1.43	6.84	0.87	4.16
2	0.26	1.24	0.16	0.76	11½	1.49	7.2	0.91	4.35
2¼	0.29	1.41	0.18	0.85	12	1.56	7.46	0.95	4.54
2½	0.32	1.56	0.2	0.96	12½	1.62	7.78	0.99	4.73
2¾	0.36	1.71	0.22	1.04	13	1.69	8.09	1.02	4.92
3	0.39	1.87	0.24	1.13	13½	1.75	8.4	1.06	5.11
3¼	0.42	2.02	0.26	1.23	14	1.82	8.71	1.1	5.29
3½	0.45	2.18	0.28	1.32	14½	1.88	9.02	1.14	5.49
3¾	0.49	2.35	0.3	1.42	15	1.94	9.33	1.18	5.67
4	0.52	2.49	0.32	1.51	15½	2.01	9.65	1.22	5.86
4¼	0.55	2.64	0.34	1.61	16	2.07	9.94	1.26	6.06
4½	0.58	2.8	0.35	1.7	16½	2.14	10.27	1.3	6.24
4¾	0.62	2.96	0.37	1.8	17	2.2	10.51	1.34	6.43
5	0.65	3.11	0.39	1.89	17½	2.27	10.87	1.38	6.62
5½	0.71	3.42	0.43	2.09	18	2.33	11.2	1.42	6.81
6	0.78	3.73	0.47	2.27	18½	2.4	11.51	1.46	7.
6½	0.84	4.05	0.51	2.46	19	2.46	11.82	1.5	7.19
7	0.91	4.35	0.55	2.62	19½	2.53	12.13	1.54	7.38
7½	0.97	4.67	0.59	2.84	20	2.59	12.44	1.58	7.56

(64a)

TABLE
FOR PHOSPHATES IN THE URINE,

showing the bulk percentages of uranyl phosphate ($H[UO_2]PO_4$) and the corresponding gravimetric percentages and grains per ounce of phosphoric acid (P_2O_5).

BULK PERCENTAGE OF $H(UO_2)PO_4$.	PERCENTAGE, P_2O_5.	GR. PER OZ., P_2O_5.	BULK PERCENTAGE OF $H(UO_2)PO_4$.	PERCENTAGE, P_2O_5.	GR. PER OZ., P_2O_5.
½	0.02	0.1	11	0.14	0.67
1	0.04	0.19	12	0.15	0.72
1½	0.045	0.22	13	0.16	0.77
2	0.05	0.24	14	0.17	0.82
2½	0.055	0.26	15	0.18	0.86
3	0.06	0.29	16	0.19	0.91
3½	0.065	0.31	17	0.2	0.96
4	0.07	0.34	18	0.21	1.
4½	0.075	0.36	19	0.22	1.06
5	0.08	0.38	20	0.23	1.1
6	0.09	0.43	21	0.24	1.15
7	0.1	0.48	22	0.25	1.2
8	0.11	0.53	23	0.26	1.25
9	0.12	0.58	24	0.27	1.3
10	0.13	0.62	25	0.28	1.35

TABLE
FOR SULPHATES IN THE URINE,

showing the bulk percentages of barium sulphate ($BaSO_4$) and the corresponding gravimetric percentages and grains per fluidounce of sulphuric acid (SO_3).

BULK PERCENTAGE OF $BaSO_4$.	PERCENTAGE, SO_3.	GR. PER OZ., SO_3.	BULK PERCENTAGE OF $BaSO_4$.	PERCENTAGE, SO_3.	GR. PER OZ., SO_3.
⅛	0.04	0.19	2¼	0.55	2.64
¼	0.07	0.34	2½	0.61	2.93
⅜	0.1	0.48	2¾	0.67	3.22
½	0.13	0.62	3	0.73	3.5
⅝	0.16	0.77	3¼	0.79	3.79
¾	0.19	0.91	3½	0.85	4.08
⅞	0.22	1.06	3¾	0.91	4.37
1	0.25	1.1	4	0.97	4.66
1¼	0.31	1.49	4¼	1.03	4.94
1½	0.37	1.78	4½	1.09	5.23
1¾	0.43	2.06	4¾	1.15	5.52
2	0.49	2.35	5	1.21	5.21

of these sediments, because the method was then new and untried, save in the author's laboratory, and it seemed a radical departure from methods better known and considered more accurate, such as titration, weighing, etc. Moreover, only bulk percentages had then been worked out without any attempt having been made to give corresponding gravimetric values, much less corresponding values in Cl., P_2O_5, and SO_3 from the bulk percentages of these combined as salts in the sediment. Since its introduction, however, it has been demonstrated in the author's laboratory that centrifugal analysis of the urine, if carried out by refined methods and improved apparatus, may readily reach results that are entitled to rank with the older standard methods, the gravimetric and volumetric included; that bulk percentages of sediments may be worked out in their equivalent values of their elements, not only with precision, but also with a rapidity and facility that at once renders this method of the greatest practical value in clinical work.

The essentials for securing accurate results in centrifugal analysis of urine are in no way complex or difficult of comprehension, much less to put into practice in the most ordinary laboratory. The equipment consists of an efficient motor, capable of the standard speed, possessing a standard radius of arm and tube (6¾ inches) accurately-graduated percentage tubes, and a gauge to regulate the speed. The author's improved electric motor (described in full at page 149) fulfills all requirements for accurate work. Very recently a further improvement in the author's percentage tubes has been adopted as follows: The points have been drawn out finer, and the first 5 cubic centimetres have been more minutely graduated so as to indicate measurements in 0.25 per cent. (¼ percentages) instead of 1 per cent. (one per cent.) as before.[1]

For the determination of chlorides, phosphates, and sulphates in the urine by the centrifugal method, the following standard and tables are now adopted in the author's laboratory.

For determination of albumin see page 80.

Process.—The double arm of the motor is employed, carrying

---

[1] Messrs. Elmer & Amend, of 205 and 211 Third Avenue, New York, manufacture and supply the author's improved standard percentage tubes.

four tubes. Three percentage tubes are filled to the 10-cubic-centimetre mark with the urine (the urine having been previously filtered if not perfectly clear). To the first tube is added 1 cubic centimetre of strong nitric acid and 4 cubic centimetres of standard solution of silver nitrate.[1] To the second tube is added 2 cubic centimetres of 50-per-cent. acetic acid and 3 cubic centimetres of uranium-nitrate solution (5 per cent.). To the third tube is added 5 cubic centimetres of the standard barium-chloride mixture.[2] The tubes are next inverted three times to insure mingling of the urine and reagents and then *allowed to stand for three (3) minutes* to secure complete precipitation. In order to balance the arm of the motor, the fourth tube is filled to the 15-cubic-centimetre mark with water. The centrifugal is next operated at a speed of 1200 revolutions per minute for three (3) minutes. The tubes are then removed and the percentages of precipitates are read off on the scale. No. 1 gives the bulk percentage of silver chloride (AgCl), No. 2 the bulk percentage of uranyl phosphate ($H[UO_2]PO_4$), and No. 3 the percentage of barium sulphate ($BaSO_4$). The bulk percentages are converted into their equivalent values in gravimetric percentages by means of the subjoined tables, and from these the grains or grammes of total chlorine (Cl), phosphoric acid (P_2O_5), and sulphuric acid (SO_3) are readily calculated by a glance at the tables. The results are more accurate if the urine be diluted in the cases of chlorides and phosphates if the bulk percentage of these exceed 15 per cent. The time required to carry out these quantitative determinations should not exceed ten minutes. As a rule, the more rapid and ready processes in uranalysis are comparatively few, and, for the most part, limited to qualitative rather than to quantitative data. The author, therefore, hopes that the above contribution of centrifugal analysis to our resources, which he has worked out with great care and pains, will prove of equal value to others in practical urinary work to that found in his own laboratory. Indeed, the amount of practical information

---

[1] Standard nitrate-of-silver solution consists of silver nitrate, 3j; distilled water, ℥j.

[2] Standard barium-chloride mixture consists of barium chloride, 4 parts; strong hydrochloric acid, 1 part; distilled water, 16 parts.

that this method is capable of laying before the clinician without loss of time cannot fail to prove of inestimable value in practical work. Thus, the time required to carry out these quantitative determinations centrifugally as described above should not exceed ten minutes. It has, indeed, been repeatedly demonstrated in the author's laboratory that the use of modern centrifugal methods has made it possible to make a fairly complete analysis of urine, both qualitative and quantitative, in from twenty minutes to half an hour which formerly required twenty-four hours' time.

SECTION III.

ABNORMAL URINE.

PROTEIDS.

The four proteids of the blood—viz., serum-albumin, serum-globulin, fibrin, and hæmoglobin—are met with in the urine in various pathological conditions of the kidneys, the blood, or the system at large. Other proteids are sometimes met with in the urine which do not exist in the blood, such as egg-albumen upon the liberal ingestion of eggs as food, and, under certain conditions, also peptone. Finally, certain proteoses are met with in the urine in pathological conditions, the more prominent of which are pro-albumose, deutero-albumose, and hetero-albumose.

ALBUMINURIA.

The chief clinical interest with regard to proteids in the urine will probably always centre about serum-albumin. While albumin is doubtless the most common of all the constituents of morbid urine, it still remains a debated question if it be present in the urine in health. No doubts can further exist that the urine occasionally contains a variable—usually small—but distinct amount of albumin when the kidneys present no appreciable alterations of structure; but, as will be shown, albuminuria often arises from causes aside from the kidneys themselves. Albuminuria, therefore, cannot be proved to be a condition of health, so long as the kidneys alone are considered; yet, the absence of renal lesions would seem to be the chief, if not indeed the only, condition sought to be established by many advocates of a so-called physiological albuminuria.

Albumin belongs to the class of colloids which do not crystallize, and under ordinary conditions do not penetrate animal membranes; but alterations from the normal conditions, as in the integrity of the basement membrane, in the quality of the albumin itself, or in the pressure to which they are both subjected,

(67)

may result in transudation. It is altogether probable that most forms of albuminuria are referable to causes corresponding to one or more of the above-named conditions. In other words, albuminuria may be due (1) to changes in the kidneys themselves, which impair the integrity of the structures between the vessels and the excretory channels of the organs; (2) alterations in the quality of the blood which render its serum-albumin more diffusible; (3) alterations in the degree of blood-pressure. Albuminuria may depend upon one or, indeed, all three of the above conditions.

Clinical Significance.—1. The more common form of albuminuria, as well as the most serious in its clinical significance, is that depending upon pathological conditions of the kidneys. The most frequent of these are inflammatory and degenerative changes in the renal structure, and include the whole class of disorders commonly grouped together under the term of Bright's disease.

It is impossible *always* to estimate the gravity or progress of renal changes by the quantity of albumin present in the urine. Sometimes, however, as in acute inflammatory conditions, when the amount of albumin ranges high,—1 per cent. by actual weight or more,—the quantity may be taken as a rough gauge of the extent of the lesions as well as the progress of the same from day to day. The same may be said of certain degenerative changes in the kidneys, notably of amyloid disease. This, however, by no means applies to all diseases of the kidneys, for, indeed, in certain renal diseases of the most serious character,—*interstitial nephritis,*—not only is the quantity of albumin in the urine usually small, but it is often temporarily absent and even, occasionally, throughout. The quantity of albumin in the urine, therefore, is not a safe guide as to the gravity of the situation in diseases of the kidneys, especially in cases attended by moderate or even very slight grades of albuminuria.

2. The second class of albuminurias depend upon changes in the constitution of the blood, which so alters the diffusibility of its albumin as to permit it to pass into the renal tubules. The hæmatogenic causes of albuminuria have been most ably expounded by Semola, and, although he probably claims too wide a range for these causes, there remains no just reason to doubt

their existence. We often meet with such albuminuria in anæmia, and in strumous and enfeebled individuals, when no lesions of the kidneys can be made out. The effect of certain poisons upon the blood probably so alters that fluid as to permit of transudation of its albumin into the renal tubules. The effects, also, of some infectious fevers—micro-organisms in the blood—no doubt seriously alters the constitution of the circulating fluid, so that transudation of albumin is the rule, while the kidneys do not always become damaged.

3. The third form, disturbances of the circulation, may bring about albuminuria without inducing structural changes in the kidneys, provided they be not too long continued. Circulatory disturbances, in order to induce albuminuria, must include the renal vessels. In nature they must consist of acceleration of the arterial current or slowing of the venous current, in either case resulting in increased blood-pressure. Probably this cause is responsible for the majority of that large class of cases of so-called " *physiological* or *functional albuminurias.*" This is most marked upon prolonged or fatiguing muscular exercise. Leube found albumin in the urine in 16 per cent. of soldiers after prolonged march, and Chateaubourg gives the percentage as even higher. A similar result sometimes occurs after the application of cold to the surface of the body ; the blood being driven to the interior, the renal vessels become overfilled and albuminuria often results. Again, in some derangements of the nervous system, which interfere with the vasomotor-nerve regulation of the renal vessels, temporary albuminuria is not an uncommon result. Albuminuria from increased blood-pressure is readily demonstrable by experimentation in the following ways : 1. By pressure upon the renal veins. 2. Ligature of the aorta below one kidney, and extirpation of the other. 3. Compression of the trachea. The quantity of albumin in the urine from disturbances of the circulation is for the most part small. It may be but temporarily present, or it may become a permanent condition, depending upon the continuance of the cause. Thus we may have temporary albuminuria after a seizure of epilepsy which soon after the attack subsides, or when depending upon organic disease of the heart it becomes permanent.

Finally, albuminuria often owes its origin to two or even all three of the causes just considered. In fevers, for instance, all the described causes of albuminuria are sometimes present. We have, for instance, accompanying changes in blood-pressure, and when long continued the febrile state is apt to induce structural changes in the renal epithelium, while profound changes in the constitution of the blood are often induced by fevers, more especially the acute infectious ones, which without doubt are the active cause of albuminuria.

It remains to consider the significance of a form of albuminuria which is often intermittent in character, to which Pavy's *"Cyclic Albuminuria"* and Moxon's *"Albuminuria of Adolescence"* doubtless belong. In a large percentage of these cases the albuminuria is intermittent; if not, usually it is remittent; the intermission or remission occurring during rest, as at night. On rising in the morning the urine is often free from albumin, but soon after rising, and especially upon exercising, the urine contains albumin, which may or may not wear away toward evening. A large percentage of these cases is observed in youths and young adults.

In another class of these cases the albuminuria is more constant, and if an intermission occur it is usually measured by weeks or months instead of hours. In these cases the age of the patient is less constant, although the albuminuria is still most common before middle age.

For the most part all these cases possess certain features in common: 1. The quantity of albumin in the urine is small, usually ranging from one-half to one-tenth or two-tenths of one per cent. 2. The urine either contains no renal casts or very few perfectly hyaline ones. 3. The specific gravity of the urine is somewhat above the normal standard,—1.024 to 1.030. 4. Evidences of cardiac and general vascular changes of a permanent nature are absent. 5. Close observation will usually reveal evidence of some local or general impairment of the functions, as measured by the standard of vigorous health.

The causes of this group of albuminurias are identical with those already considered, only operating perhaps in milder degrees. To changes in the renal structures, alterations in the quality of the blood, or abnormal increase of blood-pressure,—to one or more, or all of these combined, we may with great probability refer every case of albuminuria as to its essential cause or causes.

It will be seen, from the preceding considerations, that albuminuria is a symptom of the most variable clinical significance, and therefore, in itself, should never be accepted as proof of the presence of renal disease. As has been truthfully said, " this was the error of former times." It can only be positively asserted that albuminuria is the result of renal changes when it is accompanied by those products in the urine which are a consequence of renal lesions, such as casts, epithelium, etc. On the other hand, it must be remembered that albumin in notable quantity is not present in healthy urine. On the whole, it will be safer to accept albuminuria as an evidence of an existing abnormal state, the gravity of which must be determined by its accompanying symptoms. The author holds that so-called functional albuminuria forms no exception to the above rule, inasmuch as he has never met with a case of albuminuria in which the patient did not present more or less evidence of departure from the normal balance of perfect health, either local or general.

It is only necessary here to allude to the occurrence of albumin in the urine derived from sources other than the kidneys. Such albuminuria has been variously termed *adventitious, false,* or *accidental.* In such cases the urine on leaving the kidneys is perfectly normal; but, meeting with the products of inflammatory changes in the urinary passages,—the renal pelvis, ureters, bladder, or urethra,—it becomes albuminous. As a rule, in such cases, the source of the albumin may be determined by chemical and microscopical investigation, together with local symptoms.

Detection of Albumin in the Urine.—1. *Heat:* Boiling the urine constituted the first test employed to detect albumin, by Contugno (1770). The more common method of application of this test is in conjunction with nitric acid. A test-tube of ordinary size is filled half full of the suspected urine, and heat is applied until boiling occurs throughout the whole. If a precipitate occur, it consists either of albumin or earthy phosphates. A few drops of nitric acid are next added, and if the precipitate remain undissolved it is due to the presence of albumin. If, on the other hand, the precipitate disappear upon the addition of the acid, it consists of the earthy phosphates, and the urine is free from albumin. In testing, the acid should be added in small quantity, at first,—say, 2 to 5 drops,—and the urine should then be re-boiled. If now no precipitate occur, acidulation should be continued until precipitation occur or a limit of acidification be reached of about 15 to 20 drops. Some prefer to reverse this order, and first acidify the urine before applying the heat.

The heat and nitric-acid test is subject to certain errors. Thus, if there be little albumin present and the acid be in excess, the albumin may combine with the acid, forming a soluble acid albumin,—*syntonine*,—which is not precipitated by boiling. If, on the other hand, the acid be insufficient to distinctly acidify the urine, and if the phosphates be in excess, a part only of the basic phosphates may be acidified, while the albumin may combine with the remainder, forming a soluble alkali albuminate, which will not be precipitated by boiling. The heat and acid test throws down albumin, globulin, and mucin, and upon cooling albumose separates, if present. No reaction occurs with peptone, vegetable alkaloids in the urine, or with the urates. If the urine contain the pine acids, as sometimes occurs after the use of cubeb or copaibæ, these may cause slight precipitation by this test, which may be mistaken for albumin. The ready solubility of this precipitate in alcohol distinguishes it from albumin.

Various modifications have been suggested, with the view of avoiding the mucin reaction which sometimes undoubtedly occurs with this test, such as first boiling the urine and then very faintly acidifying with nitric acid, or by employing acetic instead of nitric acid in quantity not to exceed 2 drops. Such modifications, however, are not to be depended upon in eliminating the occasional mucin error, as will appear by a study of the chemistry of mucin reactions.

2. *The Author's Method.*—Have on hand a saturated aqueous solution of chemically-pure sodium chloride. Fill a clean test-tube about two-thirds full of the previously-filtered urine, and add to this about one-sixth of its volume of the sodium-chloride solution. Next add 5 to 10 drops of acetic acid (50 per cent.) and gently

boil the upper inch or so of the contents of the test-tube for about half a minute. If albumin be present, even in the minutest traces, it will appear in the upper, boiled portion of the test if examined in a good light. This test possesses all the sensitiveness of the heat-and-acid reaction with albumin, while it avoids faulty reactions. After repeated and crucial investigations the author confidently recommends this test as superior to all others for distinguishing minute quantities of albumin from other proteids in the urine—mucin or nucleo-albumin included.

3. *Nitric-Acid Test.*—This test is applied according to Heller's method, as follows: Upon a column of pure nitric acid in a test-tube the suspected urine is gently floated, so that the column of urine and that of acid are about an inch in depth. In order to accomplish the above without mixing the acid and urine, the test-tube should be held in an inclined position and the urine slowly delivered along the inside of the tube, so that the urine may flow gently down and overlie the acid. If albumin be present, an opalescent zone will be observed *at the point of contact* between the acid and the urine, which becomes more or less pronounced according to the quantity of albumin present. If no change be perceptible upon careful examination in a good light, the tube should be set aside and re-examined in half an hour, because, when only a trace of albumin is present, twenty to thirty minutes may elapse before the zone of coagulated albumin becomes visible.

In concentrated urines with this test the acid is apt to precipitate the amorphous urates in the form of a light, rather brownish cloud, which may be taken for albumin. The cloud of precipitated urates, however, does not appear *at the point of contact* between the acid and the urine, but *higher up*, within the urine itself; moreover, it is more diffused than the albuminous zone, and spreads downward through the urine. The precipitated urates disappear upon the application of gentle heat.

If the urine contain mucin in excess, a light cloud may come into view *toward the surface* of the urine with this test. It will be remembered that mucin is soluble in strong nitric acid, but is precipitated by the same in dilute form, and therefore the mucin reaction always occurs high up in the strata of urine which contain the acid well diluted. If the urine contain the pine acids a reaction may occur with this test somewhat similar to that of albumin, though usually less defined. The precipitate due to oleo-resins is soluble in alcohol, which distinguishes it from that due to albumin.

The nitric-acid test precipitates all modifications of albumin, acid and alkaline, as well as albumose. On the other hand, it gives no reaction with peptone or the vegetable alkaloids.

4. *The Ferrocyanic Test.*—This test is very simple and rapid in application, as follows :—

1. Fill an ordinary test-tube half-full of urine and add a half-drachm or more of potassium-ferrocyanide solution (1 to 20). After thoroughly mingling the urine and the reagent, add a few drops of acetic acid (50 per cent.) ; then pause for a half-minute and note any change. If albumin be present, it will come plainly into view, within half a minute to a minute, in the form of a white, milk-like opacity, diffused throughout the whole contents of the tube.

2. Into the bottom of a clean test-tube is poured a half-drachm of acetic acid ; then about a drachm to a drachm and a half of potassium ferrocyanide (1 to 20) is added and the two thoroughly mingled. The suspected urine is next allowed to gently flow down the side of the tube and overlie the reagents to the depth of about an inch. If albumin be present, a sharply-defined white zone or band will come plainly into view.[1] The ferrocyanic test applied by either of the above methods precipitates all modifications of albumin. On the other hand, it gives no reaction with phosphates, peptones, mucin, the alkaloids, urates, or the pine acids.

---

[1] The reaction that sometimes occurs on long-standing between the acid and potassium ferrocyanide should not be mistaken for albumin. The albuminous reaction appears within half a minute or so, while the other occurs only after ten minutes to half an hour, and is mingled with more or less blue coloration.

OTHER TESTS FOR ALBUMIN.

Within the past twenty-five years a number of additional tests for albumin in the urine have been brought forward. While few, if any, of these may yet be said to have become standard, yet the author will endeavor to here present the more prominent ones, together with a brief account of their special individual claims for recognition.

5. *Tanret's Test.*—This test was first proposed by Tanret in 1872 and was subsequently pronounced by the Clinical Society (London) the most delicate of a series of reagents for the detection of albumin in the urine at that time investigated. The formula is as follows: Potassium iodide, 3.32 grammes; mercuric chloride, 1.35 grammes; acetic acid, 20 cubic centimetres; distilled water, to 100 cubic centimetres. The potassium iodide and mercuric chloride should be separately dissolved in water, and the solutions mixed; the resulting reagent is the double iodide of mercury and potassium, to which the acetic acid is added and the whole made up to 100 cubic centimetres with distilled water. Thus prepared, the test is applied by the contact method of Heller; the reagent, being the heavier, is first introduced into the test-tube, and the urine is allowed to overlie it. The reaction with albumin consists of the development of a sharply-opaque white ring, or band, at the junction of the reagent and urine. The test responds to all modifications of albumin, to peptones, the vegetable alkaloids, and the pine acids. It is claimed, however, that all reactions other than with albumin, the pine acids, and nucleo-albumin are dissipated by heat.

6. *Picric Acid.*—This test was strongly advocated by Sir George Johnson as a most delicate reagent for albumin in the urine. The solution is prepared by simply saturating distilled water with picric acid (6 or 7 grains per ounce). While this test is applied by the contact method (the urine below), yet there must be an actual mingling of the urine and reagent in the upper stratum of the latter. This is, no doubt, a delicate reagent for albumin, but it also reacts with creatinin, copaiba, peptone, nucleo-albumins, as well as with alkaloids. Those save with albumin are claimed to disappear by heating, but they reappear upon cooling.

7. *Sodium Tungstate.* — Sonneschin in 1874, and subsequently Oliver, advised the use of this agent as a delicate test for albumin. The solution of this salt (1 to 4) must be acidulated with acetic or phosphoric acid. The test is applied by the usual contact method, and reacts with albumin, nucleo-albumin, peptones, urates, and the vegetable alkaloids; all save albumin and nucleo-albumin are probably cleared up by heat.

8. *Trichloracetic Acid.*—This agent was first suggested by Raab as a test possessing special advantages in that it is claimed not to precipitate peptones or nucleo-albumin, while it is exceedingly sensitive to albumin. Trichloracetic acid is applied in saturated solution by the contact method, the reagent below. It precipitates albumoses, alkaloids, and sometimes uric acid in addition to albumin, all save the latter disappearing on the application of heat.

9. *Metaphosphoric acid* was suggested by Hindenlang, in 1881, as a delicate reagent for albumin in the urine. The application is simple, viz.: in the test-tube is placed a little of the solid metaphosphoric acid, and upon this the urine is filtered, and agitated, when, if albumin be present, a turbidity results. This test reacts with albumoses, and sometimes with uric acid, but these are dissipated by heat.

10. *Spiegler's Test.*—This is composed of a solution of 8 grammes of mercuric chloride, 4 grammes of tartaric acid, and 20 grammes of sugar in 200 cubic centimetres of distilled water. One-third of a test-tube may be filled with the reagent, and the urine allowed to overlie this an inch or more in thickness; if albumin is present a sharply-defined white ring or zone appears at the line of contact of the two fluids. Globulin and albumoses react with this test, but peptones produce no change.

11. *Salicyl-sulphonic acid*, or sulphosalicylic acid, was first suggested by Roch, in 1891, as a delicate reagent for albumin in the urine; subsequently Macwilliam modified the test and advised its use as follows : About 20 drops of the urine are placed in a test-tube (small size) and 1 or 2 drops of a saturated aqueous solution of the reagent added, or more if the urine is alkaline. The tube is next agitated and inspected. Opalescence or turbidity occurring immediately is claimed to indicate greater

delicacy than Heller's method. Turbidity occurring slowly— one and a half to two minutes—implies minute traces of albumin. The test is lastly boiled and the precipitate remaining is due to albumin or globulin.

This test is often more simply applied as follows: To the suspected urine merely add a few crystals of the reagent, agitate, and, if turbidity results, correct by heat. An acid urine is necessary for this test, and, therefore, if the urine be alkaline, it must be treated with acetic acid before adding the reagent.

12. *Phenic-Acetic Acid.*—This agent, in conjunction with alcohol, was first employed by Méhu for determining the percentage of albumin. Subsequently Millard modified the formula for qualitative purposes as follows : Acid. phenic. glacial., 3ij ; acid. acetic. pur., 3vj ; liquor potassæ, 3vj, 3ij. As indicated, the quantity of liquor potassæ may be varied. The test is applied by the usual contact method. It reacts with peptone, nucleo-albumin, albumoses, and alkaloids, all of which save that with albumin are claimed to disappear with heat.

13. *Stutz's Test.*—This test consists of a mixture of mercuric chloride, sodium chloride, and citric acid. When added to albuminous urine, this solution causes an abundant precipitate of albuminate of mercury in the form of a dense, white opacity.

14. *Resorcin* was proposed by Carrez as a test for albumin. One gramme of resorcin is dissolved in 2 cubic centimetres of distilled water in a test-tube, and the urine is poured upon its surface. If albumin is present a white ring develops at the junction of the two fluids. Peptone is indicated also by a white ring, the latter disappearing if the tube is immersed in boiling water.

15. *Nitric-Magnesium Test.*—This test was proposed by Sir William Roberts. Its composition is 1 ounce of strong nitric acid and 5 ounces of saturated solution of magnesium sulphate. This is applied by the usual contact method, and is claimed to be more sensitive than the cold-nitric acid method of Heller.

16. *Sodium Nitroprussiate.*—Nya recommends this salt as a reagent for albumin. The test is applied as follows : The urine previously acidulated with acetic acid is treated with a concentrated solution of sodium nitroprussiate. The reagent must be kept from the light to avoid decomposition.

6

Color Reactions.—The albumins yield certain color reactions, but they are not suitable for direct qualitative testing, especially in cases of deeply-colored urines containing only small amounts of proteid matter. They are more useful as confirmatory or furnishing more positive proof that a given precipitate when considerable is really proteid. Thus, the supposed albuminous precipitate may be tested with Millon's reagent as follows:—

17. *Millon's reagent* is prepared by dissolving 1 part of mercury in 2 parts of nitric acid of specific gravity of 1.42 and diluting with two volumes of distilled water. The characteristic reaction is as follows: To 1 drachm of the albuminous solution or urine add 10 minims of Millon's reagent and heat to boiling. The presence of proteid is indicated by the liquid turning red, which color will include the precipitate, if any. The test also reacts with numerous derivatives of the aromatic series.

18. *Biuret Test.*—The urine is first treated with a solution of potassium or sodium hydroxid and subsequently drop by drop with a dilute solution of copper sulphate. In the presence of proteid first a reddish, then a reddish-violet, and lastly a violet-blue color is obtained. If albumin is absent from the urine the presence of albumoses and peptone may be tested by this method.

19. *Xanthoproteic Reaction.*—Add to the urine concentrated nitric acid and boil. Let the liquid cool and then add ammonia. If albumin is present, an orange color is produced.

Many of the tests just considered have many admirers and some strong advocates, largely upon their claims for exceeding delicacy. Doubtless in some cases at least this is really true. The question naturally arises, however: are these delicate reactions trustworthy? Upon this point the profession, as well as authorities, at present seem divided. Some claim that, for the most part, the tests for which unusual delicacy of reaction is claimed also react with substances found in many normal urines as well as with substances in pathological urines other than albumin. Nucleo-albumin is most often considered responsible for the doubtful reactions. Mitchell recently contends, after a review of all these tests, that a number of them—including Spiegler's, Tanret's, picric acid, and trichloracetic acid—give reactions with urines containing alkaline carbonates when albumin is absent.

It will be noted that almost unexceptionally the newer tests appeal to the influence of heat as their chief corrective: a circumstance that speaks strongly in favor of heat as the crucial distinguishing agent. The author believes that, while heat may not be the most delicate test for albumin, yet, when properly applied, it is, in all probability, the most trustworthy we yet possess.

While, doubtless, it is desirable that we should possess tests for albumin of somewhat greater sensitiveness than the old method of boiling the urine, yet, after all, extreme delicacy of reaction is altogether a matter of secondary consideration, as compared with accuracy, because, when the quantity of albumin in the urine is *very slight*, resort must necessarily be had to other means than the presence of *traces of albumin* in the urine, in order to be able to establish a positive diagnosis of renal disease. Notwithstanding the above facts, much unnecessary confusion and uncertainty in our present methods have been caused by the multiplication of tests for albumin in the urine whose chief claim for recognition is that of great sensitiveness of reaction rather than that of trustworthiness.

20. *Tests in Paper Form.*—According to the suggestion of Dr. George Oliver, of Harrogate,[1] a number of the tests named have been prepared and used in paper form. This is accomplished by saturating chemically inert filtering-paper with solutions of the albumin reagents, and with citric acid and then drying. The papers are then cut into slips of convenient size for testing, and may be carried about in the pocket-case for use at the bedside of the patient. In testing, the following method is advised: Into a small test-tube containing one drachm of distilled water are dropped a reagent paper and one charged with citric acid. After agitation for a minute or so the test-papers are removed and the solution is ready for testing. The urine is now added, and the test may be conducted either by a mixture of the two or by the contact method, of which Dr. Oliver prefers the latter.

Dr. Oliver now advises the use of two reagents only for albumin, viz.: the ferrocyanic and potassio-mercuric papers.

---

[1] Bedside Urine Testing. London, 1885.

These will be found very convenient in clinical work, or at the bedside in visiting practice. The ferrocyanic paper is recommended as the more ready and trustworthy, since in all cases the reaction with the mercuric test must be corrected by heating, otherwise it is liable to be misleading.

Quantitative Estimation of Albumin in Urine.—The number of tests proposed—methods for estimating the quantity of albumin in the urine—is scarcely less than those for qualitative testing, but they are far less satisfactory, because without exception they consume too much time for practical clinical work. Fully realizing the necessity for some more ready and rapid process, the author introduced the centrifugal method in the first edition of this work as at least a ready approximate process. More recently the centrifugal method has been worked out with great pains and accuracy. The author here presents the improved centrifugal method in detail, for which he claims something more than approximate results.

THE AUTHOR'S CENTRIFUGAL METHOD.—The process, in brief, consists of the following steps: Precipitation of the albumin in improved percentage tubes of 15 cubic centimetres' capacity. To 10 cubic centimetres of the urine, 3 cubic centimetres of 1 to 10 aqueous solution of potassium ferrocyanide are added and 2 cubic centimetres of 50-per-cent. acetic acid are added. After mingling the reagents and urine, the tube should stand for 10 minutes to insure entire precipitation of the albumin. At the end of 10 minutes the percentage tubes are placed in a centrifugal machine, the radius of which, with tubes extended, must be exactly six and three-quarter ($6\frac{3}{4}$) inches. The tubes are next revolved for exactly three (3) minutes at a uniform speed of fifteen hundred (1500) revolutions per minute. Lastly, the tubes are removed and the amount of albumin is read off in bulk percentage, which, by consulting the accompanying table, can be readily converted into percentage by weight and grains per fluidounce. It will be noted that the time necessary to carry out this test does not exceed 15 minutes, and it has been found that the results carefully compared with the gravimetric method need not amount to errors exceeding 0.01 per cent. More accurate results than the above are not ordinarily claimed

PURDY'S QUANTITATIVE METHOD FOR ALBUMIN IN URINE (CENTRIFUGAL).

Table showing the relation between the volumetric and gravimetric percentage of albumin obtained by means of the centrifuge with radius of six and three-quarter inches; rate of speed, 1500 revolutions per minute; time, three minutes.

VOLUMETRIC PERCENTAGE BY CENTRIFUGE	PERCENTAGE BY WEIGHT OF DRY ALBUMIN	GRAINS PER FLUIDOUNCE DRY ALBUMIN	VOLUMETRIC PERCENTAGE BY CENTRIFUGE	PERCENTAGE BY WEIGHT OF DRY ALBUMIN	GRAINS PER FLUIDOUNCE DRY ALBUMIN	VOLUMETRIC PERCENTAGE BY CENTRIFUGE	PERCENTAGE BY WEIGHT OF DRY ALBUMIN	GRAINS PER FLUIDOUNCE DRY ALBUMIN
¼	0.005	0.025	13½	0.281	1.35	31½	0.656	3.15
½	0.01	0.05	14	0.292	1.4	32	0.667	3.2
¾	0.016	0.075	14½	0.302	1.45	32½	0.677	3.25
1	0.021	0.1	15	0.313	1.5	33	0.687	3.3
1¼	0.026	0.125	15½	0.323	1.55	33½	0.698	3.35
1½	0.031	0.15	16	0.333	1.6	34	0.708	3.4
1¾	0.036	0.175	16½	0.344	1.65	34½	0.719	3.45
2	0.042	0.2	17	0.354	1.7	35	0.729	3.5
2¼	0.047	0.225	17½	0.365	1.75	35½	0.74	3.55
2½	0.052	0.25	18	0.375	1.8	36	0.75	3.6
2¾	0.057	0.275	18½	0.385	1.85	36½	0.76	3.65
3	0.063	0.3	19	0.396	1.9	37	0.771	3.7
3¼	0.068	0.325	19½	0.406	1.95	37½	0.781	3.75
3½	0.073	0.35	20	0.417	2.	38	0.792	3.8
3¾	0.078	0.375	20½	0.427	2.05	38½	0.801	3.85
4	0.083	0.4	21	0.438	2.1	39	0.813	3.9
4¼	0.089	0.425	21½	0.448	2.15	39½	0.823	3.95
4½	0.094	0 45	22	0.458	2.2	40	0.833	4.
4¾	0.099	0.475	22½	0.469	2.25	40½	0.844	4.05
5	0.104	0.5	23	0.479	2.3	41	0.854	4.1
5½	0.111	0.55	23½	0.49	2.35	41½	0.865	4.15
6	0.125	0.6	24	0.5	2.4	42	0.875	4.2
6½	0.135	0.65	24½	0.51	2.45	42½	0.885	4.25
7	0.146	0.7	25	0.521	2.5	43	0.896	4.3
7½	0.156	0.75	25½	0.531	2.55	43½	0.906	4.35
8	0.167	0.8	26	0.542	2.6	44	0.917	4.4
8½	0.177	0.85	26½	0.552	2.65	44½	0.927	4.45
9	0.187	0.9	27	0.563	2.7	45	0.938	4.5
9½	0.198	0.95	27½	0.573	2.75	45½	0.948	4.55
10	0.208	1.	28	0.583	2.8	46	0.958	4.6
10½	0.219	1.05	28½	0.594	2.85	46½	0.969	4.65
11	0.229	1.1	29	0.604	2.9	47	0.979	4.7
11½	0.24	1.15	29½	0.615	2.95	47½	0.99	4.75
12	0.25	1.2	30	0.625	3.	48	1.	4.8
12½	0.26	1.25	30½	0.635	3.05			
13	0.271	1.3	31	0.646	3.1			

Test. — Three cubic centimetres of 10-per-cent. solution of ferrocyanide of potassium and 2 cubic centimetres of 50-per-cent. acetic acid are added to 10 cubic centimetres of the urine in the percentage tube and *stood aside for ten minutes*, then placed in the centrifuge and revolved at rate of speed and time as stated at head of the table. If albumin is excessive, dilute the urine with water till volume of albumin falls below 10 per cent. Multiply result by the number of dilutions employed before using the table.

(80a)

for the gravimetric method itself. In order to insure accurate results, the following conditions should be complied with: (a) If the albumin be excessive, the urine should be diluted with one or more volumes of water, until the volumetric percentage does not materially exceed 10 per cent. Observations conducted in the author's laboratory have demonstrated the fact that accurate and uniform volumetric measurements of albumin in the urine by this method are only possible when the percentage does not materially exceed 10 per cent. (b) The reagents and the urine must stand after mingling 10 minutes to insure entire precipitation of the albumin. (c) The centrifuge must possess the following essentials or be capable of such modification as to include them, viz.: The arm should possess a radius of exactly 6¾ inches; that is to say, the linear distance from the centre of the axle to the tip of either tube must be just 6¾ inches. The motor must be capable of an even and sustained speed of 1500 revolutions per minute, with the stated radius and carrying 30 cubic centimetres of urine. Lastly, some trustworthy method of gauging the exact speed of the motor must be employed.

The advantages claimed for this method over those hitherto in use are its rapidity, simplicity, accuracy, and comprehensiveness in expression of results—the volumetric percentage, its corresponding percentage of dry albumin, the number of grains of dry albumin per ounce, and from these the total weight of dry albumin in 24 hours, all being apparent by a glance at the accompanying table.

THE GRAVIMETRIC METHOD.—The process consists in coagulating the albumin, which may be accomplished either by (a) boiling or (b) by means of a chemical agent. The succeeding steps are filtering out the albumin, collecting, drying, and weighing.

1. *Coagulation by Heat.*—To 100 cubic centimetres of urine acetic acid is added until the urine is distinctly acid, after which it is filtered, and gradually heated to boiling and the boiling continued for half a minute. The urine is next passed through a filter, the weight of the filter having first been ascertained and noted. The flask in which the urine was boiled is next washed with distilled water to secure all particles of albumin, and the

contents are again thrown on the filter. Next the albumin on the filter is washed with boiling distilled water, the washing being continued until the albumin is perfectly clean and white.

The filter is next placed in an oven the temperature being 100°C. (212°F.), and there left until drying is complete. Drying is known to be complete when two weighings at an interval of an hour are identical. From the whole weight that of the filter is deducted, and the difference represents the weight of albumin in 100 cubic centimetres of urine, from which the whole amount may be readily calculated.

2. *Coagulation by Chemical Agent.*—There are a number of processes of which Méhu's will serve as an example : 2 or 3 drops of acetic acid are added to the urine and the latter is filtered. One hundred cubic centimetres of the filtered urine are next taken and 2 cubic centimetres of nitric acid are added, with 10 cubic centimetres of the following solution : Crystallized carbolic acid, 10 grammes ; acetic acid, 10 cubic centimetres ; alcohol (90 per cent.), 20 cubic centimetres. The albumin immediately coagulates upon the addition of the above, and the whole is thrown on the filter. Drying and weighing are to be conducted as already described.

The tediousness of the gravimetric process is its chief drawback for clinical use, since the process cannot be carried out in less time than five or six hours.

ESBACH'S METHOD.—This test is conducted by means of a standard graduated glass tube or albuminometer, shown in the cut (Fig. 8). The following standard solution is required : Picric acid, 10 grammes ; citric acid, 20 grammes ; distilled water, to 1000 cubic centimetres (1 litre).

FIG. 8.—ESBACH'S
ALBUMINOMETER.

Process.—Fill the albuminometer tube with the urine to the letter *U*, then add the test solution to *R*; close the tube with the stopper and invert several times, until the urine and the test solution are thoroughly mingled. Stand the tube in a rack for twenty-four hours and then read off the number of grammes of albumin per litre, as will be indicated on the side of the tube on a level where the albumin settles. If it be desired to know the percentage of albumin instead of the number of grammes per litre, remove the decimal point one figure to the left: thus 5 grammes per litre would be 0.5 per cent. If the urine be highly albuminous, it should be diluted with one or more volumes of water, before testing, and the result multiplied by the number of dilutions employed. This test has attained some popularity largely on account of its extreme simplicity. Its disadvantages for clinical use are that it takes twenty-four hours to complete the process, and the results are only claimed to be approximately accurate in the end.

TITRATION METHOD OF TANRET.—Tanret recommends for the volumetric estimation of albumin its precipitation by the following solution: Potassium iodide, 3.22 grammes; mercuric chloride, 1.35 grammes; distilled water, to 100 cubic centimetres. For the confirmatory solution: Mercuric chloride, 1 gramme; distilled water, 100 cubic centimetres. One drop of the precipitating solution given by a pipette of standard size precipitates 0.005 gramme of albumin; so that as many drops as it takes to precipitate all the albumin so many times 0.005 gramme of albumin must be contained in the urine. To save trouble in calculation, a certain quantity of urine should always be employed, a convenient quantity being 10 cubic centimetres, since then the number of drops of the solution that it takes to precipitate all the albumin in this quantity of urine represents so many half-grammes per litre.

Process.—Take 10 cubic centimetres of the urine and add 2 cubic centimetres of acetic acid, and stir with a glass rod; add the precipitating solution drop by drop, stirring carefully after each drop, until the albumin is no longer affected by the reagent, as ascertained as follows: After adding each drop of the reagent place a drop of the urine on a porcelain dish and note if a yellowish-red color appears on adding a drop of the confirmatory solution. As soon as it does all the albumin is precipitated and the process is completed. The amount of albumin per litre will be arrived at by taking the number of drops of the reagent employed, subtracting 3 as having been employed in excess to make the yellow color perfectly apparent, then considering the remainder as so many half-grammes.

PROTEOSES.

These substances are the intermediate products in the hydration of proteids, the final products being peptones. In the body they are formed by the action of the gastric and pancreatic juices, and they may be formed artificially by heating albumin with water,—more readily by dilute mineral acids or sulphuretted steam. They correspond to the propeptone of Schmidt-Mulheim, and to the A-peptone of Meissner. They are uncoagulable by heat, are precipitated but not coagulated by alcohol; they all respond to the biuret reaction, and are precipitated by nitric acid, the precipitate thus formed being dissolved by heat, but re-appearing upon cooling.

These substances may be subdivided into albumoses, globuloses, vitelloses, caseoses, myosenoses, depending upon the proteid from which they are formed. The albumoses and globuloses are absent from normal urine, but appear in the urine under a number of abnormal conditions.

Albumosuria.—The albumoses are of two varieties: *hemi-albumoses*, or those convertible by further digestive action into *hemi-peptone*, and *anti-albumoses*, or those similarly converted into *anti-peptone*. Albumoses have also been classed according to their solubilities, as follows: (*a*) Proto-albumose ; soluble in cold and hot water, and in saline solutions. They are precipitated as are globulins, by saturation with sodium chloride and magnesium sulphate. (*b*) Hetero-albumose ; insoluble in water ; soluble in 0.5 to 15 per cent. sodium-chloride solutions in the cold, but precipitable by heat at 65° C., the precipitate being readily soluble in dilute acid or alkali. It is precipitated by alcohol, as are other albumoses, but, unlike them, it is partly converted into an insoluble *dys-albumose*. Hetero-albumose is precipitated by dialyzing out the salines from its solutions, and, like proto-albumose, it is precipitated by saturation with salines. Proto- and hetero- albumoses constitute the first products of the hydration of proteids, and have hence been called primary albumoses. (c) Deutero-albumose is soluble in cold and hot water, is not precipitated from its solutions by saturation with sodium chloride or magnesium sulphate, but it is precipitated by strong solutions of ammonium sulphate. It is not precipitated by copper sulphate, and only gives the nitric-acid reaction (characteristic of albumoses) in the presence of excess of saline. It is therefore, in reactions, nearest to peptones of the albumoses; it is an intermediate stage in the conversion of primary albumoses into peptone.

Clinical Significance.—Albumose, like peptone, is found in pus; but, unlike the latter, it is also present in the blood, most notably so during digestion. An albumose was first discovered in the urine by Bence Jones in a case of osteomalacia, and in a like case since by Kuhn. Virchow has found this albumose in the medulla of bones in osteomalacia, while Fleischer found it in the medulla of normal bone. Senator has found albumoses in a number of cases in the urine,—viz., tertiary syphilis, hemiplegia, double pneumonia, diphtheria, carcinoma, and muscular atrophy. Hoppe-Seyler has found it in a number of cases of atrophy of the kidneys. Lassar has found it in the urine of people rubbed with petroleum, while Oertel has met with it in a few cases after severe exertion. Albumose and peptone are both found in the urine of animals when injected into their circulation. Deutero-albumose so closely resembles peptone in its reactions that it is often mistaken for the latter substance. Its distinction from peptone will be considered in connection with peptonuria. The clinical significance of albumosuria, so far as our present knowledge extends, is very indefinite; so that practical conclusions in connection with its presence in the urine are as yet lacking. In the few diseases in which it has been noted, its appearance, under similar circumstances, has subsequently been found inconstant.

Detection.—The proteoses are known by their solubilities and reactions, as follows: *Proteo-albumose*, soluble in both hot and cold water, in both hot and cold saline solution (10 per cent.

NaCl), is precipitated with strong solutions of sodium chloride and magnesium sulphate ; also by saturated solution of Am_2SO_4 ; is precipitated by nitric acid in the cold ; not soluble on heating, or only slightly so ; it is precipitated by copper sulphate.

Hetero-albumose, insoluble in hot and cold water, is precipitated by dialysis from saline solutions ; soluble in both hot and cold solutions of sodium chloride of 10-per-cent. strength,— *i.e.*, is partly precipitated, but not coagulated, on heating to 65° C. ; is precipitated by saturation with sodium chloride or magnesium sulphate ; also with saturated ammonium sulphate ; and is precipitated by nitric acid in the cold, the precipitate being dissolved with heat and re-appearing on cooling.

Deutero-albumose is soluble in hot and cold water ; soluble in hot and cold solutions of sodium chloride of 10-per-cent. strength ; *is not precipitated* by saturation with sodium chloride or magnesium sulphate, but is precipitated by saturation with ammonium sulphate ; while with nitric acid it is only precipitated in the presence of excess of salt.

PEPTONURIA.

Peptones are best known as the final products of gastric and pancreatic digestion. They are also products of retrogressive changes in albuminoids and of the corpuscular elements of the blood, and as such assume importance in their clinical relations. It has just been stated that peptones are the final products of gastric and pancreatic digestion. If hydration were continued a step farther, peptone would be split up into simpler substances, and would no longer constitute a proteid. Peptones are soluble in water, uncoagulable by heat, and are not precipitated by nitric acid, copper sulphate, ammonium sulphate, potassium ferrocyanide, and a number of other precipitants of proteids. They are precipitated by tannin, potassio-mercuric iodide, phosphomolybdic acid, phosphotungstic acid, and picric acid.

Peptones are divisible into two forms : (*a*) *Hemipeptone,* which by further action of the pancreatic juice is split up into leucin and tyrosin and such simpler products. (*b*) *Antipeptone,* which is not decomposed further by pancreatic juice. It, furthermore, does not yield tyrosin on treatment with sulphuric acid, and does not respond to Millon's reagent.

Both forms of peptone are readily diffusible through animal membranes, albumoses being only slightly so, and albumin not so at all, under ordinary circumstances. The peptones are not precipitated by saturation with ammonium sulphate, in which respect they differ from albumoses.

Clinical Significance.—Peptone is absent from normal urine, but has been described in the urine in connection with numerous pathological conditions. Since the recent publication of Kühne and Chittenden's work on proteoses and peptones, it has been made evident that many of the cases formerly described as peptonuria were, in reality, albumosuria, the real proteid present being deutero-albumose, and, therefore, much of our supposed knowledge of peptonuria needs revision.

Peptonuria has been frequently described as associated with the following conditions : In phosphorus poisoning, in suppurative diseases, in croupous pneumonia, acute rheumatism, typhoid fever, typhus, small-pox, scarlet fever, mumps, tuberculosis, erysipelas, empyema, cancer of the viscera (notably of the liver and intestines), catarrhal jaundice, apoplexy, etc. The local—sometimes termed the pyogenic—causes seem to be connected with resorption of exudations so situated as to favor the products of disorganization—the peptone constituent of leucocytes—being absorbed into the circulation, from whence it is eliminated by the kidneys. This form of peptonuria is met with in the declining stages of pneumonia, in purulent pleuritis, suppurating tuberculosis, chronic bronchial catarrh, psoas abscess, purulent meningitis, and acute articular rheumatism.

In acute inflammatory affections the appearance of peptonuria may be taken as an evidence that suppurative changes have been established, other known causes of peptonuria being excluded. Jaksch insists that, as a means of distinguishing between tubercular and epidemic cerebro-spinal meningitis, peptonuria is of crucial significance, being absent in the former and characteristic of the latter. In this connection, however, ulcerative changes in the lungs must be excluded to render this sign trustworthy.

In like manner, peptonuria appearing as is usual in septicæmia may serve to distinguish it from latent disseminated
sarcoma when, as often happens, the clinical symptoms are very
similar.   In short, Maixner has declared it the law that peptone
is always present in the urine when pus is forming in the
organism.

In addition to the pyogenic causes of peptonuria, extensive
destruction of the corpuscular elements of the blood seems to
constitute a prominent cause, and hence the frequent appearance
of peptonuria in acute infectious diseases and toxic conditions
already enumerated.

Peptonuria also arises from a few additional causes.   It is
almost invariably associated with cancer of the liver, and this
fact has led Pecancowski to the conclusion that the liver in
health is concerned in the conversion of peptone into albumin.
Maixner has shown that in ulceration of the intestines the
peptic products of the stomach pass directly into the blood
through the ulcerated surfaces, and give rise to peptonuria.
Then, again, we have puerperal peptonuria, Fischel having
shown that peptone is a normal constituent of the urine in the
puerperal condition.   Finally, when injected into the blood,
peptone quickly appears in the urine.

Detection.—Peptone may be recognized by the following
method : Saturate the urine (slightly acidified first with acetic
acid) with ammonium sulphate, and filter out any precipitate
formed which may consist of albumin, globulin, proto-albumose,
hetero-albumose, or deutero-albumose.   Any proteid remaining
may be precipitated by potassio-mercuric iodide or picric acid,
and can only be peptone.   This is, in fact, the only certain
method of identification of peptone.

Differentiation.—Since peptone and deutero-albumose so
closely resemble each other in reactions, it is well to be able
to distinguish them, more particularly since the frequency with
which they have been confounded has undoubtedly led to
numerous clinical errors.   Halliburton contrasts them as follows :—

PEPTONE.	DEUTERO-ALBUMOSE.
1. Gives no precipitate with nitric acid.	1. Gives no precipitate with nitric acid unless a considerable amount of salt be added. This precipitate disappears on heating and re-appears on cooling.
2. Is not precipitated by saturation with ammonium sulphate.	2. Is precipitated by saturation with ammonium sulphate.

In all other respects these two substances, as far as known, behave similarly.

Recent researches by Kühne[1] have shown that, in order to effect complete separation of the albumoses from peptones, the mixture containing these substances should be saturated *whilst boiling* with ammonium sulphate. Furthermore, a single saturation with ammonium sulphate should not be depended upon to remove all the deutero-albumose, but saturation should be repeated till precipitation no longer occurs.

GLOBULINURIA.

Globulin is insoluble in water, but dissolves in dilute neutral salt solutions. From these solutions it is precipitated by sufficient dilution with water, and on heating it coagulates. Globulin dissolves in water on the addition of very little acid or alkali, and on neutralizing the solvent it re-precipitates. Solutions of globulin in a minimum amount of alkali are precipitated by carbon dioxide, but the precipitate may be dissolved by excess of the precipitant. The neutral solutions of globulin containing salts are precipitated on saturation with sodium chloride and magnesium sulphate at normal temperatures. Normal urine is free from globulin, but this proteid appears in the urine in a number of pathological conditions.

Clinical Significance.—Globulinuria is nearly always associated with albuminuria, and, indeed, globulin may greatly exceed the quantity of albumin present in some cases, although the proportion of globulin in the blood is only as 1 to albumin 1.5. Globulin is, however, a more diffusible form of proteid than albumin, which may account for its proportional excess in the

---

[1] W. Kühne, Erfahrungen über Albumosen und Peptone, I. Reinigung der Peptone von Albumosen. Separatabdruck aus der Zeitschrift f. Biologie, 1893.

urine at times. From the fact already stated, that globulin is nearly always associated with albumin in the urine, its clinical significance is nearly identical with that of albuminuria. In a few cases, however, its presence in the urine seems to imply a special significance. Thus, globulin is noted in unusual quantities in the urine in catarrhal inflammations of the bladder, in acute nephritis, and especially in amyloid degeneration of the kidneys. The same is said to be the case in albuminuria associated with digestive disorders. On the other hand, in chronic Bright's disease globulin is said to be present in very small amount, or even at times absent.

Detection.—1. Exactly neutralize the urine, filter, and treat with magnesium sulphate in substance until it be completely saturated at an ordinary temperature, or with a saturated solution of ammonium sulphate. In both cases a white precipitate is formed if globulin be present. In using ammonium sulphate with urines rich in urates, precipitation of ammonium urate may appear. These, however, do not immediately appear, but only after some time, and they may thus be distinguished from the globulin precipitate.

In detecting serum-albumin in the same urine, heat the filtrate after precipitation of the globulin to boiling, after the addition of a few drops of acetic acid.

2. Globulin falls out of solution when the urine is diluted until the specific gravity is about 1.002, and upon the above fact Roberts suggested the following simple test: Fill a wineglass or test-tube with water and let fall into it several drops of albuminous urine. If globulin be present in any quantity, each drop as it falls is followed by a milky streak, and when a number of drops have been added the water assumes a milky opalescence throughout. The addition of acetic acid causes the opalescence to disappear.

Determination.—The separate determination of globulin and albumin may be accomplished by carefully neutralizing the urine and precipitating with magnesium sulphate added to saturation, or by simply adding an equal volume of saturated solution (neutral) of ammonium sulphate. The precipitated globulin is thoroughly washed with saturated magnesium sulphate, or half saturated ammonium-sulphate solution, dried at 110° C., boiled with water, extracted with alcohol and ether; then dried, weighed, and ashed; then weighed again, and the weight is the amount of globulin.

DIFFERENTIAL TESTING.

Serum-albumin, serum-globulin, hetero-proteose, deutero-proteose, and peptone may all be present in the urine simultaneously. This is very unusual, but in doubtful cases the only certain method is to test for each one in the list. The best method of doing this is that proposed by Halliburton, as follows :—

1. If the urine give no precipitate on boiling after acidulation, albumin and globulin are absent. If a precipitate occur, albumin or globulin, or both, are present.

2. If the urine after neutralization give no precipitate on saturation with magnesium sulphate, globulin and hetero-proteose are absent. If such precipitate occur, one or the other is present.

3. If the urine be saturated with ammonium sulphate and filtered, and the filtrate gives no xanthoproteic or biuret reaction (a large excess of potash must always be added), peptone is absent.

4. If the urine give no precipitate on boiling after acidulation, no precipitate with nitric acid, and no precipitate on adding ammonium sulphate to saturation, peptone can be the only proteid present. Confirm this by the biuret reaction.

5. If all proteids are present, they may be separated as follows :—

Saturate the urine (faintly acidified with acetic acid) with ammonium sulphate. A precipitate is produced. Filter.

(a) PRECIPITATE.	(b) FILTRATE.
Contains albumin, globulin, hetero- and deutero- proteose.	Contains peptone.

Collect the precipitate on a filter, wash it with saturated solution of ammonium sulphate, and redissolve it by adding a small quantity of water. To this solution add ten times its volume of alcohol; a precipitate is formed; collect this, and let it stand in absolute alcohol for from seven to fourteen days. Then filter off the alcohol, dry the precipitate at 40° C., extract with water and filter. An insoluble residue is left.

7

(a) RESIDUE.	(b) EXTRACT.
This consists of albumin and globulin coagulated by alcohol.	This contains the proteoses in solution.

Hetero-proteose is precipitated by heating the solution to 65° C., or by saturating a portion of the extract with magnesium sulphate. Deutero-proteose remains in solution.

Take another portion of the urine, neutralize it, and saturate with magnesium sulphate. A precipitate is produced. Filter.

(a) PRECIPITATE.	(b) FILTRATE.
This consists of globulin and hetero-proteose, which may be separated by the prolonged use of alcohol, as above.	This contains albumin, deutero-proteose, and peptone. Add alcohol, as above; albumin is rendered, in seven days, insoluble in water. The deutero-proteose and peptone are soluble, and may then be separated by ammonium sulphate.

The reactions of the several proteids in the urine already considered may be seen at a glance in the following table, after Halliburton. (See next page.)

HÆMOGLOBINURIA.

Hæmoglobin, the red pigment of the blood, is a somewhat remarkable compound in that it contains iron, is intimately associated with a proteid, and gives the proteid reaction; it is hence non-diffusible, but yet is crystalline. It exists in the blood in two conditions,—in arterial blood it is termed oxyhæmoglobin, being charged with oxygen; in venous blood it is deoxygenated or reduced hæmoglobin. Hæmoglobin belongs to the group of blood-proteids, which yields as cleavage products small amounts of volatile fatty acids with about 96 per cent. albumin and about 4 per cent. hæmochromogen, containing iron, which, in the presence of oxygen, is readily oxidized into hæmatin. Hæmoglobin prepared from different kinds of blood has not always the same constitution, indicating the probable presence of different hæmoglobins. This is further shown by the facts that the products obtained from different kinds of blood in many cases differ in solubility and crystalline form and possess a varying quantity of water of crystallization.

The Clinical Significance.—The blood-pigment, hæmoglobin, sometimes appears in the urine without the appearance of any

TABLE OF REACTIONS OF PROTEIDS IN THE URINE.

Variety of Proteid.	Hot and cold water.	Hot and cold saline solution, e.g., 10 per cent. NaCl.	Saturation with NaCl or MgSO₄.	Saturation with Am₂SO₄.	Nitric acid.	Copper sulphate.	Copper sulphate and ammonia.	Copper sulphate and caustic soda, or potash.
Albumin.	Soluble in cold and coagulated in hot water.	Soluble in cold and coagulated in hot solutions.	Not precipitated.	Precipitated.	Precipitated in cold; not dissolved by heating.	Precipitated.	Blue solution.	Violet solution.
Globulin.	Not soluble in either.	The same as albumins.	Precipitated.	Precipitated.	The same as albumins.			
Proto-albumose.	Soluble in both.	Soluble in both.	Precipitated.	Precipitated.	Precipitated in cold; precipitate dissolves with heat and returns in cold.	Precipitated.	Violet solution.	Rose-red solution.
Hetero-albumose.	Insoluble, i.e., like globulins, precipitated by dialysis from saline solutions.	Soluble in both; partly precipitated but not coagulated by heat to 65° C.	Precipitated.	Precipitated.	"	Precipitated.	"	"
Deutero-albumose.	Soluble.	Soluble.	Not precipitated.	Precipitated.	Only reacts in presence of excess of salt.	Not precipitated.	"	"
Peptone.	Soluble.	Soluble.	Not precipitated.	Not precipitated.	Not precipitated.	Not precipitated.	"	"

corpuscles associated therewith, as was first shown by **Pavy.** This is always the result of the destruction of the blood-corpuscles in the circulating stream. The hæmoglobin thus liberated is thrown out by the kidneys and appears in the urine. This may be produced by injection into the circulation of substances which act as solvents of the corpuscles, such as glycerin, solutions of the bile-salts, distilled water, and the injection of the blood of one animal into another. Similar results follow in cases of poisoning with arseniuretted hydrogen ; hydrochloric, sulphuric, carbolic, and pyrogallic acids ; phosphorus, and potassium chlorate. In certain diseases, notably pyæmia, typhus, scurvy, fat-embolism, in some cases of jaundice, and after extensive burns, hæmoglobinuria often results. The most interesting and marked pathological condition in which hæmoglobin appears in the urine is the disease known as *paroxysmal hæmoglobinuria,* which will form the subject of special consideration in a subsequent section of this work.

Detection.—1. Solutions of hæmoglobin may be determined by spectroscopic examination with great precision. They strongly absorb the rays lying between D and E. In a proper dilution the solution shows a spectrum with one broad, not sharply-defined, band. This band does not lie in the middle, between D and E, but is toward the red end of the spectrum, a little over the line D.

2. *Guaiacum Test.*—Mix in a test-tube equal volumes of tincture of guaiacum and old turpentine which has become strongly ozonized by the action of air under the influence of light. To this mixture, which must not have any blue color, add the urine to be tested. In the presence of hæmoglobin, first a bluish green, and then a beautiful blue ring appears where the two liquids meet. On shaking, the mixture becomes blue. Normal urine, and albuminous urine, do not give this reaction. Urine containing pus also gives a blue color with the above reagents ; but in this case the tincture of guaiacum alone, without the turpentine, is colored blue by the urine. The blue color produced by pus differs from that produced by hæmoglobin by disappearing on heating the urine to the boiling-point.

Urines, if alkaline from decomposition, must first be made

faintly acid before applying this test. The turpentine should be kept exposed to the light, while the guaiacum should be kept in a dark-glass bottle.

3. *Heller's Test.*—If a neutral or faintly-acid urine containing hæmoglobin be heated to boiling, there is obtained a mottled precipitate of albumin and hæmatin. If caustic soda be added to the boiling-hot test, the liquid becomes clear and turns green when examined in thin layers, and a red precipitate, appearing green by reflected light, re-forms, which consists of earthy phosphates and hæmatin.

FIBRINURIA.

Fibrin is a whitish, stringy solid when fresh, but upon drying it becomes of a grayish color. It is feebly soluble in 6-per-cent. solutions of potassium nitrate, in 5- to 15-per-cent. solutions of sodium chloride, in 5- to 10-per-cent. solutions of magnesium sulphate; and in solutions of other neutral salts, such as sodium sulphate and ammonium sulphate, most readily at a temperature of 40° C. It is insoluble in water, alcohol, and ether. It swells up in weak hydrochloric-acid solutions (0.2 per cent.), and also in sodium or potassium hydrate solutions (0.2 per cent.), into a gelatinous mass. Stronger acids slowly dissolve fibrin with the formation of acid albumin (syntonin) and albumoses.

Digestive ferments act readily on fibrin. Thus, pepsin in an acid solution and trypsin (from the pancreas) in an alkaline solution cause, in the first place, a splitting up of the fibrin into two globulins, one coagulating at 56° C., the other at 75° C. Then the formation of albumoses and peptones follows.

Fibrin separates on the so-called spontaneous coagulation of the blood, lymph, and transudations.

Clinical Significance.—Fibrin may appear in the urine from various causes, either in solution or in coagulated flakes or masses. In the urine fibrin constitutes the basis of the so-called coagulable urine, which, upon standing some time, forms the fibrinous coagula. The quantity of fibrin present in the urine determines the extent of coagulation; sometimes only a sticky sediment forms at the bottom of the vessel; more rarely the coagulation involves the whole volume of urine, converting it into a sticky, gelatinous mass. This form of fibrinuria is seldom met with in the United States, but it occurs frequently as a special form of disease in Brazil and the Isle of France.

Fibrinuria, as it occurs with us, uniformly indicates that an exudation of fibrinous fluid—blood-plasma—has gained access to some part of the urinary tract. In most cases it comes from the kidneys, although it may result from intense inflammation of the lower urinary passages. It is often associated with hæmorrhages into the urinary tract, and it has been observed frequently in cases of villous tumors of the bladder.

It is important to distinguish between fibrinuria and those cases of pyuria in which large quantities of pus are present in highly-alkaline urine. In cystitis of long standing the ammonium carbonate resulting from bacterial decomposition of the urine may form with pus a gelatinous, sticky mass, which may be readily mistaken for fibrin coagulum. Such a urine, however, is thinned by the addition of water, and, moreover, if treated with acetic acid, a white precipitate falls, consisting of alkaline albuminate.

Detection.—The coagulum should be separated by filtration and washed with water, when it will show the following characters:—

1. It does not dissolve in water, but it swells up in 1-per-cent. solutions of hydrochloric acid, and is only dissolved when pepsin is added.

2. Separate the coagula from the urine by filtration through fine muslin, and wash with cold water. Treat a part of the mass with dilute solution of sodium hydroxide; if insoluble upon long standing, the indication is that it is fibrin, since albuminous bodies dissolve in sodium-hydroxide solutions. Treat another portion of the mass with a 1-per-cent. solution of sodium carbonate. Fibrin dissolves completely in this solution if warmed gently several hours on a water-bath. This solution is then filtered and treated with Millon's reagent, when a deep-red color is produced.

NUCLEO-ALBUMIN (MUCIN ?).

It has been well known that a substance exists in nearly every normal urine that is precipitated by acetic acid, insoluble in the latter in excess, but soluble in strong mineral acids. The mucoid appearance of this urinary proteid led to its long being known as mucin. More recent investigations, however, have demonstrated that it does not yield a reducing substance on hydrolysis, while, on the other hand, it is rich in phosphorus. It has, therefore, of late been customary to look upon this body as nucleo-albumin. While it is probably true that minute quantities of nucleo-proteid derived from the cells of the urinary passages are seldom, if ever, absent from the urine, it has, on the other hand, been shown by recent researches made upon large quantities of urine, that the precipitate given by acetic acid contains small quantities of ordinary mucin, or phosphorous-free mucoid, as well as nucleo-proteid. The truth appears to be, therefore, with regard to nucleo-albumin and mucin, that they are both constituents of normal urine. In the majority of cases the amount of these substances in the urine is very minute—so minute, indeed, that it is difficult to demonstrate their presence. In cases of irritation of the urinary passages, catarrhal inflammation attended by free exfoliation of epithelial cells, and in certain febrile conditions of the system there may be great increase of one or both of these substances in the urine. The mucin of the urine has its origin in the urinary passages, notably the lower tract and especially in the bladder; it never originates from the kidneys (which are devoid of muciparous glands), much less from the blood. Nucleo-albumin of the urine, for the most part, originates from the epithelial cells throughout the urinary tract, through some loss of integrity of the cells whereby the protoplasm yields to the macerating influences of the salines contained in the urine, thus effecting more or less solution. Nucleo-albumin and mucin have doubtless lent considerable confusion to a number of the methods of testing for albumin in the urine, notably in those tests requiring previous acidification of the urine with acetic acid. This is evident from the fact that both mucin and nucleo-albumin are precipitated by acetic acid. The author has previously called attention to the fact, however, that by charging the urine with a concentrated saline (sodium chloride) before the addition of the acid, both these bodies are subsequently held in solution in the presence of both acetic acid and heat. Likewise with the ferrocyanic test; if the reagent be added first to the urine it acts as does the saline, and holds these bodies in solution so that they are unaffected by subsequent addition of the acid.

Detection and Distinction.—To detect mucin in the urine, the urine should first be diluted with water, otherwise uric acid may be precipitated upon addition of the acid. The dilution also reduces the solvent influence of the salts of the urine over the mucin. After dilution add an excess of acetic acid. The precipitate may be purified by dissolving in water with a little sodium hydrate and reprecipitated by acetic acid. To distinguish nucleo-albumin from mucin the precipitate must be boiled with a dilute mineral acid, and, if no reducing substance is found by this means, mucin is absent. *Ott's* method of detecting nucleo-albumin in the urine is as follows : To the urine is added an equal quantity of saturated salt solution, and Alméns's tannin solution[1] is slowly added. If nucleo-albumin is present, even in small amount, an abundant precipitate will appear.

---

[1]Alméns's tannin solution is composed of tannin, 5 grammes; 25-per-cent. acetic acid, 10 cubic centimetres; 40-per-cent. methylated spirit, 240 cubic centimetres.

SECTION IV.

CARBOHYDRATES.

THE carbohydrates resemble one another in their chemical composition, in all containing 6 atoms of carbon, or a multiple thereof. They also resemble one another in their chemical characters, being neutral in reaction, not prone to enter into combinations, and, with the exception of inosite, they all possess a strong rotatory power over polarized light.

A number of carbohydrates are met with in the urine,—viz., glucose, levulose, lactose, inosite, etc.,—but the chief clinical interest with regard to this class of compounds at present belongs to grape-sugar.

GLYCOSURIA.

Grape-sugar, or dextrose, in its pure form crystallizes in rhombic tablets, is soluble in its own weight of water, and gives a dextro-rotatory power over polarized light of $= + 56°$ (Hoppe-Seyler). Its solutions become brown when boiled with liquor potassæ, but with picric acid a deep mahogany red. In alkaline solutions it reduces salts of silver, bismuth, mercury, and copper; with the first three the metal is precipitated, while with the last cupric are reduced to cuprous compounds with separation of cuprous oxide. Faintly-alkaline solutions of grape-sugar, colored blue by indigo, when boiled exhibit a beautiful color reaction, beginning with violet and ending in yellow. With solutions of sodium acetate it reduces phenyl-hydrazin hydrochloride to phenylglucosazone, forming highly characteristic and beautiful, golden-yellow, acicular crystals.

Grape-sugar exists in minute quantity in normal blood, varying chiefly with the functional activity of the liver. In some abnormal states of the system the amount of sugar in the blood becomes markedly increased, reaching its maximum— about $\frac{1}{10}$ per cent.—in the more pronounced diabetic conditions.

The ardently-disputed question if sugar be present in normal urine seems to have been conclusively settled in the affirmative

(99)

through the researches of Wedenski (see Section II), although this is by no means universally conceded. Clinically speaking, this question is one of minor importance, since, if sugar be present in normal urine, it exists in such exceedingly minute quantity that it is unrecognizable by the ordinary methods of testing, and such quantities give rise to no marked clinical symptoms.

Clinical Significance.—Glycosuria may appear as a *temporary* condition in the course of a number of diseases, as cholera, intermittent fever, scarlatina, gout, cerebro-spinal meningitis, diseases of the lungs, liver, and the brain, especially if involving the fourth ventricle or vicinity. The administration of certain drugs and toxic substances causes the appearance of sugar in the urine, such as curare, carbonic oxide, amyl nitrite, methyl-delphinin, morphine, chloral, hydrocyanic acid, sulphuric acid, mercury, alcohol, strychnine, salicylic acid, turpentine, uranium nitrate, benzol, acetone, and phloridzin. Recent investigations have shown, however, in some of these cases, that the toxic agent merely causes the appearance of reducing agents[1] in the urine other than sugar.

Glycosuria may be produced experimentally by various lesions, as follows: By puncture of the floor of the fourth ventricle of the brain; injury to the vermiform process of the cerebellum; section of the spinal cord in different locations; destruction of various sympathetic ganglia; section of the splanchnic nerves; irritation of the vagus; section and stimulation of the central end of large motor nerves, as the sciatic; and by total extirpation of the pancreas in dogs. In addition to these, experimental glycosuria may be induced by measures which cause increased determination of blood to the liver, as by tying the accessory branch of the portal vein in animals, irritation of the liver by needle-punctures, compression of the aorta, etc.

Glycosuria in mild form may occur from disorders of the stomach, and, in some cases, the overingestion of starchy and saccharine foods will cause the appearance of sugar in the urine.

Glycosuria, however, of pronounced and *persistent form*

---

[1] Usually glycuronic acid.

belongs to the province of diabetes mellitus, and may always be regarded as symptomatic of grave defects either in the brain, liver, or pancreas. In young subjects the disease is almost uniformly progressive toward a fatal termination in from a few months to four or five years, notably so with patients under 20 years of age. During the middle period of life the disease is often less severe and less fatal. After the age of 50 it is often mild and amenable to treatment. Persistent glycosuria is the most constant and certain of all the symptoms of diabetes mellitus, being, in fact, sometimes the only symptom of the disease, and herein lies its great value for diagnostic purposes. It, therefore, becomes highly essential to be able to readily detect the presence of sugar in the urine, since diagnostic data of the most positive nature often hinges upon this point alone.

Detection of Sugar in the Urine.—The most popular method of searching for sugar in the urine has, heretofore, been by means of the copper tests. These all depend upon the fact, already noted, that in strongly-alkaline solutions grape-sugar reduces cupric oxide to lower grades of oxidation.

1. *Trommer's Test.*—This test may be conveniently performed as follows : About 1 drachm of urine, in an ordinary test-tube, is first treated with sufficient cupric-sulphate solution to render the urine a light-green color ; then an equal volume of liquor potassæ is added. At first a blue precipitate of hydrated cupric oxide results, which dissolves upon shaking the tube, forming a beautiful, clear, blue solution. If allowed to stand half an hour or so, reduction gradually takes place, especially if much sugar be present, resulting in precipitation of yellow or yellowish-red suboxide of copper (cuprous oxide). If, instead of standing half an hour, gentle heat be applied, this test becomes more delicate, and reduction, moreover, occurs at once.

This test is open to two objections : (a) If it be not submitted to boiling it is not very sensitive, only detecting sugar when present in considerable quantity. (b) If boiled especially long the test is rendered oversensitive, so that reaction may occur with substances in the urine other than sugar. The power of reducing cupric oxide in alkaline solution is possessed, to a feeble degree, by a number of substances[1] in both normal and abnormal urine, and iu Trommer's test the quantity of urine submitted to the copper solution is relatively large, which greatly increases the chances of reduction by non-saccharine agents.

2. *Fehling's Solution.*—This solution is best prepared as follows : (1) 34.64 grammes of pure crystallized copper sulphate is dissolved in about 300 cubic centimetres of distilled water by the aid of gentle heat, and (2) 180 grammes of crystallized potassium sodium tartrate (Rochelle salt) together with 70 grammes of caustic soda, or 100 grammes of caustic potash, is likewise dissolved in about 300 cubic centimetres of warm distilled water. When cold the two solutions are mixed and the resulting dark-blue liquid made up to 1000 cubic centimetres with distilled water ; or, what is generally better, the two solutions are each made up to 500 cubic centimetres and kept separate until the test-liquid is needed for use, when carefully-measured equal volumes of the two are mixed. Whether prepared in one solution or two, the liquids should be kept in well-stoppered bottles in a dark, cool place, in order to reduce as far as possible the spontaneous decomposition they are liable to undergo. Apply by placing about 1 drachm of the solution in an ordinary test-tube and boiling. If the solution remain clear, add the suspected urine a few drops at a time, continuing the boiling. If sugar be present the solution assumes an opaque-yellow color, and shortly after a dense yellowish-red sediment falls to the bottom. Should no change occur, the addition of urine may be continued until its volume equals, but *must not exceed*, the volume of the test. In addition to the fact that Fehling's solution slowly undergoes spontaneous decomposition, Seekamp[2] has shown the instability of tartaric acid in aqueous solution on exposure to light ; and, moreover, Fenton[3] has demonstrated that its oxidation takes place in the presence of iron and alkali. Again, very recently M. L. Jovitschitsch has shown[4] that alkaline copper solution freshly prepared according to Fehling's formula deposits cuprous oxide,

---

[1] Chiefly uric acid, creatin, creatinin, hippuric acid, hypoxanthin, tannin, carbolic acid, alkaloids, and glycuronic acid.

[2] Ann. Chem. (Liebig), 278, 373 ; also Rotto, Berichti, 1894, 27, 799.

[3] Jour. Chem. Soc. (London), 1894, 899 ; 1896, 546 ; 1897, 375.

[4] Berichti, 16, 33, 2431.

either at ordinary temperature or when heated, if it has been partially neutralized with sulphuric acid, hydrochloric acid, or nitric acid. This result he ascribes to decomposition of the tartrates present in the liquid. Finally, Tingle has reviewed and confirmed these observations,[1] and, moreover, adds: "Purdy's formula is not thus affected by the presence of mineral acids. Specimens treated in the same manner described by Jovitschitsch, both at ordinary temperature and at the boiling-point, gave no sign of reduction. The reduction of Purdy's formula is shown by the weakening or complete discharge of the blue color, and not by the production of a precipitate, as in the case of Fehling's solution; and it may be kept indefinitely without undergoing change."

The author desires it to be distinctly understood that he does not advise the use of his formula for qualitative work, but only in quantitative determination of sugar, for which it was designed.

3. *Haines's Test.*—The best form of copper test for sugar, qualitatively, is that devised by Prof. Walter S. Haines, of Chicago. The original formula for this test is as follows: Pure cupric sulphate, 30 grains; pure glycerin, 4 drachms; caustic potash (in sticks), 3 drachms; distilled water, to 6 ounces. The solution is prepared by dissolving the cupric sulphate and glycerin in part of the water and the caustic potash in the remainder. Mix the two solutions.

For purposes of greater convenience, Professor Haines has recently simplified this formula, as follows: Take pure copper sulphate, 30 grains; distilled water, ½ ounce; make a perfect solution, and add pure glycerin, ½ ounce; mix thoroughly, and add 5 ounces of liquor potassæ.

In testing with this solution, take about 1 drachm and gently boil it in an ordinary test-tube. Next add from 6 to 8 drops—*not more*—of the suspected urine, and again gently boil. If sugar be present, a copious yellow or yellowish-red precipitate is thrown down. If no such precipitate appear, sugar is absent. This test is stable, and, though kept on hand indefinitely, it may always be depended upon to be in order for testing.

With regard to the copper tests in general: It is important to bear in mind that, if boiling be continued too long, slight

---

reduction is apt to occur with the urine, even when free from sugar, as evidenced by a slight *greenish* (not yellow) opacity; about half a minute should constitute the usual limit of boiling. It should also be borne in mind that strongly alkaline solutions are apt to precipitate the earthy phosphates of calcium and magnesium of normal urine in the form of a grayish cloud, which should not be mistaken for sugar. Albumin, if present in more than traces, reduces the delicacy of the copper tests, and should, therefore, first be removed by slightly acidulating with acetic acid, boiling, and filtering.

4. *The Fermentation Test.*—This test depends upon the fact that grape-sugar is decomposed in the fermentation set up by yeast, yielding alcohol, carbon dioxide, succinic acid, and a number of other products, with resulting decrease in the specific gravity of the urine. The following is the method employed: Fill an ordinary test-tube half full of mercury and the remaining half with the urine to be tested, and introduce into the urine a small piece of German yeast. Next close the mouth of the test-tube with the thumb and invert over a small vessel of mercury, and set aside in a warm room for several hours. If sugar be present fermentation proceeds at once, liberating carbonic-acid gas, which collects in the upper end of the tube, displacing the urine and mercury more or less, according to the quantity of sugar present. One precaution should be observed. Some specimens of yeast spontaneously evolve gas, and it is, therefore, best to perform a parallel experiment with yeast mixed with water, so that the spontaneously evolved gas may be estimated.

This test will detect sugar, if present in considerable quantity,—1 per cent. and over,—but the capacity of the urine itself for absorbing carbonic gas renders this test uncertain in detecting smaller quantities of sugar. The more serious disadvantage of this test for practical work is the fact that it requires several hours to complete the analysis.

5. *The Bismuth Test.*—Sugar possesses the power of reducing bismuth salts with resulting black precipitate, and upon this fact Böttger first suggested the following simple test: First add to the urine an equal volume of liquor potassæ and then a

PLATE IV.

CRYSTALS OF PHENYLGLUCOSAZONE.
(After v. Jaksch.)

little basic nitrate of bismuth. Gently boil the whole, and if sugar be present the test turns gray or black, according to the amount of sugar present.

Traces of sulphur, which are often present in the urine, cause the same reaction with this test as does sugar. For the above reason, albumin, if present, must be removed before applying this test.

In order to eliminate the errors due to the presence of sulphur, Brucke suggested the following improvement of the bismuth test: Fill a test-tube half full of the suspected urine and another with a similar volume of water, and stand them side by side. To the one containing water add hydrochloric acid until a drop of Frohm's reagent[1] no longer produces a cloudiness. By this means an approximate estimation may be made of the quantity of hydrochloric acid which should be added to the urine. Acidify the urine with such quantity of hydrochloric acid, add the reagent, and filter. The filtrate, which should now remain clear upon adding hydrochloric acid, or the reagent, should be boiled with an excess of potassium or sodium hydrate, as in Böttger's method; and, if a gray or black color result, sugar is present.

6. *Phenyl-hydrazin Test.*—This test, suggested by Fischer, depends upon the power, possessed by phenyl-hydrazin hydrochloride, of forming with grape-sugar a highly-characteristic crystalline compound, termed *phenylglucosazone.* The best method of applying this test is as follows: To 25 cubic centimetres of suspected urine add 1 gramme of phenyl-hydrazin hydrochloride, 0.75 gramme of sodium acetate, and 10 cubic centimetres of-distilled water in a capsule. The capsule should be placed in a water-bath and warmed at least an hour, then removed and allowed to cool; and if sugar be present even in minute quantity, there forms a yellowish deposit, which may appear amorphous to the naked eye, but which, when examined under the microscope, is seen to contain fine, bright-yellow, needle-like crystals, either single or in stars—*phenylglucosazone,*

---

[1] Frohm's reagent is prepared by mixing 1.5 grammes of freshly-precipitated bismuth subnitrate with 20 grammes of water, heated to boiling; then 7 grammes of potassium iodide and 20 drops of hydrochloric acid are added.

—which melt at 204° C. (Plate IV). The presence of small or large yellow scales or powerfully-refracting brown spherules must not be taken for evidences of sugar, as only the bright-yellow, needle-like crystals are conclusive.[1]

V. Jaksch applies this test as follows: 2 parts of phenyl-hydrazin hydrochloride and 3 parts of sodic acetate are placed in a test-tube with 6 to 8 cubic centimetres of urine. If the salts do not dissolve on warming, a little water is added and the test-tube is placed in boiling water. After twenty minutes it is removed to cold water, and, if sugar be present, the characteristic crystals are soon deposited.

This test gives very trustworthy results with every variety of morbid urine, and is therefore equally applicable whether albumin be present or not. It gives no reaction with uric acid, urates, creatin, creatinin, oxybutyric acid, urochloralic acid, uroxanthic acid, tannin, morphine, salicylic acid, or carbolic acid.

A number of other tests have been proposed for the detection of sugar in the urine, among which may be mentioned: Boiling with sodium or potassium hydroxid (Moore), picric acid (Johnson), acetate of lead and ammonia (Rubner), alpha-naphthol and thymol (Molisch), indigo carmine (Mulder), bichloride of tin (Maumene), chromic acid (Hunefeld), diazo-benzol-sulphonicacid (Penzoldt), and sulphuric acid (Runge). None of the above tests possess special advantages over those already described, while, on the contrary, most of them are greatly inferior.

In addition to these, tests for sugar are prepared in paper form, by charging chemically-inert filtering paper with sodium carbonate and indigo carmine. The chief merit of these papers is their portability, since they are neither sensitive nor trust-worthy in detecting sugar in the urine.

Testing for Sugar in the Urine.—In searching for sugar in the urine, a test should, if possible, be selected which is simple in application, reasonably trustworthy, and perfectly stable, so

---

[1] In manipulating the phenyl-hydrazin hydrochloride caution should be observed not to get it on the hands, as it often causes troublesome eczema of very acute grade.

that it may be depended upon when required, in routine work. In these respects Haines's test is very satisfactory.

Before submitting the urine to reagents always thoroughly cleanse all of the utensils to be employed in the analysis. In the use of the copper tests, *always employ a minimal amount of urine at first, gradually increasing until reaction is obtained or the stated limit of urine be reached.* This method greatly diminishes the chances of reduction by other substances in the urine than sugar, and, moreover, it gives a rough idea of the quantity of sugar when present. Thus, if Haines's test be selected, after gently boiling a drachm of the solution add 2 or 3 drops of urine; then wait a moment, to see if reduction occur, before adding more urine, meantime continuing the boiling; if no reaction occur after a few seconds, add 2 or 3 more drops of urine, and so on until 8 drops be added, *but no more.*

If Fehling's solution be employed, gently boil a drachm of the test and, if it remain clear, add but a few drops of the suspected urine; then pause for a moment, continuing the boiling, and, if no reaction occur, add a few more drops of urine, and so on. If no reaction occur, continue the addition of urine until the volume thereof equals that of the test-solution, *but not more.*

If any doubts arise as to the presence of sugar in the suspected urine, after the application of a routine test, an appeal should be made to one or more of the others described. For such purpose the phenyl-hydrazin test is desirable above all others, both because of its exceeding delicacy and its property of reacting with practically no substances in the urine other than grape-sugar.

Determination of Sugar in the Urine.—Having detected the presence of sugar in the urine, it becomes all important to determine its quantity in all cases, because (*a*) such knowledge furnishes the most precise evidence of the grade and severity of the diseased state upon which it depends; (*b*) it furnishes the most solid basis upon which to construct the diagnosis; (*c*) it gives the most trustworthy evidence as to the results of treatment.

1. *The Fermentation Method.*—The best results are reached

with this test after the method suggested by Roberts.[1] It has already been stated that grape-sugar is decomposed in the fermentation set up by yeast, yielding alcohol, carbon dioxide, and a number of other products, with resulting *decrease in the specific gravity* of the urine. Each degree of specific gravity lost in fermentation corresponds to 1 grain of sugar per fluidounce. Thus, if before fermentation the specific gravity of the urine be 1.040, and after fermentation it be 1.020, it will have contained 20 grains of sugar per ounce. The method advised by Roberts is as follows: About 4 ounces of saccharine urine are put into a 12-ounce bottle, and a lump of German yeast is added to it. The bottle is then corked with a nicked cork (which permits the escape of the carbon dioxide) and set aside, in a warm place, to ferment. Beside this is placed a tightly-corked 4-ounce bottle, filled with the same urine, but without any contained yeast. In about twenty-four hours the fermentation will have ceased. The fermented urine is then decanted into a urine glass and its specific gravity is taken; at the same time the density of the unfermented urine in the companion bottle is observed and the *density lost* ascertained. The degrees of density lost represent the number of grains of sugar in each fluidounce of the urine tested. The percentage may be approximately ascertained by multiplying the number of degrees lost by 0.23.

The chief objection to this method for clinical work is the fact that it requires from eighteen to twenty-four hours to complete the analysis.

2. *The Author's Method.*—The formula for the author's standard solution is as follows: Pure cupric sulphate, 4.752 grammes; potassium hydroxid, 23.50 grammes; strong ammonia (U. S. P.; sp. gr., 0.9), 350 cubic centimetres; glycerol, 38 cubic centimetres; distilled water, to 1000 cubic centimetres (1 litre). Prepare by dissolving the cupric sulphate and glycerin in 200 cubic centimetres of distilled water with the aid of gentle heat. In another 200 cubic centimetres of distilled water dissolve the potassium hydroxid. Mix the two solutions, and when cooled add the ammonia. Finally, with distilled water bring the

---

[1] Edinburgh Monthly Journal, October, 1861.

TABLE.

By means of this table the amount of sugar in the urine may be readily calculated with the author's formula (metric), both in percentage amount and in grains per fluidounce, from one titration. The table indicates all ratios of reduction in fractions of hundreths per cent. from 5 per cent. down to 0.2 of 1 per cent. when undiluted urine is titrated with 35 cubic centimetres of the test-solution.

Degrees of the burette C. C.	Percentage of sugar.	Grains of sugar per fluid-ounce.	Degrees of the burette C. C.	Percentage of sugar.	Grains of sugar per fluid-ounce.	Degrees of the burette C. C.	Percentage of sugar.	Grains of sugar per fluid-ounce.
0.4	5.	24.	0.76	2.63	12.6	1.65	1.21	5.8
0.41	4.88	23.4	0.77	2.6	12.5	1.7	1.18	5.7
0.42	4.76	22.8	0.78	2.56	12.3	1.75	1.14	5.5
0.43	4.65	22.3	0.79	2.53	12.1	1.8	1.11	5.3
0.44	4.55	21.8	0.8	2.5	12.	1.85	1.08	5.2
0.45	4.44	21.3	0.81	2.47	11.9	1.9	1.05	5.
0.46	4.35	20.9	0.82	2.45	11.8	1.95	1.03	4.9
0.47	4.26	20.4	0.83	2.41	11.6	2.	1.	4.8
0.48	4.17	20.	0.84	2.38	11.4	2.1	0.95	4.6
0.49	4.08	19.6	0.85	2.35	11.3	2.2	0.9	4.3
0.5	4.	19.2	0.86	2.33	11.2	2.3	0.87	4.2
0.51	3.92	18.8	0.87	2.3	11.	2.4	0.83	4.
0.52	3.85	18.5	0.88	2.27	10.9	2.5	0.8	3.8
0.53	3.77	18.1	0.89	2.25	10.8	2.6	0.77	3.7
0.54	3.7	17.8	0.9	2.22	10.7	2.7	0.74	3.6
0.55	3.64	17.5	0.91	2.2	10.6	2.8	0.72	3.5
0.56	3.57	17.1	0.92	2.17	10.4	2.9	0.7	3.4
0.57	3.5	16.8	0.93	2.15	10.3	3.	0.66	3.2
0.58	3.45	16.6	0.94	2.13	10.2	3.25	0.61	2.9
0.59	3.4	16.3	0.95	2.1	10.1	3.5	0.57	2.7
0.6	3.33	16.	0.96	2.08	1C.	3.75	0.53	2.5
0.61	3.28	15.7	0.97	2.06	9.9	4.	0.5	2.4
0.62	3.23	15.5	0.98	2.04	9.8	4.25	0.47	2.3
0.63	3.17	15.2	0.99	2.02	9.7	4.5	0.44	2.1
0.64	3.12	15.	1.	2.	9.6	4.75	0.42	2.
0.65	3.06	14.8	1.1	1.82	8.7	5.	0.4	1.9
0.66	3.03	14.5	1.15	1.74	8.4	5.5	0.36	1.7
0.67	3.	14.4	1.2	1.66	8.	6.	0.33	1.6
0.68	2.94	14.1	1.25	1.6	7.7	6.5	0.31	1.5
0.69	2.9	13.9	1.3	1.54	7.4	7.	0.29	1.4
0.7	2.86	13.7	1.35	1.5	7.2	7.5	0.27	1.3
0.71	2.82	13.5	1.4	1.43	6.9	8.	0.25	1.2
0.72	2.78	13.3	1.45	1.38	6.6	9.	0.22	1.1
0.73	2.74	13.1	1.5	1.33	6.4	10.	0.2	1.
0.74	2.7	13.	1.55	1.29	6.2			
0.75	2.67	12.8	1.6	1.25	6.			

Example.—If 35 cubic centimetres of the standard test-solution be reduced (decolorized) upon boiling by 0.4 cubic centimetre of the undiluted urine, the latter contains exactly 5 per cent. of sugar, or 24 grains per ounce; if it require 1 cubic centimetre of urine to reduce the 35 cubic centimetres of test-solution, there is just 2 per cent. of sugar, or 9.6 grains per ounce; if it require 4 cubic centimetres of urine to reduce the test, there is just 0.5 per cent., or 2.4 grains of sugar per ounce, etc.

Note.—If the percentage of sugar be high (above 5 per cent.), the urine should be diluted by three volumes of water, in which case the product should be multiplied by 4 after using the table.

volume of the whole to exactly 1000 cubic centimetres (1 litre). The principle of the test depends upon the fact that, in the reduction of cupric oxide in solutions of definite strength by grape-sugar, the blue coloration disappears by addition of a definite quantity of grape-sugar—35 cubic centimetres to 0.02 gramme of sugar—without any attendant precipitate, but leaving the reduced solution perfectly transparent and colorless. Thirty-five cubic centimetres of this solution are reduced by exactly 2 centigrammes (0.02 gramme) of grape-sugar.

The analysis is best conducted as follows: Have on hand a glass flask (150 or 200 cubic centimetres' capacity), an ordinary retort-stand, a 10-cubic-centimetre finely-graduated burette, and a large spirit-lamp or Bunsen burner.   Proceed by measuring accurately 35 cubic centimetres of the test-solution into the flask, add water to nearly half-fill the flask, and bring the contents thereof to the boiling-point.   Fill the burette with the urine to be tested and *slowly discharge* the urine into the boiling test-solution, drop by drop, until the blue color *begins to fade;* then *still more slowly,* three to five seconds elapsing after each drop, until the blue color completely disappears and leaves the test-solution perfectly *transparent and colorless.* The number of cubic centimetres required to discharge the blue coloration in 35 cubic centimetres of the test contain exactly 2 centigrammes (0 02 gramme) of sugar.[1]   The following, therefore, is the percentage relationship of reduction of the test: If it require 2 cubic centimetres of urine to reduce 35 cubic centimetres of the test-solution, there is present exactly 1 per cent. of sugar; if it require but 1 cubic centimetre of urine to reduce the 35 cubic centimetres of test-solution there is present 2 per cent. of sugar; if it require $\frac{1}{2}$ cubic centimetre of urine to reduce the test, there is 4 per cent. of sugar present; but if it only require $\frac{1}{4}$ cubic centimetre of urine to reduce the 35 cubic centimetres of test-solution, 8 per cent. of sugar is present.

The best results are reached by first diluting the urine before

---

[1] It will be noted, after testing, that upon standing some time the test-solution slowly resumes the blue color again. This is due to re-oxidation, which is somewhat rapid, and should not be mistaken for imperfect reduction or defect in the test-solution.

titration.   Of course, any dilution may be employed and the result multiplied by the number of dilutions.   If the quantity of sugar be very large (over 5 per cent.), 3 volumes of water and 1 of urine will prove best, and multiply the product by 4. If the quantity of sugar be small (2 per cent. or less), 1 volume of•urine to 1 of distilled water is better, and multiply the result by 2.

Those who find it more convenient to work with the English weights and measures may obtain the same rapid and accurate results by using the following formula and methods : Formula—pure cupric sulphate, 44 grains ; potassium hydroxid, 214 grains ; strong ammonia (U. S. P. ; sp. gr., 0.9), 9 fluidounces ; glycerol (pure), 6 drachms ; distilled water, to 20 ounces.

Prepare by dissolving the copper sulphate and glycerin in 4 ounces of distilled water, with the aid of gentle heat.   In another 4 ounces of distilled water dissolve the potassium hydroxid. Mix the two solutions and cool to 65° F. ; then add the ammonia.   Finally, with distilled water bring the volume of the whole to exactly 20 ounces.   Ten drachms of the above solution are reduced upon boiling by exactly ½ grain of sugar.   ·The test is conducted in the same manner as described in the metric system, save that a minim burette is substituted for the cubic-centimetre burette, and the result is reckoned in grains instead of grammes.[1] The best results are obtained by diluting the urine with 2 volumes of distilled water before titration ; then the number of minims in 1 ounce (480) are divided by the number of minims of diluted urine it takes to reduce the 10 drachms of the test, and the product is the number of grains of sugar present in each ounce of urine tested.

Example.—If 48 minims of diluted urine (1 to 2) reduce 10 drachms of the test-solution, there are just 10 grains of sugar per ounce :—

$$\frac{480}{48} = 10 \text{ grains of sugar.}$$

---

[1] Messrs. Richards & Co., of Chicago and New York, prepare and keep in stock this test, both in metric and English weights.   They also manufacture and keep in stock the burettes in both systems, for conducting titration with this test.

If it be desired to know the percentage-amount of sugar present, divide the number of grains of sugar per ounce by 4.8.

The detection and determination of sugar by means of the copper tests is a matter of some historical interest. Trommer first proposed in 1841[1] to detect the presence of sugar in the urine by means of its reducing power over cupric oxide in strongly-alkaline solutions. About seventeen years later Fehling proposed[2] the formula that bears his name as an improvement on Trommer's test, and subsequently this became a means of quantitative determination by applying the titration method thereto, and this became almost universally standard for the purpose for many years. Ultimately, however, it became well known that Fehling's formula was faultily constructed, owing to the instability of its contained tartaric salts. Later on, Schmiedeburg proposed to remedy this defect by substituting for the tartrate pure mannite. This somewhat improved, but by no means corrected, the instability of the test-solution, and accordingly Pavy again modified Fehling's original formula, substituting potassium hydroxid for sodium hydroxid, and furthermore divided the test into two solutions, keeping each solution separately until required, as follows :—

1. Neutral potassium tartrate, 640 grains; potassium hydroxid, 1280 grains ; water, 10 ounces.

2. Cupric sulphate, 320 grains; water, 10 ounces.

Later on Pavy[3] proposed to keep the elements of the two solutions in the form of pellets. The latter, though convenient, were found, like the original formula, liable to undergo change.

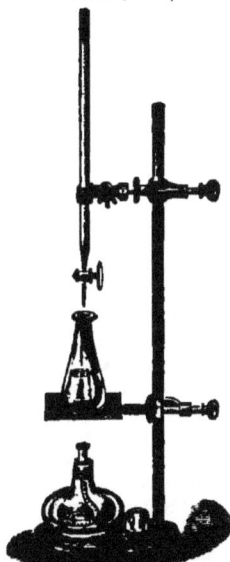

FIG. 9.—THE AUTHOR'S APPARATUS FOR QUANTITATIVE DETERMINATION OF SUGAR IN URINE.

The difficulties encountered in qualitative testing for sugar with Fehling's formula were slight, however, as compared with those in quantitative work. The reduction of cupric oxide being attended by a precipitate greatly obscures the end-reaction, entailing tedious delay and uncertainty. In order to remedy this Pavy brought forward his ammonio-cupric test[4] for quantitative work, as fol-

[1] Ann. Chem. und Phar., xxxix, 360, 1841.
[2] Ann. Chem. und Phar., lxxii, 106, 1848 ; cvi, 75, 1858.
[3] Clinical Soc. London, January, 1880.
[4] Chemical News, xl, 77.

lows : Cupric sulphate, 4.158 grammes ; potassium sodium tartrate, 20.4 grammes; strong ammonia-water (sp. gr., 0.880), 1 litre. Ten cubic centimetres of this is equal in oxidizing power to 0.005 gramme of sugar, the reduction being attended by disappearance of the blue color without any resultant precipitate, the latter being dissolved by the ammonia. This undoubtedly constituted the greatest advance in the copper test for quantitative work yet brought forward, for which the profession is under perpetual obligations to Pavy. One unfortunate draw-back remained, however, in the fact that Pavy still retained the treacherous tartaric salt.

In the autumn of 1888, while in London, Dr. Pavy kindly demonstrated his ammonio-cupric test to the author in his private laboratory. Being especially struck with the beautiful end-reaction thereof, the test was put in use in the author's laboratory. But a brief experience was necessary to prove the im-mense convenience of the improved end-reaction ; but it also demonstrated the instability of the test-solution, and this the author set about to remedy if possible.

As the result, a year or so later the author brought forward his formula after considerable experimentation. The ammonia was retained to secure the striking and beautiful end-reaction, but the tartrate was substituted by glycerol as the organic element, which has since proved entirely satisfactory in preserving the stability of the solution however long it be kept on hand. The test has since been carefully worked out, both in the metric and English systems, and its accuracy checked by thousands of practical determinations. Theoretically, a few objections have been raised against the test, chiefly by those who have not used it, while one or two have objected apparently on general principles, not having given any reasons for their objections. Thus it has been claimed, since ammonio-cupric solutions are so highly oxidizable, that, unless measures be adopted to exclude the atmosphere while testing, reduction is sufficiently inter-fered with to prevent accurate results. This objection is more theoretical than practical. Repeated experiments in the author's laboratory in testing, in which measures were adopted to exclude the atmosphere during titration, such as passing a current of coal-gas through the flask, and even overlaying the test-solution with paraffin-oil, conclusively proved that the variation of reduction was too small to be reckoned.

One distinguished chemist objected to the test on the grounds that the fumes of ammonia evolved in testing were disagreeable in the laboratory. If such were in reality so, it would be the simplest possible matter to conduct the fumes through a glass tube and perforated cork into a wash-bottle containing dilute hydrochloric acid, which would not only dispose of this, but also prac-tically of the previous objection.

These devices have all been tested in the author's laboratory, where quan-titative sugar testing is often in progress for hours at a time, and for the most part they have been abandoned as unnecessary. If the flask containing the test-solution be fitted with a piece of ordinary glass tubing about one quarter of an inch in diameter, with a bend at right angle above the mouth of the flask, it will serve both to conduct the fumes of the ammonia aside and also to retard the ingress of air sufficiently to preserve the absolute accuracy of the testing, even to the satisfaction of the most scrupulous manipulator. It has been found that when it has been thus manipulated the ammonio-cupric test corresponds in results with great accuracy to the gravimetric method.[1]

3. *Fehling's Test.*—This is conducted by the titration method with Fehling's solution, the formula of which has already been given both in one solution and also divided into two parts, to be mixed when used (see page 102). The latter form is always preferable. The principle of the process is the same as in the author's method, previously described, save that reduction is accompanied by a precipitate in addition to disappearance of the blue coloration of the test. Each 10 cubic centimetres of Fehling's solution corresponds to 0.05 gramme of sugar, or 200 grains to 1 grain of sugar. The urine should be diluted to a known degree,—usually 1 to 10,—unless the quantity of sugar be very small, in which case 1 to 5 is better. Titrate precisely as with No. 2, save that, after each few drops discharged from the burette, the test should be allowed to stand for a short time, so that the precipitate may settle and the observer may see if the mixture contain any blue color. As soon as the blue color has disappeared, the quantity of diluted urine employed is read off, and since it takes just 0.05 gramme of sugar to remove the blue coloration in the 10 cubic centimetres of the test-solution, from this the percentage of sugar in the urine may be reckoned.

4. *Optical Saccharimetry.*—As already stated, grape-sugar possesses a right-rotatory power over polarized light, and upon this fact has been based a method of quantitative testing for sugar by the polariscope. Among the more elaborately-constructed instruments for this purpose are those of Lippich, Misterlich, Soleil, Laurent, Wild, and von Fleischel.

Ultzmann has devised a polarizing saccharimeter, which possesses several important advantages over the instruments named, as follows : (*a*) No artificial light is needed, for the concave mirror of the microscope-stand brilliantly illuminates the field

---

[1] The superior stability of the ammonio-cupric test over other copper tests is largely due to the fact that the ammonia constantly maintains a high degree of oxidation of the copper salt in solution.

of vision. (*b*) The apparatus itself is small, scarcely longer than the elongated tube of the ordinary microscope, and needs no separate stand. (*c*) By means of this instrument the percentage of sugar can be directly calculated. (*d*) The entire apparatus can be had for a comparatively small cost.

FIG. 10.—ULTZMANN'S POLARIZING SACCHARIMETER ADJUSTED TO
MICROSCOPE-STAND.

In using this saccharimeter, the tube, objectives, and ocular of the microscope are removed, and in their place the saccharimeter is inserted and made fast by means of a small screw. The concave mirror is then adjusted, and, by looking through the instrument, it is determined whether or not it is properly adjusted.

In Fig. 11 *a* is the biconcave and *b* the objective lens of a small Dutch telescope, the focal distance of which extends to *p ; c* is the upper Nicol prism, with which a vernier is closely connected ; *d* is a glass tube for holding the suspected fluid, which should be filtered or otherwise cleansed before analysis ; *p* is a double plate of right and left rotating quartz, and *f* the lower Nicol prism.

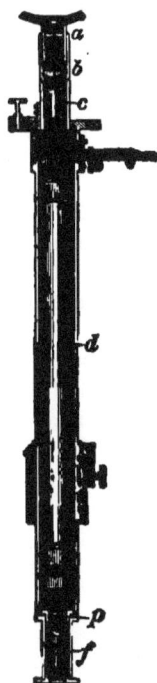

FIG. 11.—SECTIONAL VIEW OF ULTZMANN'S POLARIZING SACCHARIMETER.

The arc or fixed scale is so divided that one division of it represents 1 per cent. of grape-sugar at a temperature of 20° C. By means of the vernier, *tenths of a degree (i.e., of 1 per cent.)* can be approximately determined. Since 10 degrees of the vernier correspond exactly with 9 degrees of the arc, to the percentage of sugar found must be added as many tenths as spaces are counted on the vernier up to that division which exactly coincides with a division of the arc.

If, for example, the zero-point of the vernier does not quite reach (toward the right) the 5-point of the scale, it indicates that the percentage of sugar is more than 4 and less than 5 *per cent.* If it be desired to estimate *the tenths per cent.*, and the *sixth* division of the vernier is the first (counted from the zero-point) to coincide with a division of the arc, then 6 is the number of *tenths* required, and the apparatus would indicate, in this case, 4.6 per cent. grape-sugar present. In estimating the strength of cane-sugar solutions, it is to be borne in mind that the polarization power of cane-sugar is three-fourths that of grape-sugar.

Cane- and grape- sugar, as well as lactose, turn the polarized ray to the right, while albumin and levulose, on the other hand, turn it toward the left. If the glass tube of the saccharimeter be empty, or contain a fluid holding in solution substances having no optical influence (as normal urine), the zero-point of the vernier coincides exactly with the zero-point of the scale, and the two halves of the field of vision are exactly isochromatic. If, on the contrary, an optically-active substance be contained in the fluid,—as, for example, sugar,—the normal isochromatism of the two halves disappears, and a distinctly unequal coloring takes place. This is the more apparent the greater the amount of optically-active substance present in solution. When this unequal coloring occurs the vernier is to be moved toward the right or left (according to the presence of sugar or albumin) until the color of the two halves is again exactly the same. The percentage is then read off the scale in the way above mentioned.

If a diabetic urine be very light colored and clear, it can at once be put into the glass tube of the instrument and the determination made. If, however, it be dark and cloudy, and contain albumin, it is advantageous to first clarify it and remove all disturbing substances. This is best accomplished by means of a 10-per-cent. aqueous solution of sugar of lead. The lead acetate causes in urine a copious white precipitate, consisting of lead chloride, phosphate, and sulphate, and the precipitate carries down with it all the coloring matter of the urine and such albumin as may be present. If the urine be then passed through a dry filter, the resulting filtrate is almost as clear as water, and is particularly well adapted for the apparatus. Since, however, the amount of sugar in the mixture (after the addition of the lead-acetate solution) differs from that in the urine, the amount of dilution must be taken into account in estimating the sugar present. It is best, therefore, to take 75 cubic centimetres of urine, and to that add 25 cubic centimetres of lead-acetate solution, shake, and filter. In estimating the sugar present in the urine, one-third of the percentage of the mixture added to that percentage will give the percentage of sugar in 100 cubic centimetres of urine. In other words, the percentage of the mixture is three-quarters that of the urine. Thus, if to 75 cubic centimetres of a dark, albuminous, saccharine urine have been added 25 cubic centimetres of lead-acetate solution, the mixture filtered, and found to contain 4.8 per cent. sugar, then 1.6 per cent. must be added to give the percentage in 100 cubic centimetres of urine, which would, therefore, contain 4.8 + 1.6 per cent. = 6.4 per cent. sugar. In filling the glass tube care must be observed that no air-bubbles are included in the fluid. It is well, therefore, to fill the tube as full as possible and push the glass cover on from one side before screwing down the cover (Ultzmann).

The author uses this instrument, obtaining fairly rapid and satisfactory results if the quantity of sugar be over 1 per cent., but in quantities much less than this the results are uncertain.

LEVULOSURIA.

Fruit-sugar, or levulose, has been found in the urine of persons whose symptoms correspond closely with those of diabetes mellitus. In such cases the levulose may be associated with grape-sugar, or it may appear alone, but usually the former is the case.

Levulose turns the plane of polarization to the left, and this fact enables us to distinguish it from grape-sugar, which turns it to the right. Levulose reduces copper salts as does grape-sugar, although more feebly than the latter. It also yields the characteristic reaction of yellow crystallization with phenyl-hydrazin hydrochloride, and the crystals thus obtained are identical with those formed from grape-sugar. Levulose does not crystallize, and does not melt so readily as does grape-sugar. When cane-sugar is treated with dilute mineral acids, it undergoes a process known as inversion,—i.e., it takes up water and is converted into a mixture of equal parts of dextrose and levulose.

Clinical Significance.—Aside from the fact that levulose is sometimes found in the urine in diabetic conditions, either alone or, as is more common, in association with grape-sugar, little else is known of its clinical relations. It has been stated that excessive ingestion of cane-sugar, as well as the sugars of certain kinds of fruits, may cause the appearance of levulose in the urine, more especially in conditions of disturbed digestion. This, however, is rather conjectural than the result of observation, although cane-sugar is converted in the intestines into glucose and levulose.

Detection.—If saccharine urine deflect polarized light strongly to the left, we may infer that the saccharine substance is levulose. If other known substances which turn polarized light to the left be excluded, it may be regarded as certain that levulose is present.

LACTOSURIA.

Lactose, or milk-sugar, crystallizes in white, rhombic prisms, which are soluble in 6 parts of cold and $2\frac{1}{2}$ parts of hot water. It has only a faintly-sweet taste, and is insoluble in alcohol and ether. Aqueous solutions of lactose possess a right, or dextro-rotatory, power over polarized light of $+ 59.3°$, and do not readily undergo alcoholic fermentation. It reduces the copper salts upon boiling in alkaline solutions, but about one-third less powerfully than does grape-sugar. If long boiled with dilute acids, it forms *galactose*, which, treated with nitric acid, yields *mucic acid*.

Clinical Significance.—Lactose occurs frequently in the urine of women who are nursing, the quantity usually being small, although it may reach as high as 3 per cent., and be attended by all the usual symptoms of diabetes, as in a case reported by Ralfe, at the London Hospital. In this case the woman was suffering from debility, and lactosuria occurred after three successive confinements, the urine being free from sugar during the gestation.

Lactose is nearly always present in the urine of women two or three days after confinement, and just before milk appears in the mammary glands (during milk fever); and the same may be said of women within a day or two after weaning their children. Lactosuria may also arise from any cause that prevents the milk escaping from the mammary glands during lactation, such as inflammations involving the mammary ducts.

Detection.—If urine give the characteristic reaction of grape-sugar with alkaline solutions of cupric salts, and if it also cause extreme deflection of the polarized ray to the right, it is probable that lactose is present. Confirm by treating the urine with an excess of lead acetate, filter, and to the filtrate add ammonia until a permanent precipitate forms. The fluid is then heated, but not boiled ; and if a rose-red color gradually develop, which slowly vanishes on standing, grape sugar is present; if no such reaction occur, lactose is present (Rubner).

INOSITURIA.

Inosite, or muscle-sugar, crystallizes in large, colorless, monoclinic prisms, sometimes grouped in rosettes. They are soluble in 6 parts of water at 20° C., insoluble in alcohol and ether. Inosite does not undergo alcoholic fermentation, and possesses no rotatory power over polarized light. It does not reduce alkaline solutions of cupric salts, although it gives with them a greenish tint upon boiling, which clears up on standing, and again turns green on boiling.

Although termed muscle-sugar, inosite has been found in the lungs, spleen, liver, kidneys, and brain, and it has been found in the urine in a number of pathological conditions.

Clinical Significance.—Inosituria has frequently been noted in association with diabetic conditions. It has also been observed in typhus, phthisis, syphilitic cachexia, and in diseases of the medulla. In a number of cases of inosituria Ralfe observed moderate polyuria, loss of flesh, general malaise, and aching in the limbs, although no tangible disease could be made out. Inosituria not infrequently takes the place of glycosuria, especially in the milder grades of diabetes or in convalescence from the latter. Inosituria is also occasionally associated with albuminuria in Bright's disease. Gallois found inosite in the urine of 7 out of 102 patients examined. Of these, it occurred 5 times in 30 cases of diabetes and in 25 cases of albuminuria.

Detection.—1. If a solution of inosite be evaporated with a little nitric acid on platinum almost to dryness, and the residue be moistened with a little ammonia and solution of calcium chloride, and the mixture again be evaporated carefully to dryness, a vivid rose-red or violet color arises, which is apparent with even 1 milligramme of inosite (Scherer). Other sugars do not give this reaction.

2. Add a little mercuric nitrate solution[1] to a solution of inosite on a porcelain dish; a yellow precipitate is produced. On heating this gently it will become red; on cooling the color vanishes, but re-appears again on gently heating. Uric acid, urea, starch, lactose, mannite, glycocoll, tannin, cystin, and glycogen do not give this reaction. Albumin is colored red and grape-sugar black, and therefore, if present, these should be removed.

3. Inosite may be isolated from the urine, as follows: After first removing the albumin, if present, the urine is treated with neutral lead acetate until precipitation ceases. It is then filtered and the warmed filtrate treated with subacetate of lead as long as any precipitate occurs. It is well to concentrate the urine to one-fourth of its bulk before precipitation. The lead precipitate containing the inosite combined with lead oxide is collected after twelve hours, and, after washing, is suspended in water and decomposed with sulphuretted hydrogen. The lead sulphide produced is removed by filtration, and the filtrate, upon standing awhile, usually deposits a little uric acid. This is separated by filtering, and the filtrate, after concentration by boiling, is treated with three or four volumes of alcohol while boiling. A heavy precipitate results, and the hot alcoholic solution is poured off, unless the precipitate be flocculent and non-adhesive, in which case it may be filtered through a heated funnel and allowed to cool. If, after twenty-four hours, inosite crystals have deposited, as is usual, in groups, they are filtered and washed with cold alcohol. If, however, no crystals of inosite have separated, ether is added to the clear, cold alcoholic filtrate until a milky cloudiness results upon standing, and it is then allowed to stand in the cold for twenty-four hours. If enough ether has been employed, almost all of the inosite present is separated in the form of shining, pearly leaflets.

ALLIED SUBSTANCES OCCASIONALLY FOUND IN URINE.

GLYCURONIC ACID.

This substance is closely allied to carbohydrates, its formula being $C_6H_{10}O_7$. When pure it is not crystalline, but its anhydride forms colorless, acicular crystals. It is insoluble in ether, but soluble in water and hot alcohol, crystallizing

---

[1] Prepared by dissolving 1 part of mercury in 2 parts of nitric acid, and evaporate to one-half and add 1½ parts of water. After twenty-four hours, pour the clear fluid from the basic salt.

out from the latter on cooling. Though so closely related chemically to the carbohydrates, it yields, with urea, decomposition products of an aromatic nature, as orthonitrobenzyl alcohol. It occurs in the urine as a potassium salt, $C_6H_9KO_7$. Bromine converts glycuronic acid into saccharic acid, indicating its close relationship to dextrose, as well as to the aldehyde group. It has been thought to arise from dextrose in the organism, though Kulz suggested its origin to be from inosite.

Glycuronic acid, of all substances met with in the urine, is most likely to be mistaken for grape-sugar, since it is dextro-rotatory, and in alkaline solution converts cupric into cuprous oxide and reduces the salts of bismuth, silver, and mercury. It does not, however, ferment with yeast.

Significance.—Glycuronic acid occurs in the urine abundantly after the administration of such drugs as chloral and butyl-chloral, nitrobenzol, camphor, curare, morphia, chloroform, etc., in consequence of which it was formerly thought that the use of these substances caused the appearance of grape-sugar in the urine. It sometimes, though not frequently, occurs in the urine of people who are apparently in good health and are not diabetic ; so that it is important to distinguish between this condition and glycosuria.

Detection.—1. If the urine reduce cupric oxide and is dextro-rotatory, but fails to undergo alcoholic fermentation with the yeast test, glycuronic acid is present. 2. It may be isolated from the urine by the method of Schmiedeberg and Mayer, as follows : A large quantity of urine is decolorized by animal charcoal, evaporated to a syrup, and then digested with large quantities of damp barium hydrate in the presence of gentle heat over a water-bath. It is then extracted with absolute alcohol ; glycuronic acid and other substances are left undissolved. The residue is mixed with water and filtered ; more baryta is added to the filtrate ; it is again filtered, and the filtrate evaporated down over a water-bath. An amorphous barium compound separates out ; this is washed with water, decomposed by sulphuric acid ; the barium sulphate is filtered off, the filtrate is evaporated down and dried *in vacuo*, when crystals of the anhydride will be obtained.

CANE-SUGAR.

After abundant use of cane-sugar as food, traces of it may be found in the blood and sometimes in the urine. Pure cane-sugar possesses no reducing power over cupric oxide ; but as met with commercially it sometimes contains other sugars as impurities, which may cause slight reduction of the copper tests.

Cane-sugar crystallizes in monoclinic prisms, aqueous solutions of which possess strong dextro-rotatory powers + 73.8°. By boiling with water, or more readily with dilute mineral acids, it undergoes inversion,—*i.e.*, it takes up water and separates into dextrose and levulose. Nitric acid oxidizes cane-sugar into saccharic acid.

Detection.—It has been stated that if cane-sugar be boiled some time in water it undergoes inversion, becoming glucose and levulose. With solutions of cane-sugar the polarization is dextro-rotatory, but after inversion it becomes levo-rotatory, because the left-handed action of the molecule of levulose produced = — 106° is only partly neutralized by the right-handed action of the glucose = + 56°.

9

GLYCOGEN.

This substance is found in the liver, muscles, placenta, white blood-corpuscles, cartilage-cells, pus-cells, and in embryonic tissues. It has also been found in the urine, notably in some diabetic conditions. Pure glycogen is a snow-white, amorphous powder, tasteless and odorless, soluble in water, insoluble in alcohol and ether. Glycogen is strongly dextro-rotatory $+211^O$, but does not reduce cupric oxide. Glycogen gives with iodine a port-wine red color; the color disappears on heating, and re-appears on cooling.

SECTION V.

ABNORMAL URINE (*continued*).

ACETONURIA.

Acetone—C_3H_6O—is a thin, watery, colorless liquid, of specific gravity 0.792, which boils at 56.5° C., and possesses a peculiar ethereal or fruit-like odor. It may be obtained in considerable quantity by distillation of the urine as well as the blood of diabetic patients. Acetone is also sometimes found in the urine of children apparently in good health.

Treated with iodine and an alkaline hydrate, iodoform is evolved; with solutions of nitroprusside of sodium and ammonia a rose-red color is produced.

Acetone and aceto-acetic acid are sometimes both present in the urine, sometimes only one of them. V. Jaksch holds that acetone is a normal constituent of urine, though occurring in minute quantity (0.1 gramme in twenty-four hours). On the other hand Le Noble holds that acetone only occurs in healthy urine after the use of alcohol or foods rich in proteids. Both acetone and aceto-acetic acid seem to be decomposition products of albumins.

Clinical Significance.—Acetonuria of pronounced degree often accompanies high febrile states, probably caused by blood changes which result from exalted temperatures, since acetonuria belongs to no special form of fever. Moreover, the amount of acetone in the urine in febrile conditions corresponds closely with the degree of temperature elevation, always rising and falling with the latter.

Acetonuria often occurs in diabetes mellitus, more especially in advanced cases. In such cases it sometimes precedes the more dangerous symptom of diaceturia. Acetonuria is frequently associated with certain forms of cancer, notably carcinoma. It is also observed in cases of inanition or starvation and in cerebral psychosis, especially if accompanied by great mental excite-

(123)

ment. Acetonuria may result from excessive use of animal foods, and this may, in a measure, account for its frequent appearance in diabetes, since such patients are usually restricted largely to nitrogenous diet. A condition of auto-intoxication with acetone sometimes occurs, which is accompanied by acetonuria. This state gives rise to symptoms of restlessness, excitement, and even delirium, but unless accompanied by diaceturia it tends toward a favorable termination (Jaksch), although it may end in coma and death in some cases.

Acetonuria occurs in association with the following diseases: Small-pox, typhus, pneumonia, scarlet fever, measles, Bright's disease, perityphlitis, and strangulated hernia, but it is never accompanied or followed by diabetic coma in such cases.

Detection.—1. *Lieben's Test.—*About 250 cubic centimetres of the urine is slightly acidulated with dilute sulphuric acid, placed in a flask connected with a Liebig condenser, and gently heated until about 30 cubic centimetres has distilled over. The distillate, containing any acetone present in the urine employed. is then tested by Lieben's iodoform test as follows: Compound solution of iodine is added till a light-brown color is obtained, the liquid is slightly warmed, and solution of potassium hydrate is then poured in drop by drop until the mixture is just decolorized. If acetone is present iodoform is produced, recognizable by its odor and by its characteristic hexagonal plates, rosettes, and stars under the microscope. Alcohol and a few other substances behave similarly to acetone.

2. *Chautard's Test.—*A drop of aqueous solution of magenta decolorized by sulphurous acid gives, with fluids containing over 0.01 per cent. of acetone, a violet color. This appears in dilute solutions after four or five minutes.

3. *Le Nobel's Test.—*On adding an alkaline solution of sodium nitroprusside—so dilute as to have only a slight red tint—to a fluid containing acetone, a ruby-red color is produced, which in a few minutes changes to yellow, and on boiling, after adding acid, to greenish blue or violet. A quarter of a milligramme of acetone can be thus detected.

4. *Baeyer's Indigo Test.—*A few crystals of nitrobenzaldehyde are dissolved by heat in the suspected urine; on cooling,

the aldehyde separates in the form of a white cloud. The mixture is then made alkaline with dilute sodium-hydrate solution, and, if acetone be present, first yellow, then green, and last an indigo-blue color will appear within ten minutes.

5. *Reynolds's Test.*—This test depends upon the fact that acetone promotes the solution of mercuric oxide. The test may be conducted as follows: The yellow precipitate of mercuric oxide, obtained by the reaction of mercuric chloride with an alcoholic solution of potassium hydrate, is added to a small quantity of the urine, which is shaken and filtered. To the clear filtrate ammonium sulphide is carefully added, and if acetone be present some of the mercuric oxide is dissolved, and a black ring of sulphide of mercury appears at the plane of contact between the two liquids.

DIACETURIA.

Diacetic or aceto-acetic acid ($C_4H_6O_3$) is a colorless, strongly-acid liquid which mixes with water, alcohol, and ether in all proportions. On heating to boiling with water, especially with acids, diacetic acid decomposes into carbon dioxide and acetone.

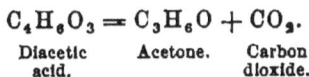

$$C_4H_6O_3 = C_3H_6O + CO_2.$$

Diacetic   Acetone.   Carbon
acid.   dioxide.

Diacetic acid has frequently been confused with and mistaken for acetone.

It differs from acetone, however, in giving a violet-red or brownish-red mahogany color with solution of ferric chloride. This color decreases at ordinary temperatures within twenty-four hours, and more rapidly upon boiling, in which respects it differs from the color produced by ferric chloride with phenol, salicylic acid, acetic acid, and sulphocyanides.

There are other substances in the urine at times which give, under certain circumstances, the ferric-chloride reaction,—viz., β-hydroxybutyric acid, sulpho- (thyo-) cyanates, acetic acid, and formic acid; and, according to Legal and Hammarsten, the urine of patients who have taken thallin, antipyrin, salicylic acid, and carbolic acid may give the reaction. If, however, as already stated, the urine be previously boiled, diacetic acid no longer gives the ferric-chloride reaction, while the other substances do. Fleischer found that the other substances which give the ferric-chloride reaction in the urine are not taken up by ether after the urine is acidulated with sulphuric acid, whereas diacetic acid is soluble in ether. Salkowski confirmed this observation, and found, as above, that urine containing diacetic acid after boiling did not give the ferric-chloride reaction. These observers hold that in most cases it is not diacetic acid from which acetone originates, but some at present unknown substance, possibly hydroxybutyric acid, already mentioned (Halliburton).

Clinical Significance.—Diaceturia is always pathological, and, for the most part, it may be regarded as a symptom of serious import. It is of least serious significance when occurring, as is not uncommon, in febrile conditions in children. Under such circumstances recovery usually follows. In the case of adults it is always of more grave significance. In diabetes the occurrence of diaceturia may be looked upon as a very probable prelude to coma, which usually terminates quickly in death. Von Jaksch, indeed, considers that diabetic coma is due to the presence of diacetic acid in the blood, and he, therefore, proposes to substitute the term *diacetic coma* for the former name.

Diaceturia is most common in the advanced stages of diabetes, and more especially in young subjects of the disease. It does not, apparently, depend upon large quantities of sugar in the urine, at least directly so, since the appearance of diaceturia is often preceded by decided diminution of sugar.

It sometimes occurs at the height of acute fevers, and in adults such occurrence is of grave significance.

Diaceturia is sometimes the index of auto-intoxication—*diacetæmia*—and is accompanied by such symptoms as vomiting, dyspnœa, jactitation, which shortly ends in coma and death without other discoverable disease or lesion.

Detection.—This is best accomplished by Gerhardt's reaction, as follows: 1. Take a recently-voided sample of urine, and add a few drops of ferric-chloride solution to it. If the phosphates be precipitated, filter them off, and to the filtrate add a few drops more of ferric-chloride solution. If a dark-red color is produced, diacetic acid is probably present. 2. The above color disappears on boiling, or is not produced if the urine be previously boiled. Salicylic acid, phenol, antipyrin, or thallin in the urine give the same color with ferric chloride, which remains, however, unchanged by boiling. 3. Acidify the urine with sulphuric acid and shake with ether. Next shake the removed ether with very dilute ferric chloride, and the watery color becomes claret-red.

CHOLURIA.

The biliary acids and pigments are the chief bile-elements of clinical interest met with in the urine.

BILIARY ACIDS.

It is an unsettled question as yet if the bile-acids occur in the urine under physiological conditions. According to Vogel, Dragendorff, Hone, and Oliver, traces of bile-acids occur in normal urine, although Hoppe-Seyler and Udransky hold the opposite view. Oliver's new and delicate method of testing for the bile-acids gives his results much weight, and it may be assumed that the evidence is in favor of the view that the bile-acids are present in minute quantities in the urine of health.

Clinical Significance.—Dr. Oliver estimates the amount of bile-salts in normal urine as about 1 part in from 10,000 to 15,000 parts. Furthermore, in a series of observations he has noted that the normal percentage quantity varies with the time of day, reaching the maximum during periods of fasting, as in the morning urine and that passed before meals; while it quickly diminishes after meals, reaching the minimum about three hours after food. Dr. Oliver also observed an increase of the bile-acids in the urine upon active muscular exertion; and he suggests that changes of temperature, atmospheric pressure, use of alcohol, etc., also influence the degree of discharge of the bile-salts by the kidneys. The bile-salts in the urine are markedly augmented in all forms of jaundice, and, moreover, according to Oliver's observations, this occurs both before the appearance of the bile-pigments and long after their disappearance from the urine.

In so-called acute bilious attacks—*i.e.*, biliary engorgement from defective bile excretion—there appears to be an overflow of the bile-elements into the blood, accompanied by such symptoms as feeble pulse, pallor, coldness of the extremities, sensations of chilliness, slow respiration, nausea and vomiting, with headache, and sometimes diarrhœa. During these symptoms the bile-salts in the urine are at first diminished, but after a time they are increased, and thereupon immediate relief from these symptoms follows.

Acute cholæmia from retention of bile-salts in the blood may pass beyond the ordinary bilious attack and produce the more serious symptoms of lowered temperature, convulsions, albuminuria, with evidences of blood dissolution such as hæmoglobinuria.

Hepatic congestion, early cirrhosis, and malarial poisoning are accompanied by increased elimination of the bile-acids in the urine, more especially if accompanied by constipation. Dr. Oliver has noted excess of the bile-acids in the urine in carcinoma, amyloid disease, enlargement of the liver, cirrhosis, and in hepatic tumors. Choluria always follows a rise of temperature, and in high fevers the increase of bile-acids may reach 400 per cent. above the normal range. In splenic leucocythæmia, anæmia, hæmoglobinuria, and scurvy a large excess of the bile-acids appears in the urine. Lastly, Dr. Oliver has noted a decided and persistent decrease of the bile-acids in the urine in cases of chronic interstitial nephritis,—granular kidney.[1]

---

[1] In view of the fact that nearly all advanced chemico-physiologists are now agreed that the liver constitutes the chief agent of destruction of those substances which we know to be auto-intoxicants to the organism, Dr. Oliver's explanation of the hepatic origin of certain symptoms seems to the author vastly more reasonable than the explanation of many of the same symptoms by Haig's uric-acid theories. The author has searched in vain among the classic experiments of Bouchard, as well as others, for proof that uric acid is in any way toxic to the organism, even when injected into the blood in the enormous dose of 0.64 gramme per kilogramme of body-weight. On the other hand, the bile-salts are toxic in almost infinitesimal doses; they are not only toxic, but in aqueous solutions of 2 per cent. *they kill* 1 kilogramme of weight; the cholate of sodium in dose of 54 centigrammes, and cholate of potassium in dose of 46 centigrammes. The enormous toxic power of bile, as a whole, may be judged from the following statement of Bouchard as a result of direct experimentation : "We must conclude that during twenty-four hours a man makes, by the activity of his liver alone, enough poison to kill three men of his own weight."

Detection.—1. Pettenkoffer's method is as follows : The urine is mixed with concentrated sulphuric acid, taking care that the temperature does not rise higher than 60° to 70° C. Then a 10-per-cent. solution of cane-sugar is added drop by drop, continually stirring with a glass rod. The presence of bile-acids is indicated by the production of a beautiful red liquid, the color not disappearing at ordinary temperature, but becoming more bluish violet in the course of a day or so. This red liquid shows a spectrum with two absorption bands, the one at F and the other between D and E near E.

This test fails if the solution be heated too high or an improper quantity—usually too much—sugar be added. In the last case the sugar carbonizes and the test becomes dark brown or brown. The reaction fails if the sulphuric acid contains sulphurous acid or the lower oxides of nitrogen. Since many other substances—such as albumin, oleic acid, amyl alcohol, and morphine—give a similar reaction, in doubtful cases the spectroscopical examination must not be omitted.

If the urine be icteric and of pronounced color, the bile-acids must first be isolated from the urine by Hoppe-Seyler's method before applying the above test, as follows: Strongly concentrate the urine and extract the residue with strong alcohol. The filtrate is freed from alcohol by evaporation and then precipitated by basic lead acetate and ammonia. The washed precipitate is treated with boiling alcohol, filtered hot, the filtrate treated with a few drops of sodium-hydrate solution, and evaporated to dryness. The dry residue is extracted with absolute alcohol, filtered, and an excess of ether added. The amorphous or, after a longer time, crystalline precipitate, consisting of alkali salts of the biliary acids, may then be submitted to the above-described test.

2. *Dr. Oliver's method* of detecting the bile-acids is most sensitive and simple. The principle of this method depends upon the physiological fact that such products of gastric digestion as peptone and propeptone are precipitated in the duodenum by contact with the bile-acids. Therefore, since peptone in an acid solution, as the urine, is precipitated by the bile-acids or their derivatives as cholate of sodium,[1] an acid solution of

---

[1] The form in which the bile-acids occur in the urine.

peptone may be used as a test for the bile-acids. The following is the formula for the test solution : Pulverized peptone (Savory and Moore's), ℥ss; salicylic acid, 4 grains; acetic acid, ℥ss; distilled water to 8 ounces. To be filtered repeatedly until rendered transparent. In testing, the urine must be perfectly cleansed, if not already so, and rendered acid if it be alkaline or neutral, and the specific gravity reduced to 1.008 if it be above it. To 60 minims of the test solution 20 minims of the urine are added. If bile-acids be present in normal amount, there will be no immediate reaction visible, but shortly a slight tinge of milkiness will be produced. If in excess, a distinct milkiness at once appears, becoming more intense in proportion to the quantity of bile-acids present. On agitation the opalescence diminishes, and may even disappear, but it returns upon the addition of more of the test solution. On this fact is based an approximate quantitative test, for which is prepared a standard solution by adding equal parts of test fluid and normal urine diluted to specific gravity 1.008. Any urine requiring 60 minims or more to bring its opacity up to that of the standard does not contain an excess of bile-acids.

DR. OLIVER'S STANDARD TABLE.

URINE.		Percentage Increase of Bile-Salts Over the Normal.		URINE.		Percentage Increase of Bile-Salts Over the Normal.
Minims.	Drops.			Minims.	Drops.	
1	or 2	=	6000	20	or 40	=	300
2	or 4	=	3000	25	or 50	=	240
3	or 6	=	2000	30	or 60	=	100
4	or 8	=	1500	35	or 70	=	83
5	or 10	=	1200	40	or 80	=	66
10	or 20	=	600	45	or 90	=	50
15	or 30	=	400				

Percentage increase over 700 above the normal is rarely encountered. With the above test Dr. Oliver detects 1 part of bile-salts in 18,000 or 20,000 parts of sodium-chloride solution. If careful attention be paid to details in preparation of the urine, nothing as yet found in urine interferes with the test.

BILIARY PIGMENTS.

Bile coloring matters appear in the urine in a number of conditions. The urine in such cases is always abnormally colored, —yellow, yellowish brown, deep brown, greenish yellow, greenish brown, or even nearly pure green.   On shaking the urine it froths and the bubbles are yellow or yellowish green in color. The morphological elements of the sediment often take the color of the abnormal pigment in the urine.

Clinical Significance.—The biliary pigments are met with in the urine in jaundice, from whatever cause it arises, but most commonly, perhaps, when due to obstruction of the bile-ducts. In such cases the bile-elements make their way into the lymphatics and the general circulation and are eliminated by the kidneys. The bile-pigments appear in the urine several days before the icteric coloration of the skin is perceptible, and therefore they may sometimes be taken as a prognostic of the approach of jaundice.

The biliary coloring matters are also found in the urine in numerous pathological conditions of the liver, in which icterus may or may not be present also.   They may also appear in the urine as a result of blood changes, and after hæmorrhage into the tissues; and in such cases they are derived from their primary source in the blood itself.   Lastly, bile-pigments in large amounts always appear in the urine in cases of phosphorus poisoning.

Detection.—1. *Gmelin's method* consists in introducing a column of strong nitric acid, containing a little yellow nitrous acid of commerce ($HNO_3 + NO_2$) into a test-tube, and upon this gently floating a column of similar depth of urine.   In the zone between the fluids appear from above downward the colors green, blue, violet, red, and yellow.   The green is most predominant, while the blue is most indistinct or sometimes absent.

(*a*) In Rosenbach's modification of this test the urine is filtered through a fine, thick filter.   After filtration a drop of nitric acid containing a little nitrous acid is applied to the inside of the filter.   A pale-yellow spot will be formed, which is surrounded by colored rings which appear yellowish red, violet, blue, and green.   This modification is very delicate, and it is hardly possible to mistake other coloring matters for the bile-pigments.

(*b*) Dragendorff has adopted still another modification of the above test, which consists in placing a little of the urine on a plaster-of-Paris disc, and, when nearly absorbed, a drop of nitric

acid is allowed to fall on the moistened spot. If the bile-pigments be present a ring forms about the acid drop, in which
green is predominant.

2. *Huppert's test* detects the faintest traces of bile-pigment
in the urine. The urine is treated with lime-water, or first with
some $CaCl_2$ solution, and then with a solution of sodium or ammonium carbonate. The precipitate, containing the bile-pigments, may be shaken out with chloroform after washing in water,
and after acidification with acetic acid. The bilirubin is taken
up by the chloroform, which is colored yellow thereby, while the
acetic-acid solution is colored green by the bilirubin.

The precipitate may also be used directly for Gmelin's test
in the following way: Spread it on a porcelain dish in a thin
layer, and add a drop of nitric acid. The reaction usually appears
very beautiful.

3. *Ultzmann's test* consists in treating 10 cubic centimetres
of the urine with 3 or 4 cubic centimetres of concentrated caustic-
potash solution and then acidifying with hydrochloric acid. The
urine will turn a beautiful green color if the bile-pigments be
present.

INDOXYL-SULPHURIC ACID.

Indoxyl-sulphuric acid ($C_8H_7NSO_4$), also called *urine indican*, and formerly known as *uroxanthin*, occurs in normal urine
as an alkali salt, whose properties have been fully considered in
Section II. Indoxyl-sulphuric acid is derived from indole, which
is first oxidized in the system into indoxyl and then is united
with sulphuric acid. Indole is formed by the putrefaction of
proteids, and hence the quantity excreted by the kidneys is
greater upon a meat than upon a vegetable diet.

Clinical Significance.—Variations in the quantity of so-called
indican in the urine occur within comparatively narrow range in
health; but in certain pathological conditions the increase becomes very marked. Clinically, therefore, an increased excretion
of this substance by the kidneys is observed in Addison's disease, cholera, carcinoma of the liver, chronic phthisis, central
and peripheral diseases of the nervous system, typhoid fever,
dysentery, acute general peritonitis, multiple lymphoma, fetid

bronchitis, ichorous pleural exudations, diabetes mellitus, as well as in a number of others.

In obstructive diseases of the small intestine its excretion is sometimes enormously increased, owing probably to the favorable conditions for the absorption of indole. The simple obstruction of the colon does not cause its increase in the urine. Obstruction of the large intestine, only when it causes considerable disturbance in the motion of the contents of the upper ileum, gives rise to its increased excretion by the kidneys.

In general, it has been considered that the appearance of large quantities of so-called indican in the urine implies that an abundant albuminous putrefaction is progressing in some part of the system. Thus, in pleurisy it indicates a copious, unhealthy exudation, and in peritonitis it may be taken as an evidence of the formation of unhealthy pus. The putrefaction of secretions rich in albumin in the intestines explains its increase in the urine during starvation.

Detection of Urine Indican.—*Jaffé's method* consists in mixing 10 cubic centimetres of strong hydrochloric acid with an equal volume of urine in a test-tube, and, while shaking, add drop by drop a perfectly fresh, saturated solution of chloride of lime, or chlorine-water, until the deepest obtainable blue color is reached. The mixture may next be shaken with chloroform, which readily takes up the indican and holds it in solution, and the quantity present may be approximately estimated according to the depth of the color. If the urine contain albumin it should be removed before applying this test, otherwise the blue color, often arising from the mixture of hydrochloric acid and albumin after standing, may prove misleading.

2. *MacMunn's Method.*—(a) Equal parts of urine and hydrochloric acid with a few drops of nitric acid are boiled together, cooled, and agitated with chloroform. The chloroform becomes violet if much indican be present, and shows an absorption band before D, due to indigo blue, and another after D, due to indigo red.

This method is more trustworthy than Jaffe's, because chloride of lime destroys small quantities of indigo.

(b) A rough, approximate method may be employed upon the

foregoing principle, as follows : Pour 4 cubic centimetres of hydrochloric acid into a small flask, and while stirring add from 10 to
20 drops of urine. If the proportion of indican be about normal
the resulting color will be rather light-yellow ; if in excess the
acid will turn violet or blue—the more intense will be the color
in proportion to the quantity of indican present. If no coloration appear after waiting a minute or two the indican is not in
excess, however deep a color may subsequently appear.

If 2 or 3 drops of nitric acid be added to the test, as in the
original method, it becomes more delicate. (See also page 43.)

THE DIAZO REACTION IN URINE.

The diazo test was suggested by Ehrlich, in 1882, as a valuable diagnostic measure in typhoid fever, although he admitted
the occurrence of this reaction in a few other conditions shortly
to be considered.

The diazo reaction depends upon the fact that if sulphanilic
acid (*amidosulphobenzol*) be acted upon by HNO_2, *diazosulphobenzol* is formed, which unites with certain aromatic substances
occasionally present in the urine to form aniline colors.

Dr. Friedenwald has recently reviewed the literature of this
reaction,[1] and shown that many of the contradictory results
obtained by some observers are due to failure in carrying out
Ehrlich's methods in performing the test, which is best accomplished as follows :—

To obtain diazosulphobenzol in a perfectly fresh condition,
sulphanilic acid is kept in solution with hydrochloric acid ; to
this sodium nitrite is added, whereupon HNO_2 is liberated and
diazosulphobenzol is formed.

Process.—Two solutions are prepared, as follows :—

1. Two grammes of sulphanilic acid ; 50 cubic centimetres of
hydrochloric acid ; 1000 cubic centimetres of distilled water.

2. A 0.5-per-cent. solution of sodium nitrite.

In performing the test 50 parts of No. 1 and 1 part of No. 2
are mixed, and equal parts of this mixture and of the urine in a
test-tube are rendered strongly alkaline with ammonia.   If the
reaction be positive the solution assumes a carmine-red color,

---

[1] New York Medical Journal, December 23, 1893.

which on shaking must also appear in the foam. Upon standing for twenty-four hours a greenish precipitate is formed.

The test must not be considered positive unless a distinct red coloration extends to and includes the foam on shaking.

Clinical Significance.—Ehrlich's original claims for the diazo reaction were as follow :—

"1. The reaction is most commonly found in typhoid fever from the fourth to the seventh day and thereafter, and if the reaction be absent the diagnosis is doubtful.

"2. Cases of typhoid fever characterized by faint reaction and occurring only for a short time may be predicted to be of very mild type.

"3. The reaction is occasionally noted in phthisis pulmonalis, but only in cases pursuing a rapid course toward a fatal termination.

"4. The reaction is sometimes, but not often, observed in cases of measles, miliary tuberculosis, pyæmia, scarlet fever, and erysipelas.

"5. In diseases unaccompanied by fever, as chlorosis, hydræmia, diabetes, diseases of the brain, spinal cord, liver, and kidneys, the reaction is always absent."

The weight of clinical evidence strongly confirms all of Ehrlich's original claims for this reaction, but more especially so with regard to typhoid fever and pulmonary tuberculosis; if present in the latter disease any length of time, the prognosis is very unfavorable.

β-Hydroxybutyric Acid ($C_4H_8O_3$).

This acid forms an odorless syrup which is readily miscible with water, alcohol, and ether. It is an optically active substance, being, in fact, levo-rotatory; so that it interferes with the estimation of sugar in the urine by polarimetry. It is non-precipitable by lead acetate and ammoniacal basic lead acetate. On boiling with water in the presence of a mineral acid, it decomposes into α-crotonic acid—which melts at 72° C.—and water. It yields acetone upon oxidation with chromate mixture.

Clinical Significance.—The appearance of hydroxybutyric acid in the urine was first demonstrated by Minkowski, Kulz, and Stadelmann. It is usually accompanied by diacetic acid in the urine, and sometimes by acetone. It is especially noted in the urine in severe or chronic cases of diabetes mellitus. It has also been observed in the urine in cases of measles, scurvy, scarlet fever, and in diseases of the brain. Aside from these, little is at present known of its clinical relations.

Detection.—1. If a urine be levo-rotatory after fermentation with yeast, it is strongly probable that hydroxybutyric acid is present. 2. *Kulz's method* consists in evaporating the fermented urine to a syrupy consistence, and, after the addition of an equal volume of concentrated sulphuric acid, distill directly without cooling; *α-crotonic acid* is produced, which is distilled, and, after strongly cooling, the distillate is collected in a glass; crystals which melt at $+ 72°$ C. separate. If no crystals be obtained, then shake the distillate with ether and test the melting-point with the residue, which has been washed with the water obtained after evaporating the ether.

PTOMAINES AND LEUCOMAINES.

The term *ptomaine* was originally used to designate those products of putrefaction which give the reaction of vegetable alkaloids and possess more or less poisonous characters. It has since been found that similar alkaloids are formed during the life of animal organisms; these are termed *leucomaines*. Ptomaines, or putrefactive bases, are transition products of decomposition; or, in other words, temporary forms through which matter is being transformed from the organic to the inorganic state by means of the action of bacterial life. They are chemical compounds of a basic nature, and their deep interest and importance in the field of modern medicine may at once be perceived from the fact that they constitute one of the chemical factors in the causation of all infectious diseases.

It has been erroneously supposed that all ptomaines are highly poisonous; but not only are many of them inert, but it may be stated that the majority of them isolated to date do not produce harmful results to the organism in ordinary quantities. On the other hand, some of them are highly toxic, and such Brieger first proposed to designate as "toxins."

Ptomaines resemble vegetable alkaloids in that they all contain nitrogen as the chief element of their basic character. Some of them also contain oxygen, corresponding to the vegetable fixed alkaloids, while those devoid of oxygen correspond to the volatile alkaloids.

Selmi was probably the first—in 1880—to claim that the basic substances formed in the organism during pathological changes often appeared in the urine, constituting an index to the pathological condition of the patient. He demonstrated, in the urine of a patient with progressive paralysis, two bases resembling *nicotine* and *coneine*. Since then Bouchard, Villiers, Lépine, Gautier, and others have demonstrated the presence of a few other basic products in the urine in other diseased conditions. It is now well known that the urine in certain diseases, as cholera and septicæmia, is decidedly toxic in character. Bouchard Chavrin, and Ruffer have proved that bacterial poisons generated in the system through disease can be excreted in the urine. They produced immunity to the action of the *bacillus pyocyaneus* upon animals by previous injections of urine of animals inoculated with that bacillus as well as with filtered cultures thereof. Unfortunately, as yet these investigations have not been pushed to sufficient completion to furnish much practical data in reference to infectious diseases, since but few bacterial ptomaines have yet been isolated from the urine. The importance, however, of this comparatively new field of uranalysis can scarcely be overestimated, since it is strongly probable that careful investigation of the urine in this direction may throw important light upon a large class of diseases.

It only remains here to refer to the few ptomaines which have been isolated from the urine, and the methods by which this has been accomplished. First, with regard to normal urine, much difference of opinion prevails as yet in reference to the presence or absence of alkaloidal toxins. When through defective renal action the leucomaines become retention products, they at once assume immense importance in the chemistry of the urine. The researches of Pouchet strongly confirm the presence of toxic alkaloids in normal urine; while, on the other hand, Villiers denied their existence, claiming that the observed physiological

10

action is wholly due to the presence of potassium salts. Since we know that toxins of an alkaloidal nature (leucomaines) are formed in the organism through tissue metabolism, and, furthermore, that the urine constitutes at least one channel of escape for similar compounds, there seems no reason to doubt their presence in urine; at least, in minute quantity. Bouchard, Guerin, and Lépine have shown that at least that which has been taken for these compounds is greatly increased in the urine in pathological states.

With regard to the isolated ptomaines, Baumann and Udransky separated several basic derivatives, amongst them *cadaverin*, *putrescin*, and a small amount of a third base from the urine of a patient suffering from catarrhal cystitis; normal urine being found free from these substances.

Putrescin—$C_4H_{12}N_2$—is closely related to *cadaverin*, since they nearly always occur together or alternately from the same source. Brieger obtained it from putrefying human viscera after exposure of from three days to three weeks at ordinary temperatures. It has been obtained from herring (twelve days' exposure), from pike (six days' exposure), from haddock (two months' exposure), and from decaying mussel (sixteen days' exposure). It is especially abundant in cultivations of comma or cholera bacillus, and hence it is believed that substances similar to *putrescin* are the true chemical poisons in cholera. *Putrescin* has been isolated from the urine in cases of cystinuria by several observers since Brieger's first discovery. It is toxic to the organism, but its tetra-methyl derivation is incomparably more so, causing symptoms of salivation, dyspnœa, contraction of the pupils, muscular paralysis of the limbs and trunk, ejaculation of semen, dribbling of urine, and violent convulsions.

Cadaverin—$C_5H_{14}N_2$—appears in decomposing tissues usually before the occurrence of *putrescin*. Brieger obtained it from putrefying heart, liver, lungs, etc., at ordinary temperatures, in three days' exposure; from putrid mussel in sixteen days. Like *putrescin*, it is a constant product of comma bacillus upon any soil upon which it may be cultivated. *Cadaverin* is a constant associate—perhaps a product—of the activity of *vibriones*, since it never occurs in cultures in which this genus is absent. It is therefore absent from both normal and typhoid stools, as well as from cultures of the bacillus of Emmerich and Eberth. Both *putrescin* and *cadaverin* may be obtained from the urine of cystinuria by precipitation with benzoyl chloride (Baumann's method).

Trimethylamine—C_3H_9N—has been found in human urine. This base occurs both in animal and vegetable tissues. It has been obtained from ergot, the blood of calves, herring-brine, in the putrefaction of yeast, in cheese, in human liver and spleen (two to seven days' exposure), in perch (six days' exposure), in cultures of *streptococcus pyogenes* on broth and blood-serum. It is not a violent . toxin, large doses being required to markedly disturb the system.

Beatin—$C_6H_{13}NO_3$—is a well-known base, the product of cotton-seed, beet-juice, turnip, vetch-seeds, etc. It has been found in the urine by Liebrich. This base is of no pathological significance, since it is non-toxic to the organism.

Jaksch, who has studied the subject of basic derivatives of the urine, both normal and pathological, finds that, while normal urine and that of some diseases hold these substances only in minute quantity, in certain other morbid states this amount is very considerable. He makes the following suggestions as a guide for investigation in this field of work :—

In the first place, he suggests that it would be well to follow the example of Brieger, Baumann, and Udrasky in withholding the name *alkaloids* from the bodies *diamines*, which are derived from the system under morbid conditions, because all that have been recognized as yet are simply *diamines*, and do not exhibit the characteristic property of alkaloids, viz., the pyridin radicle. In the second place, it is desirable to discriminate between the physiological bases of the urine creatinin, reducin, etc., which belong to normal urine, and those which are associated only with certain diseased states. It is not intended to imply by this that the physiological bases—*leucomaines*—cannot under any circumstances give rise to diseased or poisonous symptoms. On the contrary, it is highly probable that retention, and still more the increased formation of such products, under certain circumstances may induce very grave symptoms, and even greatly imperil the life of the patient. Again, it seems probable that in certain acute diseases, specific substances of a toxic nature, not present in normal urine, may be excreted with that fluid.

Jaksch makes the following suggestive classification of the subject :—

(*a*) Clinical (morbid) symptoms depending upon retention of the physiological basis. Under this head would come uræmia and certain of the symptoms of obstruction—*retention toxicosis*.

(*b*) Clinical symptoms referable to the presence of basic products which are found in the system (blood, etc.) in disease and eliminated with the urine—*noso-toxicosis*.

(*c*) Clinical symptoms which are caused by the formation of toxic basic substances from morbid matter, such as pathological fluids lodged in certain parts of the organism. Such bases being absorbed give rise to symptoms of severe poisoning. Under this head would come the collective symptoms of ammoniæmia, and others which follow the absorption of gangrenous pus—*auto-toxicosis*.

(*d*) Clinical symptoms, and consequently morbid types induced by the action of toxic bases taken into the system with the food, such as the poison of sausages, cheese, canned fruits, etc., etc.—*exogenic toxicosis*.

DETECTION.—A number of methods are in use for the detection and isolation of these bases, the more prominent of which are those of Dragendorf, Stas-Otto, Brieger, Gautier, and Etard. Since all of these methods are somewhat difficult and tedious, only the most suitable methods for uranalysis will here be described. For such purposes Brieger's method serves best ; but in some cases it is important that the urine be first concentrated *in vacuo*. Sufficient hydrochloric acid is first added to render the urine acid, and the mixture is then boiled for a few minutes and filtered. The filtrate is concentrated at first over a flame, and subsequently over a water-bath, to a syrupy consistence.

In consequence of the instability of the bodies sought, it is advisable to evaporate *in vacuum* and at the lowest possible temperature, more especially so if the urine be foul.

The thick fluid is next mixed with 96-per-cent. alcohol, filtered, and the filtrate treated with a warm alcoholic solution of lead acetate. The resulting lead precipitate is removed by filtration and the filtrate concentrated—preferably *in vacuo*—to a syrup, and again taken up in 96-per-cent. alcohol. The alcohol is next evaporated, and the residue, dissolved in water, is freed from lead by the addition of sulphuretted hydrogen and filtration. The filtrate is acidified with hydrochloric acid and evaporated to a syrupy consistence. It is then diluted with alcohol, and alcoholic solution of mercuric chloride is added. The resulting precipitate is boiled in water, and certain ptomaines may separate at this stage in consequence of different solubilities of the double salts of mercury. The better to secure this, the precipitate may be treated successively with water at various temperatures. Should it be thought that the lead precipitate may have retained some of the ptomaines, it may be suspended in water, the lead converted into sulphide, and the fluid treated in the manner just described.

The solution obtained as above is filtered, freed from mercury, and evaporated; the excess of hydrochloric acid is carefully neutralized with sodium carbonate (the reaction is kept feebly acid), then it is again extracted with alcohol to free it from inorganic salts. The alcohol is evaporated, the residue dissolved in water, the remaining traces of hydrochloric acid neutralized with sodium, the whole acidified with nitric acid and treated with phosphomolybdic acid. The phosphomolybdate double compound is separated by filtration and decomposed by neutral lead acetate or, more readily, by heating over a water-bath. The lead is next removed by means of sulphuretted hydrogen (hydrogen sulphide); the filtrate is evaporated to a syrupy consistence and taken up with alcohol. Several ptomaines are thus separated as hydrochlorates, and may be obtained in the form of double salts of gold, or platinic chloride, and of picric acid. The chloride of the base is obtained by removing the metallic base by precipitation with sulphuretted hydrogen, while the picrate is taken up with water, acidified with hydrochloric acid, and repeatedly extracted with ether to remove the picric acid. The last step is to ascertain if any ptomaines remain in the phosphomolybdic-acid filtrate after precipitation of the phosphomolybdic acid.

Brieger has obtained some of his ptomaines by a simpler modification of his above complete method. Thus he has obtained *neurodin* by treating the aqueous extract of the organic matter, after boiling and filtration, with mercuric chloride, collecting the precipitate, decomposing it with sulphuretted hydrogen, evaporating the filtrate over a water-bath, and extracting the base with alcohol.

Properties of Animal Bases.—1. They all have an alkaline reaction.

2. They are insoluble in water; soluble in acids forming compounds; precipitated from such compounds by ammonia.

3. Iodine and potassium iodide give a brown, flocculent precipitate.

4. Potassio-mercuric-iodide solution produces flocculent, yellowish-white precipitates, insoluble in acids and dilute alkalies, easily soluble in alcohol, and generally, also, in ether.

5. Iodide of bismuth and potassium give orange precipitates in solution acidulated with dilute sulphuric acid.

6. Phosphomolybdic acid gives a bright or brownish-yellow precipitate, insoluble in water and dilute mineral acids.

7. Metatungstic and phosphotungstic acids give a white, flocculent precipitate, with difficulty soluble in water and dilute acids.

8. Tannin in neutral or feebly-acid solutions give yellow or white precipitates with most ptomaines.

9. Chloride of gold gives a yellow or yellowish-white precipitate, either amorphous or crystalline.

10. They are all oxidizable and unstable, especially under the influence of an excess of mineral acid, which colors them red and then converts them into a resinous mass.

11. Picric acid precipitates most of them, the color of the precipitate usually being pale yellow.

12. The animal bases are energetic reducing agents, decomposing chromic acid, iodic acid, and silver nitrate. With potassium ferrocyanide and ferric chloride they give Prussian blue. This was formerly considered to be characteristic of the animal alkaloids, but it is now known that many vegetable alkaloids give the same reaction, and a few of the animal alkaloids (especially those containing oxygen) do not give it. As yet there is no known class reaction by which the animal bases can be separated from those of vegetable origin.

THE URINE AS A TOXIN.

It has long been generally believed that normal urine is an auto-intoxicant. The well-known fact that suppression of the urine is invariably followed by certain uniform toxic symptoms ending in death seemed to leave no further proof of the truth of this belief necessary. It was not, however, until a comparatively recent period (1881) that Feltz and Ritter first demonstrated the actual toxicity of normal urine by injecting it into the blood of animals, thereby invariably invoking symptoms which were followed by death of the animals when the dose approached a certain relative amount. These experiments were soon after repeated and confirmed by Bocchi, Schiffer, and others. Two or three years later Dupard, Lépine, and Guérin established the special toxicity of certain pathological urines which has since been confirmed by numerous observers.

Following the researches of Feltz and Ritter, Bouchard commenced the investigation of the toxicity of normal urine, and very recently Lenoir gives a complete review of this subject, as does Charrin. The method of investigation pursued by Bouchard[1] consisted of intra-venous injections of urine in animals (chiefly rabbits), and the results would seem to have established the following conclusions:—

1. That the toxic power of normal urine as a whole is such that an average of 45 cubic centimetres of urine kills 1 kilogramme of living animal; and, therefore, the urine of two days and four hours contains sufficient toxic matters to kill a man of 60 kilogrammes weight.

2. The toxic symptoms induced by intra-venous injections of urine are as follow : (a) Myosis ; contraction of the pupils beginning with the injection of 10 cubic centimetres to 15 cubic centimetres of urine per kilogramme ; the pupils contracting to pinhole size, and thus remain until after death, after which they sometimes dilate. (b) The respirations become hastened and of diminished range. (c) Somnolence and coma follow. (d) Diuresis becomes marked, micturition occurring every two or three minutes. (e) Marked lowering of temperature succeeds. (f) Diminished palpebral and corneal reflexes are present. (g) Death succeeds in coma or convulsions. (h) The heart continues to beat for some time after death.

3. The toxicity of the urine varies with certain circumstances, viz. : (a) The urine is twice more toxic during the day than during the night. (b) The night urine is strongly convulsive and but feebly narcotic, while the day urine is the reverse,— strongly narcotic, but feebly convulsive. (c) Active muscular exertion in the open air diminishes the toxic power of the urine one-third, and this diminution of toxicity continues for from twenty-four to forty-eight hours after cessation of the exercise.

4. The toxicity of the urine is not due to urea, uric acid, or

---

[1] Auto-Intoxication in Disease, Ch. Bouchard. Translated by T. Oliver. Published by The F. A. Davis Co., Philadelphia, 1894.

creatinin, since the injection of these substances into the blood in much larger proportional amounts than those in which they exist in normal urine proves non-toxic.

5. The toxicity of urine increases by permitting it to stand some time, as well as by increasing its temperature, even though fermentation be prevented. By this means a urine that ordinarily kills by coma becomes not only more toxic,—killing in smaller doses,—it also causes convulsions instead of coma.

6. The following facts are brought out regarding the isolation of the toxic elements of the urine: (a) If the urine be decolorized by charcoal it deprives it of about one-third (33 per cent.) of its toxic powers. (b) An aqueous extract of the urine (containing chiefly the mineral elements) causes contraction of the pupils, convulsions, and lowered temperature, but no coma, diuresis, or salivation. (c) An alcoholic extract of the urine produces somnolence, deep coma, and diuresis; but it does not cause contraction of the pupils or convulsions.

7. In acute uræmia the urine becomes non-toxic, and it may be injected into the blood in quantity equal to that of water (about 90 to 120 cubic centimetres per kilogramme) before it proves lethal, and then only mechanically, by interfering with the normal osmosis.

It will be seen, from these investigations, that normal urine owes its toxic properties not to any one, but to several constituents; and although Bouchard has not succeeded in completely isolating these, yet his results are suggestive in that direction. Briefly stated, his results are as follow: At least seven toxic agents are present in normal urine:—

1. A diuretic substance, which is fixed and of organic nature, non-removable by filtration through carbon, but is soluble in alcohol. This substance answers to all the features of urea, and is only toxic in enormous doses.

2. A narcotic substance, also fixed and of organic nature, non-removable by carbon, and soluble in alcohol. It is not urea, since it does not induce diuresis; but, on the other hand, it causes narcosis.

3. A sialogenous substance which produces salivation. It is only present in small amount in normal urine, and hence its

effects are unobservable in quantities of urine sufficient to kill
from other contained toxic agents. This substance is stable, of
organic nature, non-removable by carbon, and soluble in alcohol.

4. Two substances capable of causing convulsions : (a) One,
fixed, stable, of organic nature, is both retained and destroyed
by carbon, and is insoluble in alcohol. It is doubtless an alka-
loid, and is present during the day in less amount than the narcotic substance, and also of less physiological activity than the
latter. (b) A substance which causes myosis; it is fixed, organic,
and removable by carbon. It is probably a coloring substance
of normal urine.

5. A substance which reduces body-heat. It is fixed, of organic nature, and insoluble in alcohol. It may also be a urinary
pigment.

6. Another convulsive substance of mineral nature, which is
doubtless potassium.

Pathological Urine.—In pathological conditions the toxicity
of the urine may become diminished, or it may become greatly
increased. As a rule, in acute infectious diseases and fevers, if
the kidneys remain unaltered, the urine becomes more powerfully
toxic than in health. On the other hand, in pathological states
of the kidneys themselves, the toxic powers of the urine become
more or less diminished, according to the degree of functional
incapacity of the kidneys. Thus, in acute nephritis or extensive
chronic changes which greatly cripple the functional capacity of
the kidneys, the urine may become almost non-toxic. As the
condition of the kidneys improves the urine becomes more and
more toxic, and this fact may be taken advantage of as a prognostic indication in treatment. For instance, if it require 80
cubic centimetres of urine to kill a rabbit of 1 kilogramme weight,
it may be assumed that the capacity of the kidneys is crippled
about one-half (50 per cent.). If in a week later 60 cubic centimetres of urine kill a rabbit of 1 kilogramme weight, it furnishes
substantial evidence that the condition of the kidneys is much
improved.

It has already been stated that the urine in acute uræmia is
non-toxic. Under such circumstances the kidneys can no longer
eliminate the usual toxic agents from the system, and the organ-

ism becomes poisoned—uræmic—and all the phenomena described as due to intra-venous injections of urine are evoked.

Our present knowledge of this subject warrants the statement that the healthy organism is only saved from *lethal* auto-intoxication by the liver and kidneys ; the former destroys the larger proportion of the systemic toxins, and those not so destroyed are eliminated chiefly by the kidneys, if the latter be healthy.

In a large proportion of pathological urines (the kidneys remaining sound) the normal toxicity of the urine becomes increased, and, moreover, new toxic properties are developed, notably those with convulsive powers. Thus, in *tetanus* the urine is powerfully toxic, and if injected into the circulation it evokes most of the tetanic phenomena. M. Labbe injected the urine of a tetanic patient into the circulation of an animal, with the following results : After the sixth cubic centimetre (per kilogramme) mild tremors occurred ; the pupils became punctiform at 10 cubic centimetres. From 12 cubic centimetres violent tonic spasms with convulsions occurred up to 34 cubic centimetres. At the latter point death occurred from opisthotonos.

The urine in *pneumonia* is strongly toxic, the symptoms being nearly as pronouncedly convulsive as in the case of tetanus. The urine in pneumonia proves *lethal* in from 19 cubic centimetres to 38 cubic centimetres per kilogramme. In *typhoid fever*, on the other hand, Bouchard has observed that the urine produces only the toxic symptoms of normal urine ; death occurs at from 50 to 70 cubic centimetres per kilogramme with only slight myosis, coma being present, but not convulsions. In *leucocythæmia* the urine is highly toxic, causing convulsions and death at 15 to 20 cubic centimetres per kilogramme.

The urine possesses special and marked toxic powers in *cholera*. Thus, cyanosis is only produced by choleraic urine ; muscular cramps follow, unlike the convulsions produced by other urines, since the spasms begin long after the beginning of the injections, and they continue long after the injections are discontinued. Cooling of the body is more pronounced than from injections of any other urine. Albuminuria appears at once and in marked degree, while with normal urine albuminuria is rare and only occurs late. Diarrhœa always follows injections of choleraic urine; the stools become pale, watery, and devoid of bile. The albuminuria increases until complete anuria occurs, in about thirty-six hours, and death soon after occurs with a rectal temperature of 33° or 34° C.

SECTION VI.

URINARY SEDIMENTS.

URINARY sediments are most conveniently classified, for purposes of study, into two divisions, viz., *chemical substances* and *anatomical substances.*

Chemical sediments, with but few exceptions, exist in the form of solution in normal urine, and their appearance as crystalline, amorphous, or other form of sediments may result from excessive formation or excessive excretion, or alterations in the urine affecting its solvent powers. The chief chemical deposits met with in the urine are uric acid, the urates, calcium oxalate, cystin, leucin, tyrosin, xanthin, and phosphates.

The *anatomical sediments* are in most cases foreign substances, and therefore do not exist in normal urine. They consist of such structures as pus-corpuscles, blood, renal casts, spermatozoa, fragments of growths, fungi, infusoria, etc. The anatomical elements found in the urine are more or less insoluble, and, therefore, when the urine stands they fall to the bottom as sediments.

Our methods of examining urinary sediments are both chemical and microscopical. Thus, the chemical deposits may often be recognized by their characteristic reactions; or the microscope may be employed to determine the characteristic form of the deposit when crystalline, which is often of itself diagnostic. In determining the character of the anatomical deposits the microscope constitutes the chief resource, although, in some cases, chemistry materially aids the investigation.

SEDIMENTATION OF URINE.

The older method of obtaining urinary sediments for investigation consisted in letting the urine stand in conical vessels for twenty-four hours or so, when the sediment would usually be found collected in the bottom of the vessel. Much difficulty was formerly encountered by this method in securing sediments for

(147)

examination which remained unchanged, since the length of time necessary to secure the deposit almost necessarily involved alterations in the urine at ordinary temperatures. Of late years this has been in a measure overcome by the addition of preservative agents to the urine, such as chloral hydrate, salicylic acid, resorcin, etc. These, however, all interfere with the chemical examination of the urine, more especially in making examinations for sugar and urea. But the most serious objection to the old method was the necessity of waiting for several hours before a satisfactory microscopical examination of the urine could be made.

More recent experience has demonstrated the immense advantages of the centrifugal method of obtaining urinary sediments for purposes of microscopical examination. The principle of this method depends upon the fact that when the urine is placed in tubes and revolved upon horizontal rotating arms at a high speed, a centrifugal force is exerted upon all solid particles in the urine hundreds of times greater than gravity; and, consequently, the urinary sediment is deposited in the bottom of the tubes almost immediately, irrespective of the specific gravity of the urine or the character of the sediment.

Of the very large number of centrifugals at present on the market, unfortunately but very few of them are capable of efficient practical work, as the large number of discarded instruments of this order in medical offices to-day will demonstrate. In previous editions of this work the author pointed out the prime essentials of the centrifuge for urinary work. Chief of which are capability of a speed of from 1500 to 2000 revolutions per minute, with a radius of at least $6\frac{3}{4}$ inches, and a tube capacity of 15 cubic centimetres each. These requirements have not been met by the hand-centrifugals thus far. Since the early editions of this work were published the author has, with the aid of Williams, Brown & Earle,—the manufacturers,—not only greatly improved his electric centrifuge (originally designed for urinary work), but also made important additions thereto, so that it is now designed to cover the entire range of centrifugal work for medical and bacteriological purposes. Very great credit is due to the above-named gentlemen[1] for so cheerfully co-operating with the author, sparing neither pains nor time in carrying out the designs the result of which is an apparatus that the author takes pleasure in recommending as altogether efficient and satisfactory in practical work.[2]

The author's electric centrifuge, shown in Fig. 12, can be operated indefinitely on all ordinary electric currents without overheating, viz.: on the interrupted incandescent illuminating

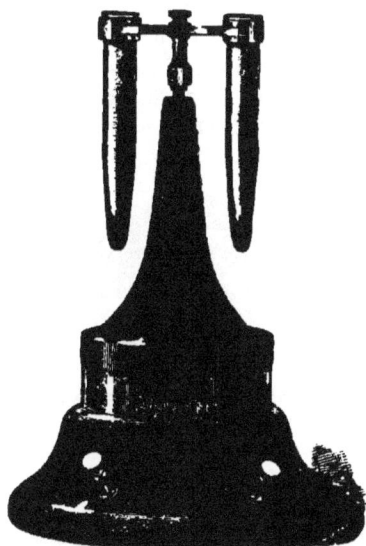

FIG. 12.—THE AUTHOR'S ELECTRIC CENTRIFUGE.

current, on the constant incandescent illuminating current, on the storage current, and on the galvanic current (sulphuric cell). The motor is furnished suitably adjusted for operation upon any of these currents at any voltage from 10 to 120 volts, if the nature and strength of the current be specified.

---

[1] The author's electric centrifuge is exclusively manufactured by Williams, Brown & Earle, 918 Chestnut Street, Philadelphia.

[2] The author has no commercial interest in the centrifugal that bears his name, neither has he had at any time; he therefore feels entirely free to speak of its merits.

This centrifugal was designed with special regard to strength, durability, efficiency, and perfect safety at the highest possible rates of speed. It is easily capable of all grades of speed from 500 to 2000 revolutions per minute, according to the strength of the current employed and the resistance of the arm. With the large urine arm, it carries 1 ounce of urine at a speed of 2000 revolutions per minute, with a radius of 6¾ inches; with the double arm for four large tubes it carries 2 ounces of urine at a sustained speed of 1200 revolutions per minute. With the new special arm for sedimenting micro-organisms, it easily carries two 1-centimetre tubes at a sustained speed of 5000 revolutions per minute, with a radius of 4½ inches. A speed-indicator is furnished for this motor (Fig. 12a) which indicates the exact rate of speed at which the motor is operating, and the speed can be

FIG. 12a.—SPEED-INDICATOR.

accurately graded on all currents of varying voltages by means of the indicator and the resister (or rheostat). In order to test the exact rate of speed of the motor, the indicator is grasped firmly between the thumb and forefinger with the dial toward the operator as shown in Fig. 12a. Next place the conical rubber tip of the indicator in the hollow depression on the top of the axle of the motor above the arm, and press rather lightly upon the indicator, when the hand on the dial will revolve more or less rapidly according to the speed attained. Care should be observed, on the one hand, to grasp the indicator firmly between the thumb and finger lest the vibrations of the motor cause the operator to lose his hold on the indicator and thus result in an accident; on the other hand, the indicator should not be pressed too firmly against the axle of the motor, as this would greatly increase the friction and correspondingly diminish the speed.

Each revolution of the hand on the dial indicates 100 revolutions of the motor. The glass tubing is provided with aluminum guards, which effectually prevent any damage from breakage.

The percentage and sediment tubes for urine and bulky fluids, shown in Fig. 12b, were specially designed for this motor, in order to avoid the defects in the old bulb-tipped Continental tubes, it having been found by experience in practical work that the latter would not hold small deposits of sediment in the tips. These tubes retain the most minute deposits of sediment intact, even though the tubes be inverted and the fluid be decanted from the sediment. The percentage-tubes are accurately graduated in fortieths of a cubic centimetre up to 0.5 cubic centimetres, then in fourths of a centimetre up to the 15-cubic-centimetre mark,—the latter to measure the reagents employed in precipitation. By means of these tubes and the methods laid down by the author accurate determination of bulk percentage may be made with this motor of the leading normal constituents of the urine, such as chlorides, phosphates, and sulphates; also such morbid elements in the urine as pus, blood, and albumin with great rapidity. Thus, with the double arm, four quantitative determinations may be made with ease in three minutes.

Finally, a new device for sedimenting and manipulating micro-organisms has been perfected and adapted to this motor. The amount of work and time that this device is capable of saving, and the ease and certainty with which it isolates micro-organisms in fluid media, are sufficient to render the apparatus an essential of the equipment of the pathological laboratory. This device consists of an arm, as shown in Fig. 12c, twenty-three centimetres in length, which carries two tubes of a little less than one cubic centimetre in capacity each. These tubes are conical at one end, which fits against a soft-rubber washer at the bottom of the

FIG. 12b.—AUTHOR'S PERCENTAGE TUBE.

slot *B*. These rubber washers are furnished in quantity, so that a new one may be, if necessary, used each time the tube is employed, or they may be readily picked out with a fine forceps and thoroughly cleaned. The large end of the tube is closed by a soft-rubber cork at *A*. This arm carries two of these tubes with perfect ease and safety at a sustained speed of 5000 revolutions per minute if desired. At a much less speed—from 3500 to 4000 revolutions per minute—practical experience has demonstrated that from 75 to 80 per cent. of the micro-organisms present in the tubes are deposited within the extreme tips in from three to five minutes.

DIRECTIONS FOR OPERATING THE MOTOR IN PRACTICAL WORK.

In sedimenting urine for ordinary microscopical examination, fill two tubes to the 15-cubic-centimetre mark with the urine, place them within the aluminums; turn on the current gradually,—never abruptly in full strength,—gauging the speed by the indicator until a speed of about 1200 revolutions per minute is

FIG. 12c.—ARM FOR SEDIMENTING MICRO-ORGANISMS.

attained. With urines of about normal specific gravity, continue this speed for two or three minutes. With urines of very low specific gravity, 1000 revolutions are sufficient if there be much sediment. With urines of very high gravity and little sediment it is well to increase the speed to about 1500 revolutions and maintain it for two or three minutes. The sediment is best manipulated by means of a nipple pipette about eight inches in length, as follows : After thoroughly cleansing the pipette slowly carry its point down the tube to within an inch and a half of the tip ; then stop and expel from 5 to 10 bubbles of air from the point by gentle pressure upon the rubber nipple. Next carry the point of the pipette firmly to the bottom of the tube and draw in about 5 drops of the sediment. Remove the pipette and expel from its point two or three drops of the sediment upon a previously cleaned glass slide, upon which is a three-fourths-inch ring of gold size (rather thickly laid on and dried) within the ring. Next cover with a cover-glass, and take up the excess of urine with a strip of filter-paper, and the slide is ready for examination. This form of slide is better than the ordinary slide with ground-out cell, because it affords a perfectly flat field. The advantage of the temporary mounting is that it can be examined at any angle of light under the microscope.

The contents of serous cavities, cysts, abscesses, and, in short, all media obtained by means of aspiration, may be dealt with in the manner just described.

Percentage determinations, by the author's standard method, of chlorides, phosphates, and sulphates have already been described in a previous section of this work (pages 63 to 66). The author's quantitative method for albumin has also been detailed on page 83. The bulk percentages of pus and blood may be determined by simply filling the percentage-tubes with the urine to the 10-cubic-centimetre marks, placing them in the aluminums, and revolving at the same speed and length of time as in the case for albumin. In order to obtain the author's uniform results in percentage determinations, care must be exercised to employ the stated speed, length of time of revolutions, and the motor must be operated with exactly 6¾-inch radius. It must be obvious to any intelligent person that, with tubes of varying capacities and form, different lengths of arms, and with no grading of speed whatever, uniform or accurate results are impossible, and that such methods are mere guess-work.

Micro-organisms.—In sedimenting micro-organisms, remove the large arm from the motor and in its place adjust the arm for the small tubes shown in Fig. 12c. In searching for micro-organisms in bulky media, such as the urine, it is best first to throw down the coarse pus-sediment in the large urine-tubes by

FIG. 12d.

about 2000 revolutions for four or five minutes. Next, with a pipette take up rather more than one centimetre of this sediment from the large tube ; remove the soft-rubber cork from the small tube at *A*, and, while holding the finger very firmly over the point *B*, transfer the sediment from the pipette to the small tube, which must be filled to overflowing in order to prevent the ingress of air-bubbles. Next, while still holding the finger firmly over the point of the small tube, press down the soft-rubber cork with a twist until it is about half-way into the large end of the tube. Next, draw back the spring in the arm of *A;* insert the point of the tube in the hollow at *B*, pushing it firmly against the rubber washer; let go the spring at *A*, and the tube will be firmly locked in the arm. Next turn on the current and increase it until the indicator shows a speed of at least 3500 to 5000 revolutions per minute, and continue this speed for about two to three minutes. Remove the tube from the arm by drawing back the spring at *A;*

11

have at hand a clean glass slide, and place the point of the small tube on the middle of the slide, and by gentle pressure with the thumb or finger (as shown in Fig. 12d) on the top of the soft-rubber cork, a drop, or, if need be, a fraction of a drop, of the now-highly-concentrated sediment may be deposited precisely where it is required to be stained and prepared for examination under a high power. It will be seen that, proceeding as above directed, the search for micro-organisms sparsely scattered through bulky media,—as, for instance, tubercle bacilli in urine,—is rendered easy and almost absolutely certain, because practically we obtain 75 to 80 per cent. of the micro-organisms in one ounce of urine concentrated within one or two minims, and manipulation results in no loss whatever.

With media of smaller bulk, as sputum, etc., it may be placed directly in the small micro-tubes after such preparation as individual preference shall determine, and sedimentation is carried out as already described. The sediment may then be transferred to the slide with the greatest possible ease and precision, as already detailed, and examined under the microscope for elastic fibres, Charcot-Leyden crystals, tubercle bacilli, etc.

In working with sputum, urine containing much mucus, fibrinous exudates, and media that are very viscid and tenacious, it will be found that the micro-

FIG. 12e.

organisms will be more quickly and more completely sedimented if the media be first thoroughly broken up and liquefied, and this can be very readily and quickly done with the author's apparatus for the purpose figured here.[1] Slight dilution of viscid media, such as sputum, with physiological salt solution greatly assists in liquefaction, and does not interfere with subsequent staining.

The Hæmatokrit Attachment.—The hæmatokrit—first suggested by Blix, for the purpose of determining the percentage and relative proportions of the red and white corpuscles of the blood—is readily adjusted to the small arm of this motor. The hæmatokrit consists of a graduated glass tube 50 millimetres in length and 0.5 millimetre bore to receive the blood. The tube is marked by a scale ranging from 0 to 100, the scale being rendered visible by a lens front (prism form). The outer end of the tube fits into a small cup-like depression at the end of the arm, the bottoms of which are covered with the rubber disks already shown, while the inner extremity is held in position by the spring at B.

---

[1] Messrs. Sharp & Smith, of 92 Wabash Avenue, Chicago, make and keep these instruments in stock.

To use the hæmatokrit in blood-examinations proceed as follows: The rubber tube with mouth-piece at one end is slipped over the end of the hæmatokrit and the latter is filled by suction on the mouth-piece from a drop of blood obtained by a prick of the finger. The blunt end of the tube is next quickly covered with the finger tip, and the tube is inserted into the arm in the same manner as adjusting the tubes for micro-organisms. The current is next turned on and the speed increased gradually to 5000 revolutions per minute, and thus steadily maintained for from two to three minutes. The hæmatokrit may next be removed and the percentage of red corpuscles is read off from the scale. In health the volume of red corpuscles is about 50 per cent. One per cent. by volume represents about 100,000 red blood-corpuscles; therefore, by adding five ciphers to the percentage of volume it gives the number of red corpuscles in one cubic millimetre of blood. Thus, in a given case, if the reading were 25; multiply that number by 100,000, and the product, 2,500,000, would represent the number of red blood-corpuscles in one cubic millimetre of blood. The amount of hæmoglobin in each corpuscle may be approximately determined, also, by dividing the quantity of hæmoglobin ascertained by Fleischl's or Gowers's instrument by the number of corpuscles determined by means of the hæmatokrit.

The white blood-corpuscles, or leucocytes, will be found to occupy a second,

FIG. 13.

but much shorter, column immediately above the column of red corpuscles, and, if leucocytosis be present, even though to a very slight degree, it is easily recognized.

The accurate regulation and determination of speed by this motor greatly improves determinations of blood, by means of the hæmatokrit, rendering the tedious and tiresome use of cytometers no longer necessary.

Further details in reference to the adjustment of this centrifugal to the various electric currents that may be available, and its establishment in working order in the laboratory or office, will be cheerfully furnished by the manufacturers upon application; and the author cannot but commend the satisfactory manner in which they have thus far met the requirements demanded.

CHEMICAL SEDIMENTS.

LITHURIA.

Uric-acid crystals occur as a sediment rarely, if ever, in other than sharply-acid urine. They differ from all other urinary deposits in possessing a deep-yellow or orange-red color; they may at times be pale yellow, but are never colorless. The crystalline nature of this deposit may usually be detected readily by the naked eye. The essential or primary form of the uric-acid crystal is that of rhombic prism, and the great variety in which it is found all constitute combinations or modifications of this form. Thus, the angles may be nearly equal, forming quadrangular plates, or sometimes nearly cubes may be seen. Again, they may be seen with rounded ends, forming ovoids or circles. Elongated crystals are sometimes observed, and these frequently join at one end, forming stars. The beauty and variety of these star-shaped clusters are very marked. (See Plate V.) Sometimes fan-shaped forms are produced by elongation of the crystals in one direction only, instead of the star form. The rough and pointed forms of uric-acid crystals are claimed by Ultzmann to be of diagnostic significance, being "almost always an accompaniment of renal calculi." For properties and tests of uric acid, see Section II, pages 30 to 36.

Uric acid possesses a strong tendency to crystallize upon contact with any solid substance, organic or inorganic. This may be observed by its behavior in a vessel upon standing, when it will often be noted that the crystals cling to the sides of the glass or to threads or specks suspended in the urine. This tendency renders it more liable than any other crystalline deposit to form around some nucleus in the urinary channels, and ultimately form gravel. This is one of the reasons why nearly 70 per cent. of the stones found in the bladder are of the uric-acid variety.

The deposit of uric-acid crystals can only be regarded as of pathological import when the deposit occurs shortly after the urine is voided,—say, within four to six hours. It has already been shown that perfectly healthy urine usually deposits uric-acid crystals after standing ten or more hours. But, on the

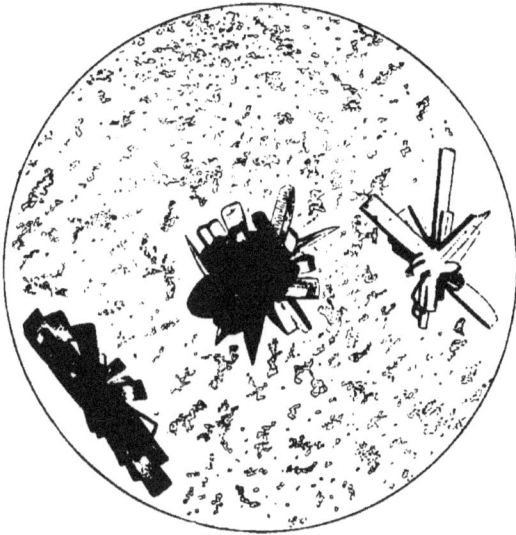

[Uric-Acid Crystals with Amorphous Urates. × 450.
(After Peyer.)]

other hand, if uric acid be precipitated from the urine shortly after it cools, it very justly forms the foundation for fear that the same may occur before it be voided, and thus give rise to the formation of calculi and gravel, with the long train of painful symptoms entailed by such conditions. Before speaking of the pathological conditions which usually attend deposits of uric acid, it may be well to allude to the conditions of the urine which favor such deposits.

Sir William Roberts, who has carefully investigated this subject recently,[1] makes the following interesting observations:—

"The presence of uric acid in human urine is somewhat anomalous, as it is not needed as a vehicle for the elimination of nitrogen. Its place is taken by urea, which, by its easy solubility, is better adapted to the liquid urine of animals. Perhaps uric acid is a vestigial remnant in mammalian descent. But, although physiologically insignificant, uric acid is pathologically the most prominent component of the urine, chiefly because of its tendency to form concretions.

"All acid urines tend inevitably to deposit their uric acid sooner or later. The time of onset of precipitation varies from a few hours to five or six days, or even longer. The inference from this is that pathological gravel is due to an exaggeration of conditions which exist in a less pronounced degree in health. To get at an explanation of this spontaneous precipitation it is necessary to examine the states of combination of uric acid in the urine.

"Uric acid ($C_5H_4N_4O_3 = H_2U$) is a bibasic acid, and forms two regular orders of salts, namely, *neutral* or *normal urates* ($M_2\bar{U}$) and *acid urates* or *biurates* ($MH\bar{U}$)[2]. But, in addition to these, it forms a series of hyperacid combinations, first discovered by Bence Jones, and termed by him *quadrurates* ($MH\bar{U}.H_2\bar{U}$). The neutral urates are never found in the animal body, and are only known as laboratory products. The biurates are only encountered pathologically as gouty concretions. The quadrurates, on the other hand, are especially the salts of uric acid. They constitute the exclusive combination in which uric acid

---

[1] Proceedings of the Medico-Chirurgical Society, 1890, p. 85.

exists in solution in normal urine, and they become visible sometimes as the amorphous urate sediment. The urinary excretion
of birds and reptiles is composed exclusively of quadrurates.
The special and characteristic reaction of quadrurates is that
they are immediately decomposed by water into free uric acid
and biurates. They exist in acid urine in the presence of water
and of superphosphates. These conditions necessarily involve
the ultimate liberation and precipitation of uric acid. The first
step is the breaking up of the quadrurate by the water of the
urine into free uric acid and biurate, according to the following
equation :—

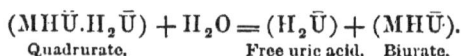

$$(\mathrm{M H \bar{U}.H_2 \bar{U}}) + \mathrm{H_2 O} = (\mathrm{H_2 \bar{U}}) + (\mathrm{M H \bar{U}}).$$
Quadrurate.   Free uric acid.   Biurate.

" This explains the liberation of half the uric acid. But the
biurate thus formed is forthwith changed in the presence of
superphosphates into quadrurates. Thus :—

$$2(\mathrm{M H \bar{U}}) + (\mathrm{M H_2 P O_4}) = (\mathrm{M H \bar{U}.H_2 \bar{U}}) + (\mathrm{M_2 H P O_4})$$
Biurate.   Superphosphate.   Quadrurate.   Dimetallic phosphate.

By these alternating reactions all the uric acid is at length set
free.[1]

" Seeing that uric acid exists in acid urine (that is, for some
sixteen hours out of the twenty-four), amid conditions which,
if the quadrurates stood alone and uncontrolled, would lead to its
immediate precipitation, and yet that in the normal course no
such early precipitation occurs, it is obvious that the urine must
contain certain ingredients which inhibit or greatly retard its
water from breaking up the quadrurates. These inhibitory ingredients consist chiefly of (1) the mineral salts and (2) the
pigments of the urine. The conditions of the urine which tend
to accelerate the precipitation of uric acid, as in the formation
of concretions and deposits, are (1) high acidity, (2) poverty in
mineral salts, (3) low pigmentation, and (4) high percentage of
uric acid. The converse conditions tend to retard precipitation.
On the interaction of these factors the occurrence or non-
occurrence of uric-acid precipitation appears to depend, and

---

[1] In these formulæ the symbol M represents a monad metal, and the symbol
$\bar{\mathrm{U}}$ the radicle $\mathrm{C_5 H_2 N_4 O_3}$.

probably the most important of these factors is the grade of acidity."

Clinical Significance.—Uric-acid sediments are perhaps most often encountered in acute fevers and inflammations attended by pronounced elevation of temperature. In such cases there is diminution of the aqueous elements of the urine, entailing increased acidity. As a consequence of increased tissue metabolism, there is also absolute increase of uric acid, as of most other urinary solids. In the so-called *uric-acid diathesis* there is often an habitual and pronounced deposit of uric-acid crystals in the urine. The causes of this state are partly defective physiological action of the liver, and partly errors in diet, coupled with sedentary habits of life; and it is often accompanied by headache, emaciation, and hypochondriasis. Since this condition is induced by faulty habits of living which entail overwork of the liver, with defective supply of oxygen, it in nowise merits the name of "*diathesis.*"

In the early stages of interstitial nephritis, uric-acid deposits are often to be observed; indeed, the urine frequently throws down this deposit habitually for some time before the interstitial defect is made known by pronounced symptoms. This is due to two causes: (*a*) the polyuria of the early stages of the disease lessens the relative amount of coloring matters in the urine, and it will be remembered that the pigments tend to hold the uric acid in solution; (*b*) both interstitial nephritis and uric-acid deposits are often the outgrowth of the same habits of living, viz., the overindulgence in animal foods. The author has repeatedly observed that people who possess generous appetites, and indulge in the use of animal foods two and three times daily, are exceedingly apt to have uric-acid deposits in their urine at middle age, and somewhat later to develop interstitial nephritis. Uric-acid deposits are frequent in cases of children convalescing from scarlatina, with or without accompanying nephritis, and concretions or gravel are very prone to arise under such circumstances.

URATES.

The acid urates of *sodium, potassium, ammonium,* and, more rarely, of *calcium* are met with as sediments in the urine.

The acid urate of sodium occurs chiefly as minute, irregular, amorphous granules, although sometimes, also, in crystalline form,—star-shaped, needle-like clusters, often of fan-shape arrangement. This deposit is more or less deeply stained brown or pink, according to the degree of pigmentation of the urine, since it possesses a great affinity for the above-named pigments.

The sodium-urate deposit occurs in acid urine, and forms a large bulk of the "brick-dust," or mixed urate, deposit found in

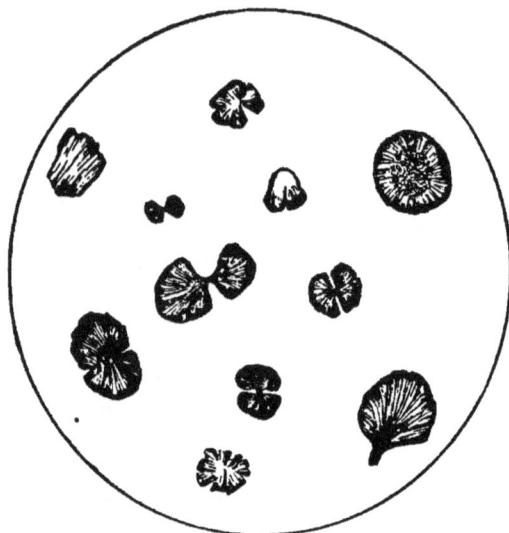

FIG. 14.—SODIUM-URATE CRYSTALS. (After Peyer.)

the bottom of the vessel after the urine has cooled. Acid sodium urate is extremely insoluble, requiring 1150 parts of cold or 124 parts of boiling water to effect its solution.

Acid potassium urate occurs only in amorphous form as a deposit, and, like sodium urate, forms a part of the mixed urate deposit met with in acid urines. It is much more soluble than the sodium urate.

Acid calcium urate occurs as a urinary deposit but rarely, and in minute quantities. It consists of a white or grayish

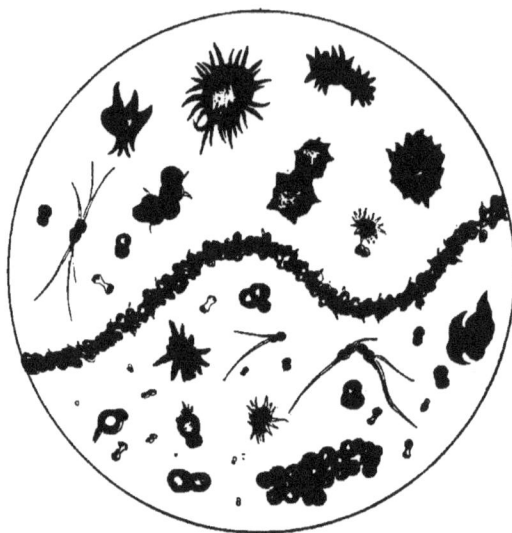

AMMONIUM URATE, SHOWING SPHERULES AND THORN-
APPLE-SHAPED CRYSTALS. (After Peyer.)

amorphous powder, highly insoluble, and on fusion leaves a white residue, consisting of calcium carbonate. The acid calcium urate, like the potassium and sodium urate deposits, occurs only in acid urine.

Ammonium urate occurs as a crystalline deposit, consisting of dark-brown, spherical masses studded with fine, sharp-pointed spicula,—"*thorn-apple crystals.*" The spicula may be long, sometimes curved, branched, or bent, forming various shapes. (See Plate VI.) The smaller crystals often closely resemble those of sodium urate, and by some they are claimed to be identical.

This sediment most frequently occurs in alkaline urine, associated with amorphous calcium phosphate and triple phosphate crystals. The ammonium-urate deposit is, in fact, the only urate sediment found in alkaline urine.

· The *mixed urate deposit* consists of a reddish, granular-looking sediment, with color always deeper than the urine from which it precipitates. It may vary all the way from a faint pinkish haze to a brick-red color. (See Plate V.) Most frequently it sinks quickly to the bottom of the quiescent vessel, but part of it may long remain suspended, imparting to the urine an opalescent turbidity; at the same time a pellicle may form on the surface or cling to the sides of the vessel, notably at the surface-line of the urine. By gently heating the urine the mixed urate sediment promptly dissolves, and this forms a ready method of its recognition, as no other urinary deposit behaves similarly. The mixed urate deposit gives the murexide reaction similar to uric acid. It is dissolved by solutions of the caustic alkalies; with mineral acids it is decomposed, with resulting precipitation of uric-acid crystals.

It is an interesting fact that urines of high density are most prone to throw down deposits of mixed urates, while those of lower density are more apt to throw out of solution uric acid. Thus, with urines of specific gravity at 1.026 to 1.030, when they cool, the excess of urates comes down because of their limited degree of solubility; while in urines of specific gravity below 1.020, the diminished pigmentation often permits the uric acid to fall out of solution and form a deposit.

Clinical Significance.—The mixed urate deposit, like that of uric acid, is most frequently encountered in febrile states; even a slight elevation of temperature. is often sufficient to cause their deposit. A more constant deposit of mixed urates may be noted in diseases of the viscera, which entail progressive emaciation, notably in the liver and in the so-called wasting diseases. Functional disorders of the stomach are frequently associated with amorphous urate deposits, due, in all probability, to incomplete transformation of proteid foods. In gout the urates are usually deposited during the attack, but disappear upon the approach of convalescence.

OXALURIA.

Oxalate of calcium is met with as a urinary sediment either in acid or alkaline urine, but most often in the former. If it occur in acid urine it is often associated with uric-acid deposits, but when occurring in alkaline urine its most frequent associated deposit is the triple phosphate.

The calcium-oxalate deposit occurs in crystalline form, consisting chiefly of two varieties of crystals, (a) and most frequent are the octahedral crystals, very beautiful and highly refracting. They are made up of four-sided pyramids, situated base to base, as seen in their long diameters. When viewed from the side they appear as squares crossed obliquely by two sharp lines, forming the characteristic " envelope-shaped " crystals. When small, the lines crossing in the centre form a bright spot, highly refractive of light,—star-like. (b) The second form of calcium-oxalate deposit is the so-called " dumb-bell " crystals. Their true form is that of ovoid or circular disc, with round margins and depressed at the centre on either side. Their variable appearance depends upon their different positions when viewed, as may be seen by causing the crystals to roll over under the cover-glass.

Calcium oxalate is insoluble in alcohol, ether, water, alkalies, and acetic acid, but readily soluble in hydrochloric or other mineral acids,—characteristics which serve to identify this salt; but in practice the microscopical appearance is the most conclusive, since the crystals are so characteristic in form that they are readily distinguishable from all other crystalline deposits; the deposits of triple phosphate and uric acid are the only ones which have the least resemblance thereto.

With regard to the triple-phosphate crystals, it is only the smaller, imperfect, and short prisms that are ever confounded with calcium-oxalate crystals. In such cases the body of the crystals, instead of forming a parallelogram, is shortened so that it becomes a square, and the prism then gives somewhat the appearance of the envelope-shaped calcium-oxalate crystal. The

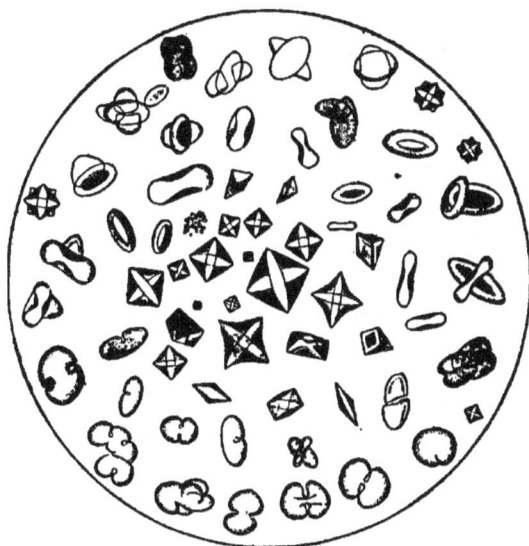

FIG. 15.—VARIOUS FORMS OF CALCIUM-OXALATE CRYSTALS. (After Peyer.)

calcium-oxalate crystals, however, are always smaller and more highly refracting. Should any doubt remain, after careful ocular examination, they may be readily distinguished by their behavior with acetic acid, which promptly dissolves the triple-phosphate crystals, while the calcium oxalate is unaffected thereby.

The "dumb-bell" form of uric acid may usually be distinguished from the calcium-oxalate crystals of similar form by the brown color of the former, as well as by their solubility in alkalies.

Clinical Significance.—The occurrence of calcium oxalate as a urinary deposit is brought about by the strong affinity which oxalic acid possesses for calcium. Oxalic acid occurs under physiological conditions in very small amounts in urine,—about 0.02 gramme in twenty-four hours. According to generally received opinion, it exists in the form of calcium oxalate, which is kept in solution by the acid phosphates of the urine. The quality of food taken often materially influences the degree of physiological oxaluria. Thus, vegetables and fruits containing much oxalic-acid combinations,—as cabbage, spinach, asparagus, sorrel, apples, grapes, tomatoes, and turnips,—when taken in excess, may cause excretion of calcium oxalate in considerable amount. Calcium oxalate is also excreted in excess upon an exclusive or excessive diet of flesh and fat, indicating its formation from proteids.

The question of so-called "*oxalic-acid diathesis*" possesses much practical interest. As early as 1842, Bird described a series of nervous and dyspeptic symptoms, which he alleged were associated with deposits of calcium oxalate in the urine. Later on, Bigbie still more minutely described the symptoms of the so-called "*oxalic-acid diathesis*," of which the following is a brief summary: "These patients are mostly males in the prime of life, ordinarily of sanguineous or melancholy temperament, of sedentary habits, and accustomed to overindulgence in the luxuries of the table. Indigestion in its varied forms is a prominent feature. These patients are often capricious, sensitive, irritable, or dull, despondent, and melancholic. The tongue is coated and the skin is dry. In inveterate cases a dirty, dingy countenance, increasing emaciation, falling out of the hair, tendency to boils, carbuncles, psoriasis, and other cutaneous disorders are frequently observable. Accompanying these are often deep pains in the back and loins, hæmorrhages from the intestines and bladder, incontinence of urine, impotence and irritation of the bladder."

Attractive though the theory be of the so-called "*oxalic-acid diathesis*," in the light of more recent and wider observation the name "*diathesis*" seems in nowise merited by any of the states associated with deposits of calcium oxalate in the urine.

It is true that oxalic acid, when taken internally in any considerable amount, exerts a poisonous action upon the organism, not only locally on the intestines, but also generally on the heart and nervous system; and this gave rise, no doubt, to the supposition that a large formation of oxalic acid or its retention in the system might produce toxic, and even dangerous, symptoms. Distinct proof, however, is yet lacking to show that the symptoms of the so-called "*oxalic-acid diathesis*" are due to an accumulation of oxalic acid in the blood. Indeed, nearly all the evidence tends in the opposite direction. In the first place, large deposits of calcium oxalate, and even the formation of oxalic calculus, is repeatedly observable in people who are otherwise in the enjoyment of the most typical good health. In the second place, the group of symptoms described as characteristic of the oxalic-acid diathesis, as Roberts has observed, "is one common to the clinician minus the deposits of calcium oxalate." Lastly, the states of the system associated with deposits of calcium oxalate are altogether too varied to admit of so narrow a classification as that of a special diathesis. It seems most reasonable to conclude that oxaluria is dependent upon a variety of conditions of the system, many of which are associated with little or no departure whatever from ordinary health.

The conclusions of Bencke, who has thoroughly investigated this subject, are as follow :—

1. Oxaluria accompanies the lighter or severe forms of illness; has its proximate cause in an impeded metamorphosis,— *i.e.*, in an insufficient activity of that stage of oxidation which changes oxalic acid into carbonic acid.

2. Oxalic acid has its chief source in the azotized constituents of the blood and food; hence, everything which retards the metamorphosis of these constituents gives rise to oxaluria.

3. Such retardation of the metamorphosis of azotized elements of the blood may be determined by the following causes : (*a*) excessive use of azotized articles of food; (*b*) excessive use of saccharine and starchy articles of food; (*c*) insufficiency of the red blood-corpuscles, entailing diminished oxidation ; (*d*) insufficient access to pure fresh air; (*e*) organic lesions which in any way impede respiration and circulation; (*f*) conditions of

the nervous system entailing depression, whether arising primarily from mental derangement or from pathological states of the blood.

4. Excess of alkaline bases in the blood.

PHOSPHATURIA.

It has already been shown in Section II that phosphorus exists in normal urine in combination with the alkalies and the earths,—the alkaline and earthy phosphates. It is only, how-

FIG. 16.—TRIPLE-PHOSPHATE CRYSTALS. (After Ultzmann.)
1. Rosette or star-shaped crystals. 2. Coffin-shaped crystals.

ever, the latter salts that are met with as urinary deposits. The earthy phosphates consist of (*a*) triple phosphate or ammonio-magnesium phosphate, and (*b*) calcium phosphate or phosphate of lime.

Ammonio-magnesium phosphate ($MgNH_4PO_4 6H_2O$), or triple phosphate, is essentially a crystalline deposit, occurring in two forms. The first—most frequent and typical—form is that of a triangular prism with beveled ends, very distinctive and often termed "*coffin-shaped*" crystals. Many modifications of this typical form are met with. Thus, the crystals may be shortened to the form of squares, instead of being oblong, or one or more corners may be absent.

The second and less frequent form in which triple phosphate appears as a urinary sediment is that of star-shaped, feathery crystals, the points appearing not unlike fern-leaves. These are often but rudiments of the prismatic form, of triple-phosphate crystals, into which latter they often become gradually trans-

FIG. 17.—CALCIUM-PHOSPHATE CRYSTALS. (After Peyer.)

formed, and therefore between these two forms numerous intermediate ones are to be observed (Fig. 16).

Calcium phosphate is met with as a urinary sediment in two forms,—(*a*) *amorphous*, (*b*) *crystalline*. The amorphous form of calcium phosphate is a whitish, flocculent deposit, often mistaken by the naked eye for pus or granular organic matter, and when precipitated from the urine by heat it is sometimes mistaken for albumin. Under the microscope this sediment appears in the form of minute, pale granules, arranged in irregular patches.

The crystalline form of calcium phosphate is a comparatively rare deposit, occurring less frequently than any other form of phosphatic deposit. Its essential or elementary form is that of crystalline rods, sometimes lying unarranged, but more often grouped in stellar or rosette form, while often they may be observed grouped in club or wedge form, but always marked by lines of crystallization (Fig. 17). A deposit of earthy phosphates is essentially a product of alkaline urine, and, with the exception of the crystalline form, they are never met with in acid urine. The above-named exception only occurs with feebly-acid urine tending to ammoniacal change.

The following are the chief conditions of the urine which lead to phosphatic sediments :—

(a) If the urine be alkaline from fixed alkali.

(b) If the earthy phosphates be in excess (the urine being alkaline or neutral).

(c) If the urine be alkaline from volatile alkali, the result of decomposition of urea into ammonium carbonate in the urinary passages, the ammonia uniting with the magnesium phosphate to form the triple phosphate of ammonium and magnesium.

Clinical Significance.—In those cases in which the phosphatic deposit occurs in alkaline urine from *fixed* alkali, the deposit is chiefly precipitated calcium phosphate, though often mixed with triple-phosphate crystals. The urine in these cases is usually of high specific gravity, alkaline in reaction *when voided*, more or less cloudy, and effervesces upon the addition of acid, after which the urine immediately clears. The clinical symptoms corresponding to the above are often those of general debility, with feeble respiration—favoring the accumulation of carbonic acid in the system. Thus, in convalescence from exhausting acute diseases, deposits of calcium phosphate are frequently to be noted. Flatulent dyspepsia is a frequent cause of alkaline urine from fixed alkali and the deposit of calcium phosphate.

As Ralfe has pointed out, the acids formed by fermentative changes being of the fatty acid series, upon entering the blood they are oxidized into carbonic acid, and, uniting with the bases of the alkaline oxides from carbonates of these bodies, increase the alkalescence of the blood and in consequence diminish the acidity of the urine or even render it alkaline, which permits the phosphates to fall out of solution. This is most frequently noted in debilitated persons with flatulence of the small intestine. It is associated with such features as loss of weight, irregular bowels, sallowness of complexion, despondency, and frequent micturition.

In those cases in which the deposit of calcium phosphate is the result of excessive elimination, very often marked systemic disturbances are associated therewith. The urine is usually alkaline, copious in volume, and the deposit is of dense whitish form. If persistent, the symptoms are usually those of nervous irritability, dyspepsia, emaciation, and backache. Sometimes symptoms akin to diabetes are observable in inveterate cases, and the condition has been called "*phosphatic diabetes.*" Indeed, it is claimed that this condition not infrequently ends in diabetes insipidus.

The deposit of the crystalline form of calcium phosphate in quantity is, in the experience of Roberts, often " an accompaniment of some grave disorder," such as cancer of the pylorus, phthisis, and exhaustion from obstinate chronic rheumatism.

In cases of phosphatic deposits in the urine resulting from the presence of volatile alkali, the sediment, as before stated, is that of triple phosphate of ammonium and magnesium. It has already been shown that ammoniacal fermentation always occurs in healthy urine upon standing sufficiently long, and then the triple phosphates are precipitated. But in the class of cases under present consideration the urine is alkaline *when voided* and precipitation of triple phosphates takes place immediately. In these cases ammoniacal decomposition of the urine occurs in the urinary passages. In addition, therefore, to the triple-phosphate deposits in such cases, the urine also contains pus and more or less mucus.

The clinical symptoms attending this state of the urine are most often those of septic inflammations of the urinary passages, such as pyelitis and cystitis. The most frequent class of causes of this condition of urine are the obstructive diseases of the lower urinary conducting channels. Whatever cause operates to retain the urine in the bladder sooner or later gives rise to cystitis and the deposit of triple-phosphate crystals. Thus, in enlarged prostate, atony and paralysis of the bladder, paraplegia, and diseases of the lower spinal cord, the urine nearly always precipitates the triple phosphates. This condition of urine nearly always precedes so-called "*surgical kidney*," and those septic conditions so dangerous to life which result from the use of instruments in the lower urinary channels. Therefore, when the urine is found to contain deposits of triple phosphate with pus, and is alkaline when voided, it constitutes a signal for the exercise of the greatest possible caution on the part of the surgeon in passing instruments into the urethra and bladder for the *first time*, more especially in the cases of elderly men.

It is well known that men addicted to exhaustive mental labor and people laboring under worry and anxiety are apt to have precipitates of earthy phosphates in their urine. If to such conditions be added habits of vegetarianism, which tend to depress the acidity of the urine, triple phosphates in addition may readily fall out of solution and form deposits.

CYSTINURIA.

Cystin ($C_3H_6NSO_2$) is comparatively rarely met with as a urinary deposit. Its origin in the economy is not clearly understood, although its highly sulphurous composition (about 25 per cent.), together with its close resemblance in composition to *laurin*, suggests the possibility that the liver may be its source. The discovery of cystin in the livers of typhus patients by Scherer, as well as the discovery of cystinuria in cases of diminished bile secretion by Marowski, would further favor the above view. Stadthagen claims that cystin is absent from normal urine, though Goldman and Baumann succeeded in separating it in very small quantities from healthy urine as a benzoyl compound.

Cystin is a crystalline compound of feeble chemical affinities, and occurs in two forms, (*a*) most commonly in six-sided tablets of variable sizes, and somewhat resembling the six-sided crystals of uric acid. These tablets possess an opalescent lustre,—"*mother-of-pearl*" appearance,—and when traced with fine lines of secondary crystallization, or formed into rosettes, they present microscopical pictures of great beauty. (*b*) The second form of cystin crystals is that of four-sided square prisms, which lie separately or in stellate form. They are highly refractive, and when their sides lie out of the direct line of vision they appear almost black, forming a strong contrast with those sides presented vertically to the light, which appear of a brilliant-white color. Cystin is soluble in the caustic alkalies, oxalic and strong

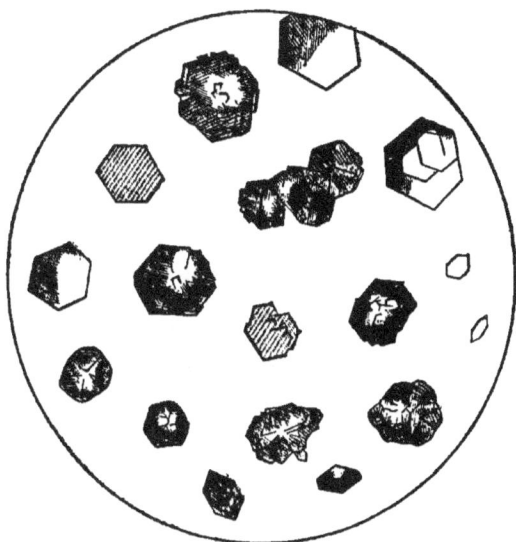

FIG. 18.—THE MORE COMMON FORM OF CYSTIN CRYSTALS. (After Peyer.)

mineral acids. It is insoluble in boiling water, acetic acid, alcohol, and ether.

Differentiation.—Cystin may be readily distinguished from the pale, lemon-colored, six-sided crystals of uric acid, as follows: Permit a drop of ammonia to mingle with the deposit on a glass slide, when either form of crystals disappears. Next, evaporate; and if cystin be present the crystals re-appear, showing that they were merely in solution. If, on the other hand, the crystals were uric acid, no re-appearance occurs upon evaporation, but.

instead, crystals of ammonium urate appear, showing chemical combination, and not solution.

Another simple method consists in treating the crystals with oxalic or hydrochloric acid, which promptly dissolves cystin crystals, but leaves uric acid unchanged.

From triple-phosphate crystals cystin is readily distinguished by its behavior with acetic acid, the former being immediately dissolved therewith, while cystin remains unchanged.

The urine containing cystin is usually pale in color, of faintly-acid reaction, and upon standing develops the odor of sulphuretted hydrogen, as well as that of ammonia. The sediment is of pale-lemon color, and often changes to green upon standing.

Clinical Significance.—Unfortunately but little at present of a positive nature is known as to the clinical relations of cystinuria. Its frequent association with hepatic disorders may be said to be established. Cystin deposits are said to be results of extensive renal degenerations. Chlorotic women and strumous children are also believed to be prone to cystic deposits in their urine. Ebstein has noted the presence of cystin deposits together with albumin in the urine in cases of acute articular rheumatism. Cystin deposits have been known to occur repeatedly in the same family by a number of independent observers, among whom are Marcet, Lenoir, Civiale, Toel, and Ebstein. Cystin calculus is well known to run in certain families. Cystinuria is said to be most common in young males, although no age or sex can be said to be exempt from it. It may be present and continue for years without any noticeable impairment of health.

The chief interest connected with cystic deposits is their proneness to form concretions of cystin gravel; and although these are comparatively rare occurrences, their consequences are none the less serious when occasionally encountered.

LEUCINURIA AND TYROSINURIA.

Leucin ($C_6H_{13}NO_2$) and tyrosin ($C_9H_{11}NO_3$), as will be seen from their formulæ, are closely related, being products of decomposition of proteid bodies or of their derivatives. Since they are nearly always found associated with each other in the urine, they will be considered together.

Leucin occurs as a urinary sediment for the most part in the form of yellowish, highly-refracting spheres, though not quite so highly refracting as oil-globules, which they somewhat resemble. In a pure state it crystallizes in scales or rosettes, often of irregular shapes, and it has a greasy feel. Leucin is insoluble in ether, which readily distinguishes it from oil-globules. It is also insoluble in mineral acids, but is partly soluble in water and alcohol and is completely soluble in caustic alkalies. For ordinary

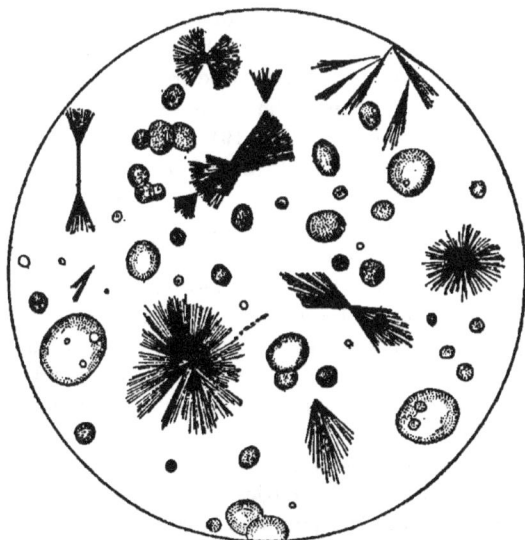

FIG. 19.—LEUCIN AND TYROSIN. (After Peyer.)

practical purposes leucin may be known by the microscopical appearance of the crystals, and in this way very minute traces may be determined with certainty. Confirmation by chemical tests may be employed if a fair amount of the material be at hand. (*a*) Thus, solutions heated with proto-nitrate of mercury give deposits of metallic mercury (Hoffmeister). (*b*) When evaporated with nitric acid on platinum-foil it leaves a colorless residue, which if heated with potassium hydrate forms drops of an oil like fluid which do not adhere to the platinum (Scherer). (*c*) Upon heating leucin in a glass tube open at both ends to about 170° C. it sublimes in feathery particles, which float about in the air within the tube. Further heat causes it to fuse and mostly disappear into carbon dioxide and amilymin. Leucin may be separated from the urine by evaporating the latter and dissolving the residue in boiling alcohol. Upon cooling the leucin present will be deposited in whitish plates or masses. Leucin is normally present in the liver, pancreas, spleen, lymph glands, salivary glands, and in the thyroid and thymus glands.

Tyrosin crystallizes in the form of very fine needles, arranged in sheaf-like collections. In masses the crystals are snow-white, tasteless, and odorless. If crystallized from an alkaline solution tyrosin often assumes the form of rosettes composed of fine needles arranged radiately (Fig. 19). Tyrosin is insoluble in alcohol and ether, feebly soluble in cold water, readily soluble in acids, alkalies, and hot water. Aside from its crystalline form and characteristic solubilities, tyrosin may be readily recognized by several pronounced reactions.

(*a*) *Hoffmann's Reaction.*—When heated with Millon's reagent, solutions of tyrosin yield a brilliant crimson or pink coloration, which, if much tyrosin be present, is accompanied finally by a similarly-colored precipitate. The test in its original form was applied by heating with a solution of mercuric nitrate in presence of nitrous acid.[1]

(*b*) *Piria's Reaction.*—If tyrosin be moistened on a watch-glass with concentrated sulphuric acid, and warmed for five or ten minutes on a water-bath, it turns pink, owing to the formation of tyrosin-sulphonic acid,—$C_9H_{10}(SO_2OH)NO^3 + 2H_2O$. This is then diluted with water, warmed, neutralized with barium carbonate, and filtered while hot. The filtrate yields a violet color on the careful addition of very dilute perchloride of iron. The color is readily destroyed by an excess of the iron salt.

(*c*) Tyrosin gives out the odor of phenol and nitro-benzol upon heating (Kuhn). Tyrosin may be separated from the urine by first precipitating the coloring matters and extractives of the urine by means of basic lead acetate, then decomposing the filtrate with sulphydric acid and again filtering. Upon

---

[1] Liebig's Annal., Bd. lxxxvii, 1853, S. 124.

evaporating the filtrate to the thickness of syrup, tyrosin crystals will be deposited upon cooling.

Clinical Significance.—Leucin and tyrosin, as already stated, usually occur together, and this applies both to the urine and the organism at large. Being products of decomposition of proteids, they form in the system in very minute quantities, if at all, during normal metamorphosis. When metamorphic changes of a retrograde nature are rapid, as in extensive suppuration and gangrene, they form in large amounts. They may in such cases pass into the urine, largely supplementing urea.

Leucin and tyrosin are found in the urine in acute atrophy of the liver and in acute phosphorus poisoning, often in very considerable amounts. They have also been observed in the urine in cases of leucocythæmia, typhoid, and small-pox.

MELANURIA.

Melanin is a black pigment which occurs pathologically in the urine under various circumstances. It is insoluble in cold alcohol, ether, acetic acid, and dilute mineral acids. It is soluble in boiling, strong mineral acids, in boiling acetic and lactic acids, and in strong solutions of sodium, potassium, and ammonium hydroxid.

Melanin contains carbon, nitrogen, iron, and sulphur,—the latter in large amount. The urine containing melanin is not usually dark when voided, but it soon becomes so upon exposure to the atmosphere, and it becomes intensely black if submitted to such oxidizing agents as nitric acid, chromic acid, and ferric chloride. Melanin occurs as a urinary deposit in the form of small, lumpy granules, much resembling carbon particles. In the urine it may be detected by several reactions: (a) By the addition of bromine-water to urine containing melanin a yellow precipitate is deposited, which gradually blackens. This is considered by Zeller the most delicate test for melanuria. (b) When melanotic urines are treated with solutions of ferric chloride, they yield, according to the concentration of the reagent, either a dark-brown cloudiness or else a black precipitate, soluble in excess of the precipitant. This test is both delicate and characteristic. (c) When melanin is present in the urine, if treated with a dilute solution of nitroprusside of sodium and some potassium hydrate be added, a pink or red coloration usually appears, which turns blue on the addition of acids, owing to the formation of Prussian blue. The latter reaction is not due to the melanotic pigment, but to some other substance simultaneously excreted.

Clinical Significance.—Melanuria is frequently observed in people who are subjects of pigmented tumors, notably melanotic cancer or sarcoma. It has also been observed in people suffering from repeated attacks of intermittent fevers. The urine of people in wasting diseases sometimes contains considerable de. posits of melanin. The practical significance of melanuria, as Jaksch has pointed out, is greatly weakened by the facts that the urine may contain a large quantity of melanin in wasting diseases, while in melanotic cancer or sarcoma the urine may be free from it. For diagnostic purposes, therefore, so far as sarcoma is concerned, it should only be regarded as adjunct.

LIPURIA.

Normal urine contains small amounts of fatty matter, *palmatin*, and *stearin*,—about 2 grains per gallon. It is probable that these neutral fats are increased upon a fatty diet, since numerous cases are recorded in which fat has been found in abnormal quantities unaccompanied by pathological conditions. Fat is soluble in hot alcohol, ether, benzol, carbon disulphide, and chloroform. When mixed with colloids in an alkaline solution, fat is broken up into fine globules, becoming white like milk,—an emulsion. Under the microscope fatty sediments appear in the urine in the form of highly-refracting globules of various sizes, with dark and somewhat irregular margins. If the urine contain fat in considerable quantity, it is usually of a milky color, but this readily clears by shaking it with ether.

Clinical Significance.—Small quantities of fat are frequently met with in the urine in chronic parenchymatous nephritis, in fatty changes in the kidneys, in phosphorus poisoning, and in diabetes mellitus. In one case of diabetes mellitus the author met with a large amount of fat in the urine, the occurrence of which was intermittent, alternating with the appearance of sugar.

Ralfe also states that he found an abundance of oil-globules in the urine of a patient who died of diabetic coma. Ebstein found a large amount of fat in the urine in a case of hydronephrosis. Lipuria is a physiological condition with pregnant women. Roberts has recorded several cases in which pure oil appeared in the urine after the administration of codliver-oil. Henderson has reported three cases of lipuria associated with heart disease.

In diseases of the pancreas lipuria is not uncommon, and in such cases lipuria has appeared before oil was to be noted in the stools. Fat is also frequently observed in the urine after fracture of bones and during the course of repair. In acute yellow atrophy of the liver, followed by fatty changes in the renal epithelium, the urine contains an excess of fatty matters.

In chyluria the urine contains a large amount of fatty matters as well as albumin and blood-corpuscles. The features of this disease, however, will be fully considered in a future section of this work.

SECTION VII.

ANATOMICAL SEDIMENTS.

HÆMATURIA.

Blood-corpuscles appear as a urinary sediment in a number of conditions, all of which are pathological. Their appearance varies according to the character of the urine in which they are found, and the location of the tract from which they exude. The typical microscopical appearance of blood-corpuscles is so characteristic that little difficulty is encountered in distinguishing them from all other urinary sediments (Fig. 20). Their original form is that of biconcave discs of yellowish color. In focusing

FIG. 20.—NORMAL BLOOD-CORPUSCLES. (After Peyer.)

with the fine adjustment of the microscope, the margins of the corpuscles undergo reversal of light and shade owing to their biconcave form. Blood-discs are distinguished from pus-corpuscles by the absence of visible cell-contents and nuclei of the former.

In acid urine blood-corpuscles long retain their characteristic features, although in time they shrivel somewhat, and become dentated at their margins—more or less stellate in form. In urine they do not, as a rule, tend to run together, or to form rouleaux as when drawn from a blood-vessel, but are for the most part distributed pretty evenly over the field of vision. Exceptions to

(178)

this rule sometimes occur in cases of pronounced hæmorrhage from the bladder or urethræ.

If the urine be concentrated the biconcave character of blood-corpuscles becomes exaggerated, but the corpuscles shrink some-what and are more apt to become indentated or jagged at their margins. On the other hand, if the urine be dilute—*i.e.*, of low specific gravity—the corpuscles swell and become biconvex, or even spherical, and at the same time they lose their optical characteristics as well as their coloring matters. This occurs the more readily if the urine be ammoniacal.

The urine containing blood is usually cloudy and more or less reddish in color, according to the quantity present. If the quan-tity of blood be considerable and the urine be acid the color is dark red, but if the urine be alkaline the color is bright red. If the quantity of blood in the urine be small the color may give no indication of its presence, especially if the urine be concen-trated. Generally speaking, if the blood come from the kidneys it is diffused evenly through the urine, imparting to the latter a reddish, hazy tint. If, on the other hand, the blood be derived from the lower urinary tract, the color is usually bright, and clots are not infrequently present. Lastly, if blood appear in the urine in quantities however small, a distinct albuminous reaction is always obtainable.

Clinical Significance.—The clinical significance of hæmaturia embraces a very wide and varied class of pathological conditions. In order, therefore, to afford any practical information it must first be determined from what source the hæmaturia arose.

In hæmorrhages *from the kidney* the urine is usually of an homogeneous, reddish-brown color, of acid reaction, of lowered specific gravity, and it often contains renal casts and renal epithelium. After standing the urine deposits more or less brown, coffee-colored sediment. If pyelitis be present, the urine may have an alkaline reaction. In hæmaturia of renal origin clots are usually absent from the urine, unless they be of the long, slender, rod-like variety, showing that they have been molded in passing through the ureters. The recognition of blood-casts in the urine forms the most conclusive proof of the renal origin of hæmaturia. The most frequent cause of renal hæmaturia is the class of renal diseases grouped together under the term of Bright's disease. In the acute forms of these lesions hæmaturia is nearly always present. The hæmorrhage is not very pronounced in these cases, being of parenchymatous origin. It usually subsides with the more acute symptoms of the disease. Of the chronic Bright's lesions the interstitial form is the most frequent cause of hæmaturia, and in such cases the hæmorrhage is the most pronounced and obstinate of all renal hæmorrhages, except that from malignant and cystic disease of the organs. This is due to the accompanying vascular changes, including cardiac enlargement and atheromatous arteries.

Amyloid disease less frequently gives rise to hæmaturia, although it is by no means rare in such cases, since the small renal vessels in this disease undergo pronounced degenerative changes. In chronic diffuse inflammatory lesions of the kidney hæmaturia is almost unknown.

Malignant growths of the kidney give rise to most trouble-some, profuse, and repeated attacks of hæmaturia. Such cases are to be recognized by renal tumor, pain, and general cachexia of the patient. Tubercular disease of the kidney not infrequently gives rise to hæmaturia. The urine in such cases contains more or less pus and broken-down tissue *débris* which do not alto-gether subside as a sediment. Hæmaturia from tuberculosis, like in cancer, is intermittent in character, and there may also be present tumor, but usually no pain. The diagnosis rests upon the general symptoms of tuberculosis, such as emaciation, elevation of temperature, etc.; but most conclusively upon the isolation and propagation of the *bacillus tuberculosis* from the urine.

Renal calculus, whether confined to the kidney or, as is more frequent, when occupying the renal pelvis, nearly always gives rise to hæmaturia sooner or later. In these cases the hæmor-rhage is most marked upon exercise, and usually diminishes or subsides upon continued rest. Pus-cells are always present in the urine in hæmaturia of calculous origin. Fixed pain in the region of the kidney, usually with tenderness upon deep pressure over a certain point, sometimes retraction of the testicle as well as reflected pain upon the affected side, often extending down the leg, serve to diagnosticate hæmaturia of calculous origin.

In endemic hæmaturia of the tropics the cause is due to a minute parasite in the kidney—*Bilharzia hæmatobi*—which will be described later on. Among the other causes of hæmaturia of renal source may be mentioned cystic disease of the kidneys, abscess, renal embolism, hydatids, acute febrile processes, purpura hæmorrhagica, uterine and crural phlebitis. Renal hæmaturia may arise from the ingestion of certain drugs, as turpentine, cantharidis, and a number of toxic substances. Lastly, hæmaturia of renal origin may arise in consequence of traumatisms involving the kidneys, either directly as by blows or wounds, or indirectly from concussion.

In *vesical hæmaturia* the urine is usually alkaline in reaction, always so if accompanied by cystitis of long standing. In such cases the urine is ammoniacal and thick from muco-pus, and crystals of triple phosphate are usually present. In vesical hæmaturia clots are more common than in other forms of hæmorrhage. These are usually of an irregular or ragged shape. The blood is brighter in color and less intimately mixed with the urine than in hæmaturia of renal origin. Stone in the bladder is perhaps the most frequent cause of hæmaturia of vesical origin, and, in such cases, the blood is almost normal in appearance and is passed mostly at the close of the act of micturition. The grade of hæmaturia in these cases depends largely upon the acuteness of the attack. In cystitis of the vesical neck the symptoms closely resemble those of stone. The most pronounced hæmaturia of vesical origin is that associated with villous growths and carcinoma of the bladder-walls. The urine is usually normal in quantity and specific gravity in such cases, and the reaction is usually feebly acid. The sediment is brownish red, flocculent, and often contains flesh-colored fibres and shreds. The quantity of blood may be so pronounced in these cases as to cause coagulation within the bladder, or, as is more frequent, shortly after the urine is voided. Vesical hæmaturia is also common with neoplasms of the organ, fibrous tumors, polypi, and varicose conditions of the vesical neck.

Hæmaturia of urethral origin may be known by the hæmorrhage preceding the flow of urine as well as between the acts of micturition, the urine itself being usually unaltered in any of its

essential characters.   It may arise from acute gonorrhœa, neoplasms, traumatisms, urethral chancre, or from surgical operations
such as cutting or divulsion of strictures.

PYURIA.

Pus may be derived from any part of the urinary tract and
appear in the urine as a sediment.   The urine containing pus is
always more or less turbid when voided, and gives the albuminous
reaction.   In their normal state pus-corpuscles appear under the

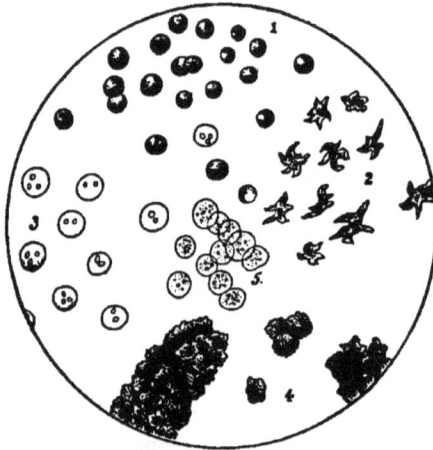

FIG. 21.—PUS-CORPUSCLES.   (After Ultzmann.)

1. Normal corpuscles.   2. Pus-corpuscles with prolongations showing amœboid movements.   3. Pus-corpuscles with nuclei rendered distinct by acetic acid.   4. Pus-corpuscles
altered by chronic pyelitis.   5. Pus-corpuscles swollen by ammonium carbonate.

microscope as circular, pale, finely-granular discs, averaging in
size nearly double that of the red blood-corpuscle.   They contain
distinct nuclei, which are often multiple,—two or three.   If pus
be diluted with water, the corpuscles may be observed to slowly
swell up and become paler, with more delicate outlines.   This
process is more quickly produced by acetic or other organic acid,
which renders the nuclei very distinct, but causes the granulated
appearance of the cell-protoplasm to disappear (Fig. 21).

Pus-corpuscles are similar to, indeed practically identical
with, mucous corpuscles, the white corpuscles of the blood and

lymph. When examined in the fresh state they exhibit the amœboid movements, and also show the usual glistening appearance of living protoplasm. As seen in the urinary sediment, pus-corpuscles are dead, the protoplasm being coagulated into coarse granules.

The chief constituents of pus-corpuscles are albuminous bodies, of which the largest proportion is nucleo-albumin, which is insoluble in water, and which expands into a tough, slimy mass when treated with sodium-chloride solution. This substance is soluble in alkalies, but quickly changed thereby into Rovida's *hyaline substance.* Besides this pus-corpuscles contain an albuminous substance which coagulates at 49° C., as well as serum-albumin and peptone. The cell-protoplasm also contains, in addition to the above, *lecithin, cholesterin, xanthin bodies, fat, soaps, and cerebrin.* In pus from congested abscesses which have stagnated some time there is *peptone, leucin,* and *tyrosin,* free *fatty acids* and *volatile fatty acids,* such as *formic acid, butyric acid, valerianic acid. Pyin* also is a specific constituent of pus,—a nucleo-albumin precipitable by acetic acid.

Pus may be derived either from the free mucous surface of the urinary tract, an ulcer, or from tissue-substance, and in each case it is likely to be mixed with elements from its place of origin which become of great diagnostic value. In addition to this pus-corpuscles themselves frequently contain micro-organisms which explain the pathological conditions of the parts from whence they are derived. Pus-corpuscles are greatly changed by contact with ammonia or other strongly-alkaline bodies; the corpuscles swell up and coalesce into an homogeneous, sticky mass, in which all but the nuclei are indistinguishable by the microscope. Ammoniacal urine containing pus deposits a vitreous-looking, slimy mass, so sticky that in decanting it from a vessel it slips out *en masse.* The peculiar behavior of pus with caustic alkalies just alluded to forms the principle of Donné's test for pus, by which the latter may usually be known without recourse to the microscope. The test is performed as follows : After the sediment has settled to the bottom of the glass or test-tube, pour off the supernatant urine and add liquor potassæ to the deposit. If the sediment be pus it is at once converted into a glairy, gelatinous-like substance which adheres to the glass or flows in a mass.

The turbidity of the urine containing pus, as well as the sediment itself, often resembles that due to the pale, granular urates. The distinction, however, is easy, since heat dissipates the turbidity due to urates, while it only serves to increase the cloud due to the presence of pus by coagulating its contained albumin. The pus-deposit also frequently resembles the earthy phosphatic sediment, but the distinction here is also easy. The addition of an acid promptly dissolves the phosphate deposit, while it only increases the turbidity due to pus by coagulating its contained albuminous elements.

Clinical Significance.—Of all sediments met with in the urine that of pus is the most common. Any affection of any part of the urinary tract, from the slightest forms of irritation up to the gravest lesions, are usually accompanied by pyuria. The clinical significance of pyuria, therefore, embraces a wide range of pathological conditions.

The first point to be determined, if possible, is to locate the particular field of the urinary tract from which the pus originated. This may often be determined by the general characters of the urine, together with the nature of the accompanying deposits. When pus originates from the kidney or renal pelvis the urine is most apt to retain its normal acidity. Round epithelium and even casts may be present, and if so it gives the most conclusive evidence of the source of the pyuria, especially the presence of casts. When the pyuria has its source in the kidney or renal pelvis the pus is intimately mingled with the urine when voided, but it quickly settles upon standing, forming a whitish, flocculent sediment. The absence of bladder symptoms in pyuria goes far toward establishing the renal source of pyuria.

Pyuria is usually associated with such renal lesions as chronic diffuse inflammations, pyonephrosis, pyelonephritis, cancer, tuberculosis, and nephritic abscess. In pyelitis, either of calculous or obstructive origin, pyuria is always a prominent accompanying symptom.

When pyuria is of vesical origin the urine is often alkaline—

ammoniacal—when voided; if not, it soon becomes so upon standing. The urine is likely also to contain considerable mucus; so that the deposit is more glairy and sticky than in renal pyuria. In addition to pus the urine is likely to contain such associated deposits as amorphous and triple phosphates and flat epithelium in excess from the bladder-walls. Local symptoms such as frequent and painful micturition aid in pointing to the source of the pus-formation.

Cystitis of all grades is always associated with pyuria, and the quantity of pus-deposit in these cases is often pronounced. Obstructive cystitis, vesical stone, ulceration, tuberculosis, and, in short, all bladder affections are ordinarily associated with pyuria.

In uncomplicated diseases of the prostate pus appears frequently in the urine, often in the form of threads long drawn out. Somewhat similar threads of muco-pus appear in the urine in chronic gonorrhœa. These are often rolled into little balls by the stream of urine as it flows down the urethral canal. The pyuria in acute gonorrhœal conditions is almost self-evident as to its source. If any doubts arise upon the question they may be readily settled by directing the urethra to be flushed, when the urine voided immediately after will be free from pus, if of urethral origin.

Determination of Blood and Pus in the Urine.—When, as is frequently the case, the urine contains a very considerable quantity of blood or pus, it is of practical importance to be able readily to determine the amount of either from day to day, in order to estimate the results of treatment. This may be rapidly accomplished by means of the author's percentage tubes and centrifuge. The process consists simply in sedimenting the urine in the percentage tubes until the urine is clear, the sediment being completely packed in the tips of the tubes, when the bulk percentage may be read off from the scale.

EPITHELIUM.

Epithelium from some part of the urinary tract usually forms a part of every urinary deposit, and, furthermore, it is usual to find scattering epithelial cells in the urine when the latter is in

13

all respects normal. Epithelium is the normal product of mucous surfaces, and may be expected to be found in small amounts in any given sample of healthy urine. But in diseased states of the urinary tract the lining epithelium is often thrown off in very considerable amount, forming plainly-visible urinary sediments.

It was formerly believed that the various divisions of the urinary tract possessed their own special forms of epithelium, and, therefore, the special forms of epithelial cells found in the urine became valuable aids in locating the seat of lesions of the urinary tract. This view is, indeed, still held by a number of prominent observers. More accurate and extensive observations, however, have shown that this can only be depended upon in a very general way. Very often the epithelium claimed to be characteristic of certain divisions of the urinary tract has been found in all its typical peculiarities in a totally different location. This, however, is the more likely to be the case in divisions most nearly located to each other. The divergent views upon this point, held even by the ablest and most experienced observers, may be illustrated by the following : Sir William Roberts describes the epithelium shed from the renal pelvis as that of " very irregular, spindle-shaped, tailed, three-cornered, elongated, rudely circular, etc." Dr. Dickinson has carefully figured the epithelium taken from the bladder, and, in reply, laconically observes : " It will be seen that these varieties of form, even to the *et cætera*, are equally characteristic of vesical disease."

The epithelium in the urine may be classed under three divisions (Fig. 22) : (*a*) Small round cells, spheroidal, finely granular, with comparatively large nuclei and nucleoli, the latter excentrically located. They occur singly or collected into groups ; in the latter case often cohering rather firmly, so that they float about in masses. They sometimes contain fatty matter, when springing from long-diseased locations. These cells may be found in their most typical form in the convoluted tubes of the kidney. They also occur in the deep layers of the mucous tract of the renal pelvis, bladder, and male urethra. These cells are to be distinguished from pus-cells by their somewhat larger size, larger and more distinct *single* nucleus, requiring no acetic acid to develop or bring the nucleus into view. No positive conclu-

sions can be drawn from the mere appearance of these cells in
the urine in reference to the precise location from which they
originated. It may be the kidney, renal pelvis, ureter, bladder,
urethra, or urethral glands. If, in a given deposit, the round
cells in their typical form greatly predominate, and if the urine
contain albumin and there be other evidences of renal disease,
it may be inferred that the cells come from the kidney. (b) The
second form of epithelium met with in the urine is the columnar
variety. These cells are of irregular, though always elongated,
form. They are described as caudate-, spindle-, and cylindrical-
shaped. They are inclined to angularity in outline, and, like the
round cells, have a well-marked nucleus, visible without the

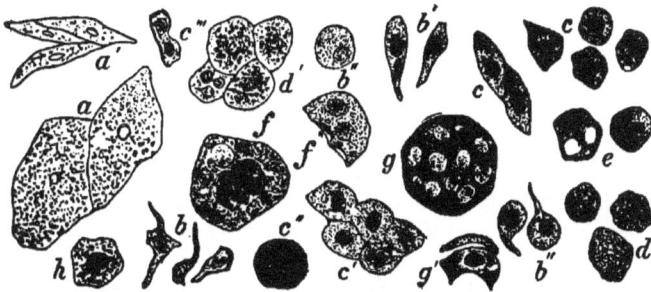

FIG. 22.—EPITHELIUM FROM VARIOUS PARTS OF THE URINARY TRACT.
(After v. Jaksch.)

a, a′, squamous epithelium ; *b, b′, b′′*, epithelium from the bladder ; *c, c′, c′′, c′′′*,
epithelium from the kidney ; *d, d′*, fatty epithelium from kidney ; *e* to *h*, epithelium from
the bladder.

action of reagents. They may occur singly or in groups. The
columnar epithelium may be derived from the superficial layer
of the mucous membrane of the renal pelvis, or from the deep
layers of the bladder, ureters, or urethra. The statement of
Ebstein,[1] that tailed epithelial cells associated with pyuria constitute the most positive evidence of pyelitis, is quite untenable.
More recent and accurate observation has amply demonstrated
that these cells exist in all their typical forms throughout the
whole urinary tract, excepting the kidney itself. (c) The third
variety of epithelium met with as a urinary sediment is the

---

[1] Von Ziemssen's Cyclopædia of Medicine, vol. xvi, p. 574.

squamous or pavement form. These cells are large, flat, some-what rounded, though irregular in outline, and have a distinct and usually central nucleus, very prominent without the aid of reagents. These cells are derived chiefly from the bladder and vagina; in the latter case the cells are usually larger than those from the bladder.

Clinical Significance.—As already stated, little more than inferences are to be drawn from the appearance of a particular form of epithelium in the urine, as to the precise location of its origin. While certain forms of epithelium predominate upon the superficial surface of the mucous tract in certain locations, the deeper layer always contains transition cells, which approach more nearly those of the surface layer in other locations. In diseased conditions, therefore, cells are thrown off from both the surface and deep layers, and the epithelium is nearly always, accordingly, of mixed varieties.

But if we are unable to locate the anatomical seat of a lesion by the character of the deposited epithelium, we may, neverthe-less, gather information of value as to the nature of the patho-logical condition present from the exfoliated cells found in the urinary sediment. With regard to renal lesions, it may be stated that practically the whole class of so-called Bright's lesions are at-tended by epithelial sediments in the urine. In the acute diffuse inflammations of the kidney the round epithelial cells from the urinary tubes are often thrown off in large quantity, so as to form a very considerable sediment. For the most part, in such cases, the epithelium is in a good state of preservation, the nuclei and outlines of the cells being sharply marked. In the more chronic lesions of the kidney the round epithelium appearing in the sediment is often fatty, the space between the nuclei and cell-walls being sometimes filled with oil-globules. The cells them-selves are often partly disintegrated or broken down, presenting a ragged appearance. These partly-disorganized cells may often be seen adhering to renal casts, or they may themselves become adhered together, forming casts.

In chronic interstitial nephritis and uncomplicated amyloid disease of the kidneys but little desquamation from the renal tubules occurs, and in these cases the fewest round cells occur in the urine. On the other hand, in acute scarlatinal nephritis the number of round epithelial cells in the urine is sometimes enormous. In acute congestive conditions of the kidney a very decided deposit of round cells are sometimes met with in the urine, without other pronounced changes in the latter save albuminuria.

In pyelitis and diseases of the renal pelvis considerable deposits of epithelium are met with, and in such cases, although round cells may be present, the tailed and spindle-shaped cells—columnar—are more apt to predominate.

In cystitis there is more or less deposit of large flat epithelium. If the cystitis be of mild grade or largely confined to the superficial surface of the mucous coat of the bladder, the large, flat, irregular-shaped epithelium predominates, but in cystitis involving the deeper layers the large flat cells are more likely to be mingled with the columnar variety.

URINARY CASTS.

Urinary casts have always and very properly been regarded of the highest diagnostic value. They were probably first seen by Vigla and Rayer, but the able investigations of Henle and Rovida gave to the profession the most complete information as to their character and significance.

Three chief views have been held as to their nature and mode of production :—

First, that they are the result of disintegration of the epithelium of the renal tubules, the resulting products becoming packed into molds by the pressure of urine, until at length they slip through the smaller convoluted into the large straight tubes and appear in the urinary sediment.

Second, that they consist of a secretion of the morbidly-irritated epithelium lining the renal tubules, which cakes into molds, and the casts thus formed are washed down with the urine.

Third, that they consist of coagulable elements of the blood which gains access to the renal tubules through pathological lesions of the latter, and that any free or partly-detached products of the tubules become entangled in this coagulable product, assisting to form the molds of the tubules, which subsequently appear in the urine as casts.

The last view is the one most generally accepted, at least so far as the nature and origin of the great majority of casts are concerned.

Although the substance forming the basis of casts is evidently closely allied to proteids, yet it is certain that it is not identical with any proteid with which we are at present familiar; perhaps it is a derivative thereof. Rovida claims for hyaline casts the characteristic of being soluble in dilute mineral acids. Renal casts have been variously classified, but the most useful division for clinical study is as follows :—

1. Those consisting of anatomical elements such as epithelial cells, blood- and pus- corpuscles.

2. Those consisting of the products or broken-down elements of anatomical substances.

3. Those clear casts often termed "*hyaline*," the nature of which, as well as is their origin, is still a disputed question.

The first division naturally includes those casts largely made up of (*a*) red blood-corpuscles, (*b*) leucocytes, (*c*) epithelial cells, (*d*) masses of bacteria.

The second division comprises (*a*) granular casts, (*b*) fatty casts.

The third division comprises (*a*) narrow hyaline casts, (*b*) broad casts, (*c*) composite casts or those largely clear, but more or less coated with the elements enumerated in the first and second divisions, such as blood, pus, epithelium, fat, etc.

Blood-Casts.—These appear in the urine under conditions which give rise to hæmorrhage within the urinary tubules. Under the microscope they often appear as very beautiful objects. The perfectly-preserved corpuscles may be observed glued together in perfect molds of the tubules, being usually short, of pretty uniform diameter throughout, and with rounded ends.

These casts are met with in the urine in hæmaturia, acute diffuse nephritis, acute renal congestion, and hæmorrhagic infarctions of the kidneys. Blood-casts do not in themselves furnish positive evidence of organic renal disease, since any hæmorrhage of the kidney may have associated therewith blood-casts in the urine. On the other hand, it may be stated that the presence of blood-casts in the urine constitutes the only positive evidence of the existence of renal hæmorrhage. Blood-casts may be considered as belonging to the rarer forms of renal casts found in the urine, and they are usually difficult to find, since a large sediment of free blood-corpuscles usually accompanies them, which greatly obscures the microscopical field.

Epithelial Casts.—These result from pathological conditions

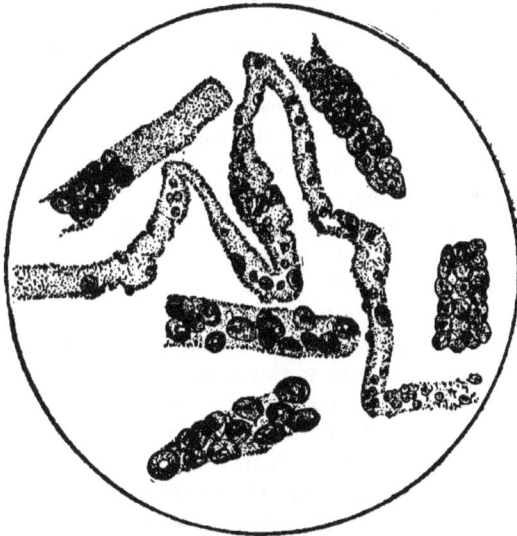

FIG. 23.—EPITHELIAL CASTS. (After Peyer.)

which cause exfoliation of the epithelium of the renal tubules. Sometimes the epithelial lining of the tubules is thrown off intact for short distances, resulting in epithelial cylinders which possess lumens. The epithelial cast also occurs in a solid form, the body being made up of hyaline substance and the surface covered with epithelial cells. These cells, as viewed under the microscope, appear more or less swollen and granular, with ill-defined margins. In some cases the epithelial cells appear in rows or in patches over the surfaces of the casts (Fig. 23). In other cases the epithelial cells have partly undergone degeneration or contain dotlets of fat. These are significant of chronic or, at least, fatty changes in progress in the kidney. Finally some casts are to be seen which are entirely composed of epithelial cells aggluti-nated together. Epithelial casts are usually of medium size and length, refracting light to a comparatively high degree, and are therefore easy to discover in the microscopical field. They resist the action of chemical reagents to a greater degree than most other casts, except those that are partly metamorphosed. The presence of epithelial casts in the urinary sediment may be taken as a positive evidence of inflammation in the anatomical struct-ures from whence they originate; and they are consequently sediments of the highest diagnostic value.

Pus-Casts.—Casts composed exclusively of pus-corpuscles are exceedingly rare. Not infrequently, however, compound casts are met with, composed of epithelium or granular matter, in which scattering pus-corpuscles may be seen dotted over their surfaces. Johnson has described and illustrated casts entirely composed of pus-corpuscles which came from subjects who sub-sequently died of multiple abscess of the kidneys. Such casts, however, have rarely been noted by other observers.

Bacterial Casts.—It is no rare occurrence to meet with casts in the urine which are composed of masses of micrococci. In appearance they closely resemble the dark granular casts, but they are readily distinguishable from the latter by their resist-ance to such chemical agents as strong mineral acids and caustic alkalies. They are more opaque than other casts, of a grayish color, and are uniform and very fine in their outlines. The use of high microscopical powers will render confusion in distinguishing these casts almost impossible. The discovery of casts in the urine made up of bacterial masses must be taken as a factor of very grave significance, since they are chiefly found in septic forms of nephritis often accompanied by embolism. They occur, therefore, in interstitial suppurative nephritis or ascending pye-lonephritis.

Granular Casts.—This form of renal cast comes under the second division named, being the result of metamorphosis of anatomical elements, usually of epithelium, pus, or blood. Granular casts are found in the urine in great variety, as is shown by the terms frequently employed to designate them, such as *finely granular, coarsely granular, granular, highly granular, moderately granular, light granular, dark granular*, etc.

Granular casts vary much in size and shape, as well as in appearance. They are most often met with in fragmentary forms, only occasionally preserving their entirety or perfect forms. They are irregular both in their coarse and fine outlines,

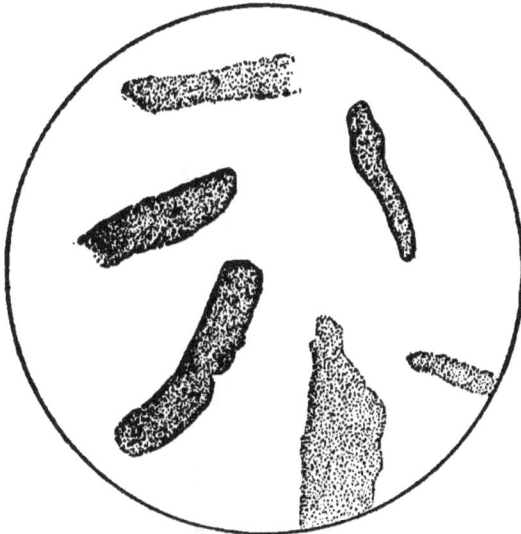

FIG. 24.—GRANULAR CASTS. (After Peyer.)

and their ends are usually ragged, as if recently fractured. The granulations are often exceedingly fine, requiring high powers to distinguish them; while, again, they are *coarsely granular*, which is apparent with comparatively low powers (Fig. 24). They are of various colors, as yellowish, white, gray, and brown. They may have scattered over their surfaces epithelium, leucocytes, fat-globules, or fatty crystals. Granular casts have generally been regarded as indicative of pathological conditions of the kidneys of chronic or degenerative character.

Fatty Casts.—It has already been stated that fatty elements are sometimes seen mingled with the elements of epithelial casts, and the same may be said with regard to granular and many other forms of casts. But, in addition to these, highly-refracting casts are often met with in the urine, whose surfaces are completely studded over with fatty globules and, less often, with fine, needle-like, fatty crystals (Fig. 25). Fatty casts are the result of a different form of transformation of anatomical elements from

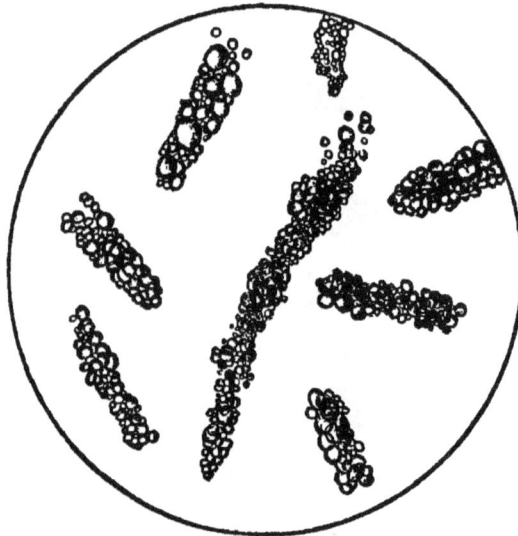

FIG. 25.—FATTY CASTS. (After Peyer.)

the last described. They constitute the index of fatty changes in the kidneys, and are found in their most typical form in "large white kidney." They may be looked upon as evidences of pathological states of the kidney, the chief feature of which is extreme chronicity, since they are probably the result of complete destruction of the cell-protoplasm, which becomes replaced by fatty elements.

Hyaline Casts.—These are pale structures of variable but usually considerable length, sometimes very difficult to detect in the sediment (Fig. 26). Sometimes they exhibit no granulation whatever upon their surfaces, being, in fact, almost transparent. Much more frequently, however, they exhibit very fine granulation of a very light color. They may exhibit here and there a dotlet of oil or a fragment of epithelium upon their surfaces, and, indeed, this is usually the case, although such casts are considered strictly of the hyaline order.

With regard to the origin of hyaline casts, much difference

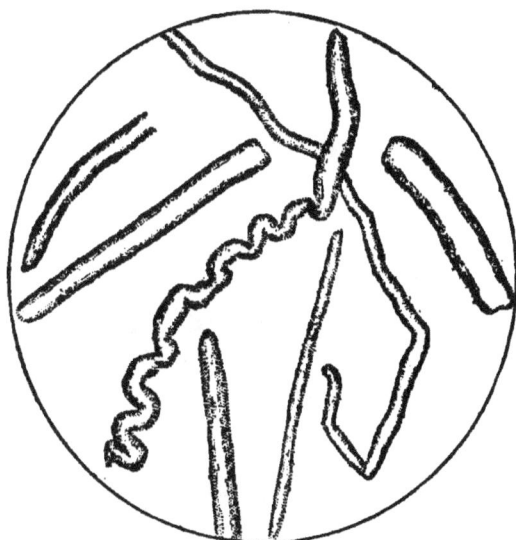

FIG. 26.—NARROW HYALINE CASTS. (After Peyer.)

of opinion has prevailed. Oertel contended that they were the result of secretion from the epithelial cells of the renal tubules, and in this opinion he was supported by Rovida. Bartels, on the other hand, holds that these casts are formed by a coagulation of the albumin or its derivatives excreted with the urine. As evidence of this, he states that they are only present in urine that is albuminous, or that has very recently been albuminous; for the occurrence of albuminuria and casts are not always simultaneous.

While the origin of the hyaline casts is not at present so clearly understood as most of the other forms, it is probable that they are formed by coagulable elements of the blood which has gained access to the renal tubules. The experiments of Ribbert upon animals indicate that hyaline casts may result directly from exudation of albumin into the tubules; and their disappearance from alkaline urine upon standing further indi-

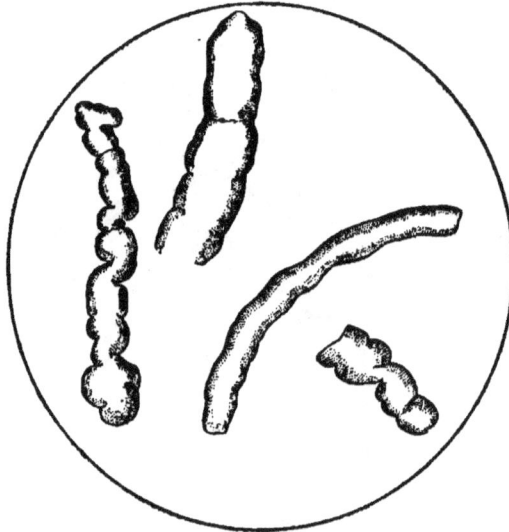

FIG. 27.—SO-CALLED WAXY CASTS. (After Peyer.)

cates their close relationship to albuminous bodies, as long since demonstrated by the author.[1]

Numerous observers have claimed to have found hyaline casts in non-albuminous urine, but the author agrees distinctly with Bartels in that he has never met with them save in albuminous urine or urine that has recently been albuminous.

The disposition by some to regard hyaline casts of the small, narrow order as of no serious import is a mistake of the gravest character, for, indeed, they are often the chief evidence, so far as

[1] Journal of the American Medical Association, September 12, 1885.

the urine is concerned, of the existence of a most serious form of renal disease, viz., interstitial nephritis; for it should be remembered that in such cases albuminuria is nearly always small in quantity.

In addition to narrow hyaline casts which doubtless come from the smaller tubules within and above the middle zone of the kidney, the urine often contains large, broad hyaline casts, evidently emanating from the large, straight tubes of the pyramids. As a rule, these larger, clear casts are more refracting, and consequently more distinct both in body and outline than the narrow casts. They are also more indented at their margins, although pretty uniform in their dimensions. There is a form of hyaline cast, usually of large size, that occasionally—not always—exhibits the characteristic amyloid reaction with methyl-violet and iodo-potassic iodine solutions. They are large and usually rather long, and their surfaces may be marked by indentations showing imperfect vertical segmentations,—"*tape-worm form*" (Fig. 27). The term *waxy* as applied to these casts is inappropriate. It was formerly thought that these casts were characteristic of amyloid changes in the kidneys, but more extended observations have shown that they are found in all forms of nephritis. "These casts may exhibit the amyloid reaction in the absence of amyloid kidneys, or they may fail to exhibit it when amyloid disease of the kidney is present, and therefore no diagnostic value can be attributed to the reaction, save, perhaps, as indicating degenerative changes in the casts themselves" (Roberts). These casts are of comparatively rare occurrence.

Cylindroids.—In addition to the casts described, the urine sometimes contains the so-called cylindroids of Thomas, who first observed them in the urine in a case of scarlatina. These are long, wavy, ribbon-like structures, which often divide and subdivide at their ends with diminishing diameters. These ends may be folded or twisted in corkscrew form. They are pale, colorless, and of greater length than the ordinary casts described, and rarely, if ever, have attached to them any cellular elements whatever. They appear flat and do not give the impression, to the eye, of being solid structures like true renal casts. It seems not improbable, however, that these cylindroids come from the renal tubules. They occur in nephritis, cystitis, and renal congestion, and may be present in urine that is free from albumin. They are not characteristic of kidney disease, but probably more often caused by irritation of the lower urinary tract, which has, in a measure, extended to the kidneys.

Lastly, it may be stated that casts are sometimes met with in the urine composed of urinary crystals or granular salts. Only those composed of urates and hæmatoidin have thus far been

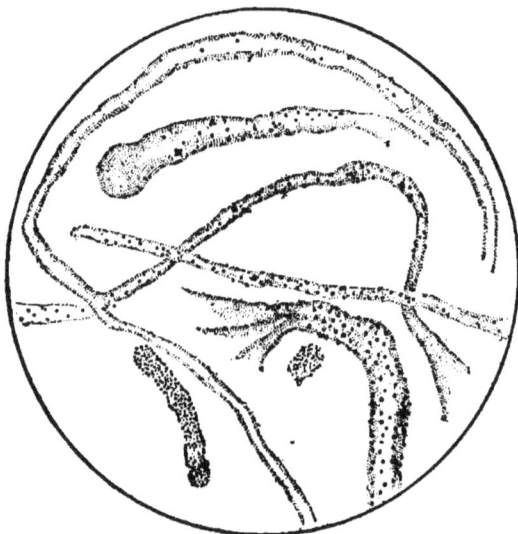

FIG. 28.—FALSE CASTS. (After Peyer.)

observed, and they are of little practical significance, being only found in the urine of infants, or in cases of gout, renal congestion, etc.

Method of Searching for Casts.—Since the more recent methods of obtaining the urinary sediment by means of the centrifuge, we no longer encounter the obstacles which were often very annoying and sometimes almost impossible to surmount in making a satisfactory microscopical inspection of the urine. The obstacles referred to especially were the following:

(*a*) The difficulty often encountered in getting casts to settle in urines of high specific gravity, because renal casts are of comparatively light weight and often float in such urines for hours without subsiding. (*b*) The changes in the urine upon standing often altered the essential features of the deposit, as before stated, sometimes rendering them unrecognizable. (*c*) In urine which stood long enough to secure the sediment, the microscopical field sometimes became so crowded with micro-organisms that the organic products were completely obscured from vision. Happily, these difficulties are now matters of the past; and it is only necessary to secure a freshly-voided sample of urine, submit it to the centrifugal apparatus for two or three minutes, and we have a true and unaltered sediment, that exactly represents the pathology of the urine as it leaves the urinary tract.

The urinary sediment obtained—always preferably by the centrifuge—is best examined in a shallow cell, upon a glass slide carefully covered with a cover-glass. This is best secured by taking up about 4 to 6 drops of the sediment in a nipple-pipette from the bottom of the sediment-tubes of the apparatus, and placing them upon a carefully-cleaned slide, then placing the cover-glass over the cell, and with the thin, freshly-torn edge of a piece of blotting-paper removing the excess of urine, which tends to spread over the cell and along the slide. By this means the sediment constitutes a temporary mount, so that it may be examined at any angle without flowing out of the cell. The slide is next placed under the microscope, which is adjusted with a $\frac{1}{4}$-inch objective for ordinary search, and examination conducted in a clear, but not too bright, light.

The greatest difficulties encountered in searching for casts in the urine will usually be met with in the case of narrow hyaline casts. These bodies are so transparent and non-refracting that they are exceedingly liable to be overlooked. This may often be avoided by proceeding as follows: After securing an accurate focus of the field by careful regulation of the fine adjustment of the instrument, gradually darken the field by the mirror-adjustment and throw the light obliquely across it, illuminating the field, in fact, but about one-half or two-thirds its extent. Next, by slowly moving the slide about, the different features of the sediment are presented to view in different lights, and the outlines or shadows of fine hyaline casts will often be brought into view, when they would be perfectly transparent and unobservable in a more direct reflection of sharp light. Once the outline is seen, by careful re-focusing the cast often stands out with distinctness. If doubts remain as to its true nature, by depressing both ends of the slide with the finger-tips currents will be created in the urine beneath the cover-glass (if mounted as directed), so that the cast will be made to move and even roll over, which will often settle any question as to its nature. The author has not found it necessary to treat these casts with staining agents to bring them into view, although this may be practiced successfully. It is, however, preferable, if possible, to view the microscopical deposit, and especially renal casts, as nearly as possible in their native state, for obvious pathological reasons.

SPERMATOZOA.

Spermatozoa are thread-like bodies provided with a head and a long, tapering, tail-like extremity. The head is of a flattened oval shape with a central depression on either side. The head and tail are united by an intermediate, cylindrical-form body or neck of uniform size. The entire length of a spermatozoön is about $\frac{1}{500}$ inch. When freshly ejected with the spermatic fluid spermatozoa exhibit active, eel-like movements, as though endowed with separate life, and under favorable circumstances—warmth and moisture—they long retain this capability. These movements may be observed in the body for several days after death, and in the uterine secretion longer than a week. The cause of the movements in spermatozoa is unknown, though it has been contended that they are mere floating cilia. Acid liquids stop these movements immediately, as do strong alkalies, especially ammoniacal liquids; also distilled water, alcohol, ether, etc. Spermatozoa show great resistance to chemical reagents. They do not dissolve completely in sulphuric acid, nitric acid, or acetic acid; but they are dissolved in boiling potassium-hydrate solution. They resist putrefaction, and after drying they may be obtained again in the original form by moistening them with a 1-percent. sodium-chloride solution. By careful heating and burning to ash the shape of these bodies is said to be seen in the ash (Hammarsten).

As found in the urine spermatozoa are nearly always in the quiescent state, and they may be found in the urine which has stood for days, their typical form being well preserved (Fig. 29).

Clinical Significance.—Spermatozoa are only found in semen or fluids mingled therewith. Their persistent absence from seminal fluid indicates sterility or incapability of procreation.

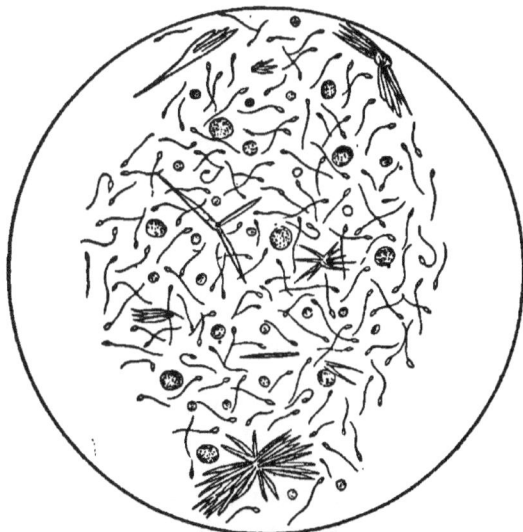

FIG. 29.—SPERMATOZOA IN URINARY SEDIMENT. (After Peyer.)

Spermatozoa may be found in the urine after every ejaculation of seminal fluid, and therefore they may aid the physician in detecting masturbation. Spermatozoa are found in the urine in some cases of severe illness, such as typhoid fever, typhus pneumonia, and after epileptic attacks. A very constant deposit of spermatozoids in the urine constitutes an essential feature of spermatorrhœa. In such cases the urine often shows the presence of white flakes, which under the microscope appear as masses of spermatozoa and finely-granular matter.

14

FRAGMENTS OF TUMORS.

Contrary perhaps to general belief, fragments of new growths
are rarely met with in the urine, and those that are occasionally
encountered nearly always originate in the bladder. Small
polypi have been found in the urine a few times, as well as fragments of villous growths. More often, perhaps, carcinoma of
the bladder parts with some of its constituents, which subsequently appear in the urine. In such cases the fragments are
usually necrotic, and therefore practically impossible to recognize.
The typical villous growth appears in its finest subdivisions as
characteristic, tree-like branches—"*fringe-like*"—which consist
of enlarged vessels covered with a single layer of epithelium; but,
owing to the necrotic changes resulting in their disintegration,
these fragments are seldom seen in the urine in typical form.
The epithelium has usually undergone molecular degeneration,
accompanied by bacterial life, and the villus itself is infiltrated
with the products of suppuration. It is only occasionally that
forms are met with which assist in diagnosticating villous
growths. Thus, crystals of hæmatoidin may be observed in the
necrotic masses, or they may be brought out frequently by
treating the masses with glycerin. They may be recognized by
their brownish-yellow color; small, rhomboidal form; or yellow,
grass-like tufts. Treated with nitric acid and observed under
the microscope, the well-known play of rainbow colors reveals
the presence of biliary coloring matters. Since hæmatoidin
occurs in the urine only in isolated crystals, it follows that, if
found imbedded in necrotic masses, it goes far to establish a
diagnosis of cancer, since such conditions are only found with
such growths (Jaksch). Hæmatoidin villus is rarely recognized
save in acid urine, because in alkaline urine the villous tissue is
more disintegrated and mixed with phosphatic deposits.

Cancer of the kidney can rarely, if ever, be diagnosticated
from the nature of the anatomical sediments in the urine. The
transitional epithelium, more or less abundantly present in all
inflammatory conditions of the urinary tract, simulates so closely
those of cancer that it is never safe to draw any positive conclusions as to the presence of cancer from this source. It is only
when considered in connection with other symptoms, such as

hæmorrhage, pain, cachexia, etc., that a positive diagnosis may usually be reached.

BACTERIURIA.

It has been generally accepted that healthy urine when freshly voided is free from bacteria, and is, in fact, an aseptic fluid. Over thirty years ago Pasteur demonstrated that such urine is sterile, and this has been repeatedly confirmed since. If, however, the ordinary normal urine be allowed to stand for some time at ordinary temperatures it becomes crowded with micro-organisms. This is due to the fact that the urine contains so large a percentage of organic matter that it practically constitutes a culture medium for many forms of these organisms.

Abnormal urine nearly always contains micro-organisms, of which nearly forty varieties have been isolated to date. They practically all belong to the class of fungi, and for purposes of study may be divided into two classes, viz.: (a) *non-pathogenic fungi*, or those which are innocuous, and (b) *pathogenic fungi*, or those possessed of pyogenic powers.

Non-pathogenic Fungi. — These include *molds*, *yeasts*, and *fission-fungi*.

FIG. 30.—YEAST-FUNGUS IN URINE. (After Harley.)

Molds are of comparatively rare occurrence in urine, even when undergoing ordinary decomposition; but if diabetic urine be allowed to undergo alcoholic fermentation, at its conclusion molds make their appearance in quantities upon the surface of the urine, which also becomes crowded with yeast-fungi.

The yeast-plants of the urine (*saccharomyces urinæ*) are single cells of about the size of blood-corpuscles. They are distinguishable from blood-cells by the irregular and occasionally large size of the cells, the presence of a nucleus in the larger sporules, and their more elongated or oval form. Usually these cells are arranged in bead-like forms, some of the beads having several small bud-like cells attached to them (Fig. 30). For their development it is necessary that the urine be distinctly acid, and they cease to multiply if the urine become alkaline. The presence of yeast-fungi in the urine in large numbers may be taken as certain evidence of the presence of sugar.

Fission-fungi are associated with urine tending toward putrefactive changes. Such urine is more or less cloudy when voided; it is never sharply acid; on the contrary, it is usually neutral in reaction or feebly alkaline. Examined under the microscope it is seen to be crowded with micro-organisms in active motion, such as the more common forms in decomposing organic fluids, the most familiar being the well-known *bacterium termo*. The urine on standing does not clear, and, moreover, it does not clear by filtration, and it tends to rapidly pass on into ammoniacal decomposition. This condition of urine is most often met with in weakly and enfeebled people, and in men who have urethral stricture, or who have frequently had catheters or bougies passed. In the process of ammoniacal fermentation of the urine, urea is transformnd into ammonium carbonate through the agency of bacterial life. To each molecule of urea two molecules of water are supplied, and the chemico-vital change wrought by bacterial activity results in two molecules of ammonium carbouate. The chief agent concerned was formerly believed to be the *micrococcus ureæ*, as almost pure cultures of this organism are often observable upon the surface of the decomposing urine. It has recently been shown, however, that nearly all microbes found in the urine possess the above powers to a greater or less extent. The micrococci ureæ are organisms of comparatively large size, and are most frequently observed in long, chain-like strings; although they also occur as free and independent, minute, round objects (Fig. 31).

FIG. 31.—MICROCOCCUS UREÆ. (After v. Jaksch.)

Ammoniacal bacteriuria is most frequently met with in cases of obstructive cystitis, in which more or less residual urine remains in the bladder, as often results from paraplegia or enlarged prostate, and urethral stricture of small calibre. The frequent use of instruments in the urethra and bladder often results in this condition.

In addition to the micrococcus ureæ, the urine may contain various other forms of non-pathogenic fungi, including rod-like bacteria of various forms and sizes. Occasionally long, spiral bacilli, with large spores and cocci, are met with in the urine. These are often grouped in various-shaped masses, usually of dark color and variable sizes. For the most part, these organisms gain access to the bladder through the urethra. This may take place through discharges or, more commonly, by the use of instruments.

Pathogenic Fungi.—The pathogenic bacteria found in the urine for the most part belong to two orders,—viz., *micrococci* and *bacilli*.

The *micrococci* found are those characteristic of suppurative diseases in general, and include the *staphylococcus pyogenes albus, aureus, citreus,* and the *streptococcus pyogenes.* The action of these germs is general, which accounts for their frequent presence in the urine. To these must be added the *gonococcus* of Neisser, which will be considered later.

The *bacilli* met with in the urine include the *urobacillus liquefaciens septicus,* the *bacillus coli communis,* the *tubercle bacillus* of Koch, besides a number of others less well known.

In all infective diseases in the healthy organism, if life be not destroyed thereby, the microbes must either be destroyed in the blood and tissues by the process termed *phagocytosis,* or be eliminated through the excretory organs, usually in an active state. The fact of the rapid disappearance of most micro-organisms from the blood, when injected into the circulation of the healthy organism, indicates that the corpuscular elements of the blood have the power of destroying these organisms,—at least, to a large extent. Passing from the blood into the tissues, these organisms meet with the same warfare in the tissue-cells. Should, however, both of these sources of *phagocytosis* prove unsuccessful in destroying these organisms, there still remains a method by which the unassisted organism may rid itself of these microbes,—viz., by means of elimination through the excretory channels; and in this process the kidneys and intestinal canal are the chief organs concerned. The frequency with which the kidneys become infected in the course of general tuberculosis furnishes proof that these organs are the source of elimination of tubercle bacilli. Furthermore, Philipowicz has succeeded in producing tuberculosis in animals by injecting urine into the peritoneal cavity which was taken from tuberculous subjects.

Neumann has demonstrated the characteristic microbes of typhus, pneumonia, and pyæmia in the urine in the course of these diseases, and he has cultivated from the urine, in acute endocarditis and osteomyelitis, the *staphylococcus pyogenes aureus*. Neumann furthermore claims that the microbes which gain access to the circulation often localize in the capillary vessels of the kidney, there causing multiple lesions without involving the whole organ; and through these lesions some of the microbes gain access to the uriniferous tubes and appear in the urine.

Schweiger has conclusively demonstrated that the urine of scarlatinal subjects is distinctly contagious; and he regards all renal lesions arising in the course of infectious fevers of microbic origin. Finally, it may be mentioned that Rimann has conclusively demonstrated the passage of bacilli through the kidneys in the following manner: The bacilli discovered in the pus of ozæna were cultivated in gelatin and agar and stained intensely green; after dilution with physiological solution of salt, the culture was injected directly into the circulation of a dog, cat, and rabbit; after a time these stained microbes appeared in the urine in large numbers.

Of the various pathogenic micro-organisms found in the urine, the recognition of the *bacillus tuberculosis* of recent years claims the greatest clinical interest. It may be stated that the presence of this bacillus in the urine, especially when arranged in S-shaped aggregations, or in colonies of irregular masses, points with unmistakable certainty to tubercular ulceration of the urinary tract. Fortunately for diagnostic purposes, if the *tubercle bacillus* appear in the urine at all, it usually occurs in abundance, and may, with recent methods, be recognized in masses often arranged as in pure cultures.

In all cases of purulent urine accompanied by anæmia, wasting, and evening temperature, the urinary sediment should be examined for the presence of tubercle bacilli. Of course, more

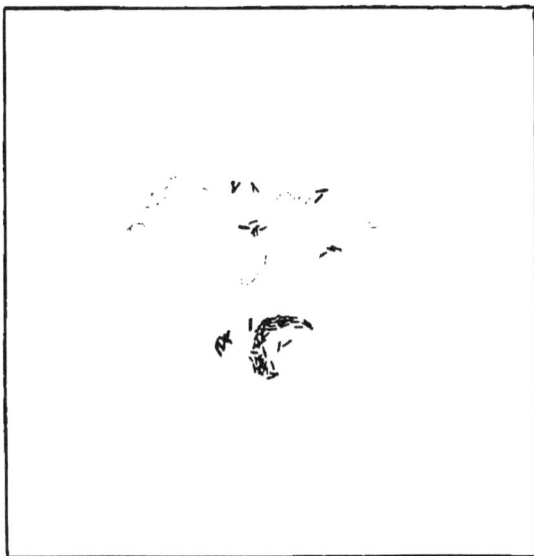

TUBERCLE BACILLI IN URINARY SEDIMENT.
(After v. Jaksch.)

patient search is necessary than in the examination of sputa, owing to the dilution of tubercular pus by the urine, and it is necessary to concentrate it as much as possible for the same reasons. This is best accomplished by the centrifuge for obvious reasons. The purulent deposit of tubercular urine should be treated precisely as the sputum in searching for tubercle bacilli. Should difficulties be encountered in finding the bacilli when the symptoms strongly point to their presence, or should doubts arise as to the true character of the bacilli found in the urine in any given case, the inoculation of animals with the sediment or plate cultures thereof should be resorted to before negative conclusions can be positively reached. Indeed, unless the time required for the latter be a matter of importance in the case (which is very seldom), the plate cultures are the easiest and most certain method of success.

Gonococci were first discovered in the pus of gonorrhœa by Neisser. They consist of minute, roll-shaped cocci. They are chiefly met with as diplococci, the individual cocci being seemingly divided by a bright, transverse band, often presenting the so-called "*roll-form*"; also termed "*kidney-*" or "*bean-*" *shape*. The cocci usually appear in pairs lying close together, their flattened surfaces usually presented to each other. They multiply by each coccus splitting in two. Bodies in all respects resembling gonococci, so far as present methods are able to determine, have been found in the genital tract under the most variable conditions, and this fact has tended greatly to diminish the diagnostic value of these organisms. On the other hand, it is pretty well established that the presence of gonococci within pus-cells in purulent urethritis is characteristic of infective gonorrhœa. In all cases of recent infection with gonorrhœa the specific gonococci are to be found within the pus-cells in abundance, although they are not limited to this location, but may be seen in the epithelia, as well as floating in the liquor puris. This cannot be said of the non-infective forms of diplococci. The gonococcus is best stained in gentian violet, methylene blue, or fuchsin, after which it may be rinsed in water and examined under the microscope.

VERMES.

Distoma Hæmatobium.—This parasite has frequently been found in the portal vein and its branches, in the splenic and mesenteric veins, and in the venous plexuses of the bladder and rectum (Fig. 33). From the investigations of Bilharz, it would seem that more than half the adult Fellaheen and Coptic population of Egypt suffer from this parasite.

The eggs of this parasite are found in numbers in the urinary passages and in the urine. They are usually accompanied by blood, pus, and sometimes by considerable fat. The eggs are oval, flask-shaped bodies, with rather sharp projections from

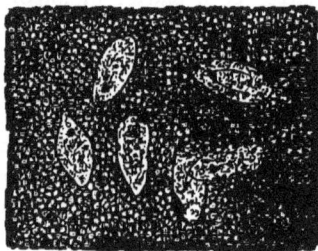

FIG. 32.—EGGS OF DISTOMA FROM URINARY SEDIMENT.

their anterior extremities. They measure rather less than $\frac{1}{800}$ inch in length.

When confined to the larger veins, these parasites do not produce much damage; but if they invade the smaller vessels, notably those of the submucous tissues of the urinary tract, they induce severe and, in some cases, fatal consequences. In the intestines the result is often that of severe dysentery.

The most serious results of the ravages of this parasite are met with in the urinary passages. The lesions produced here, according to Griesinger, consist of raised patches of injected and ecchymosed tissue, which often pass into ulcerations, giving rise to severe hæmorrhages. These patches are covered with mucus and brownish exudations of bloody matters containing masses of ova. If no lesions of the mucous membrane occur over these patches, they may remain as nodules of more or less firm and indurated tissue.

When these parasites invade the ureters and renal pelvis, the results are apt to be still more serious, as they threaten the integrity of the kidney from two sources. First, by consequent thickening they may cause occlusion of the ureters and resulting hydronephrosis or pyonephrosis, as in a case observed by Griesinger. Second, severe pyelitis is often set up by invasion of the renal pelvis, which may result in ascending pyelonephritis.

In addition to these consequences the masses of ova extruded into the urinary passages are exceedingly liable to become the nuclei for urinary calculous formation; and, indeed, this is claimed to constitute the cause of the great frequency of calculous disease in Egypt.

From the variety of lesions caused by invasion of the *Distoma*

FIG. 33.—DISTOMA HÆMATOBIUM, MALE AND FEMALE, WITH EGGS.

hæmatobium, as might be expected, a corresponding variety of clinical symptoms result. In the milder cases but little disturbance of the urinary function is to be observed. Some pain, described as of a burning character, is notable on micturition. This, however, is but momentary, and is chiefly due to the passage of the ova along the urethra, which they irritate by means of their sharp projections. At the termination of micturition a few drops of blood, or a blood-clot, are sometimes noted. The urine contains pus and blood-cells, with the eggs of the parasite as represented in Fig. 32. The symptoms become more distressing with the involvement of the bladder, entailing various grades of cystitis, always accompanied by more or less hæmaturia. To these may be added the discomforts and deterioration of the general health entailed by pyelitis when the parasite invades the renal pelvis. Septic infection, nephritis, and uræmia are among the most serious consequences entailed by extension of the morbid changes set up by the parasite in the upper urinary tract. Finally, it may be noted that Bilharz and Griesinger, during the course of their investigations in Egypt, found strong evidence that the distoma disease sometimes takes the course of an acute and rapidly fatal disorder. Griesenger observes : " We found on two occasions, in the bodies of persons who had rapidly died from an unknown acute disease, abundant recent distoma changes in the bladder, recent pyelitis, and a uniform dark-red hyperæmia of the kidneys. In other cases of supposed rapid typhus the same changes were found in the bladder and ureters."

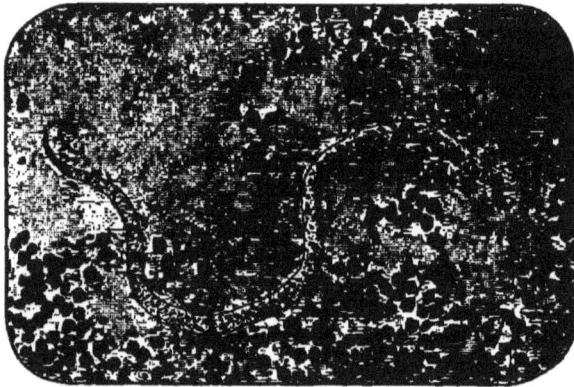

FIG. 34.—FILARIA IN HUMAN BLOOD. (After Mackenzie.)

Filaria Sanguinis Hominis.—Dr. Lewis, of Calcutta, was the first to describe this parasite as occurring in the human organism. In 1872 Lewis made the interesting announcement that, as observed in India, chylous urine always contained large numbers of these parasites. He later on discovered that these worms were present in great numbers in the blood of patients suffering from chyluria. This hæmatozoön is about the width of a red blood-corpuscle and about fifty times longer than its width. It possesses a short, rounded head, with a tongue-like protuberance, and a rather long and pointed tail. The parasite is inclosed in a loose sac, in which it moves with freedom. This covering appears structureless, but the parasite itself, as viewed under the microscope, is seen to be very granular, with transverse striations.

It was early supposed that the minute hæmatozoön described and figured (Fig. 34) was but the young of a larger, parent worm, and in 1876 Bancroft, of Brisbane, succeeded in demonstrating the mature worm in a lymphatic abscess of the arm. The mature form is a nematode worm, about the thickness of a hair and three or four inches in length. Dr. Manson has established the origin of this parasite in the mosquito, which deposits the larvæ in ponds, the water of which being swallowed, the subject becomes the seat of its further development. The most curious and interesting circumstance in reference to the habits of this parasite is the fact, pointed out by Manson, that the embryo filaria is very active at night, but quiescent during the day; it is therefore to be found abundantly in the blood of subjects of this disease during the night, but not to be found during the day. Dr. Mackenzie observed, however, that upon reversing the order of sleeping and waking, the filariæ changed their nocturnal habits,—coming out when the subject slept and remaining hidden during the waking hours. It would, therefore, seem that the activity of filariæ in the blood depends upon quiescence of the subject. What becomes of the filariæ, or where they conceal themselves temporarily during their disappearance, as yet remains a mystery.

The filariæ, though greatly obstructing the lymph-channels and literally swarming in the blood, of themselves do but comparatively little mischief. The embryos appear to pass readily through the capillary vessels without causing obstruction, irritation, or emboli. It is, however, altogether different with the ova if they remain unhatched in the blood, lymphatics, and the urine. Owing to their greater diameters, they become arrested in the smaller lymphatic vessels and, becoming impacted, accumulate until the gland becomes impervious, resulting in lymphatic congestion and even necrotic changes; *elephantiasis*, or *lymph-scrotum*, is one of the results of this condition. Chyluria is one of the most frequent consequences, also, of this condition. This is brought about through the obstruction of the larger lymphatics, the thoracic duct, or large channels between it and the urinary organs.

Echinococci.—The hooklets and scolices of the echinococcus cyst are among the rarer forms of sediments met with in the urine (Fig. 35). It is exceptional for echinococci cysts to develop in the urinary passages; more often the products of these cysts

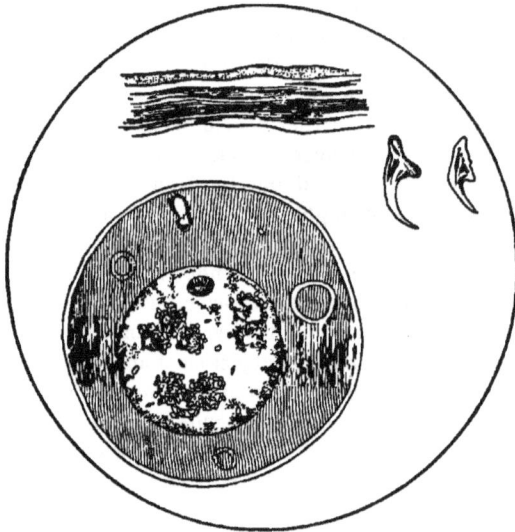

FIG. 35.—ECHINOCOCCUS, WITH TWO HOOKLETS AND SECTION OF CYSTIC MEMBRANE GREATLY MAGNIFIED. (After Peyer.)

make their way into the urinary passages from the kidney or some neighboring organ by means of rupture. If the characteristic hooklets appear in the urine, they are usually accompanied by blood, pus, and more or less cellular *débris*, as well as shreds of the membrane forming the cystic envelope.

Echinococci are minute ovoid parasites barely visible to the naked eye. They consist of a head not unlike that of the tapeworm, provided with four mouths or suckers and a double row of hooks. The echinococci are developed within a germinal membrane, and exist in groups of from six to ten. The hydatid growth consists of an outer, fibrous capsule by which it is attached to the organ in which it is developed. Within the capsule is the hydated cyst, which varies in size from that of a marble to that of a distended urinary bladder, or even larger. Within the large cyst a great number of secondary or so-called *daughter-cysts* develop and float freely in a serous or rather saline, aqueous fluid. These *daughter-cysts* also vary in size from that of an apple to minute points requiring the microscope for their recognition. The *mother-cyst* may, in rare cases, be entirely barren,—*i.e.*, contain nothing but fluid,—or it may contain thousands of secondary or *daughter-cysts*.

Siebold pointed out the interesting fact that the hydatid worm in man is the encysted form of development of a very minute tape-worm which infests several animals, notably the dog, pig, monkey, sheep, etc. This tape-worm is the *Tænia echinococcus*, and is very minute—about the size of a millet-seed. The intestines of the dog are usually infested with great numbers of these worms. The eggs are discharged with the stools, and, finding their way into food, reach the stomach in man, where they find suitable conditions for their development. The embryo, upon hatching in the stomach, burrows its way or is carried by the blood to some organ,—most often the liver or kidneys,—where it develops as the hydatid vesicle, in which are contained the echinococci described. In Iceland, where the natives live with their dogs in huts, it is said that one-seventh of the deaths are due to the ravages of these parasites.

Strongylus Gigas.—This nematode worm has been found in the renal pelvis of dogs, wolves, horses, oxen, and other animals, and less frequently in the human kidney. It somewhat resembles the *Ascaris lumbricoides*, but it has six papillæ about the mouth instead of three, and it attains a much greater length,—one to three feet,—and it is nearly a quarter of an inch in thickness. It may further be distinguished from the common round intestinal worm by its reddish color. This parasite may be said to be peculiar to the kidney, being very rarely met with elsewhere. Leuckart has, after some research, questioned the appearance of this parasite in the human organism; but there are, beyond doubt, at least half a dozen well-authenticated cases on record, in addition to a specimen in the Hunterian Museum, taken from the kidney of a patient of Sheldon's (Dickinson).

In addition to the parasites described, *ascarides* have been found in the urine in exceptional cases. In such cases the worms have usually made their way from the lower intestines into the urinary tract, usually by way of abnormal openings. In cases of women, these thread-worms have made their way into the bladder through the urethra.

SECTION VIII.

THE MICROSCOPE.

THE following general suggestions are designed for the use of beginners and as supplementary to special works on the microscope, one of which every student should possess—preferably one that deals concisely and clearly with the optical principles and mechanism of the instrument[1]: In purchasing a microscope one should buy the best instrument within one's means, and, now that good microscopes may be had at very reasonable prices, there remains no valid excuse for the physician to be without one. While it is true that the best microscopes are made chiefly by foreign manufacturers, yet in all respects excellent instruments are now turned out by a number of American firms, among which an example is shown in Fig. A. It may be added that this firm (Spencer) also furnishes admirable objectives which, indeed, are quite equal to those of European makers.

When a new microscope is received from the maker or dealer it usually arrives carefully packed in a wooden case. Before its removal from the case careful study should be given to its position therein, so that it may be readily replaced when desired. In handling the instrument it should always be seized by its solid parts, as the foot or post just above the foot, and under no circumstances should violence be employed. If the instrument is to be in daily use, it will be found convenient to keep it under a glass shade, or, after use, a silk cover may be spread over it and the instrument placed on a library shelf for the night. The microscope should not be kept in an overheated room, lest the cement about the lenses become softened ; on the other hand, it should not be kept in a very cold place, else, when the cold optical parts of the instrument are approached by the hands or face, the glasses become coated with moisture, entailing delay or unsatisfactory work. The illustration (Fig. B) will serve to

---

[1] Gage on "The Microscope," published by the Comstock Publishing Co., Ithaca, N. Y., can be highly commended to the beginner.

indicate the mechanism and different parts of the microscope, and the beginner should become thoroughly familiar with the terms employed to designate them, as well as their special uses. *The ocular*, or *eye-piece*, consists of one or more converging lenses, the combined action of which is to magnify the image formed by the objective. It is the piece at the upper end of the tube (*A*) which the eye approaches when looking through the microscope. The *objective* consists of a system of converging lenses at the distal end of the tube which give an enlarged inverted image of the object. Upon this piece depends the initial magnification of the object examined. In the cut will be observed two spring clips for the purpose of holding the object to be examined firmly on the stage. In high-class instruments a mechanical stage is furnished, and by removing the spring clips the mechanical stage may be fastened to the platform by means of which the object under examination may be moved systematically in different directions by means of milled head-screws. *Diaphragms* are perforated stops that fit under the central opening of the stage, and serve to admit the light for the purpose of illuminating the object under examination. The diaphragms vary in size in order that different amounts of light may be employed for the different degrees of illumination required. The *iris-diaphragm* (*J*) is the most convenient and useful form of stop, since by the simple movement of a small lever at its side the diaphragm opens and closes as does the iris of the eye, and thereby the light may be gradually increased or decreased as desired. The *reflector* (*S*) is the small mirror beneath the stage, and serves to illuminate the object under observation by reflecting the light upon it. One surface of the reflector is plain for the purpose of reflecting the light in parallel rays, while the other surface is concave and serves to converge, or focus, the light upon the object. The concave mirror without the condenser is usually employed with low-power dry objectives, while the plain mirror in conjunction with the condenser is usually used with high-power objectives and with immersion objectives. The *coarse adjustment* (*T*) is the rack-and-pinion mechanism which serves to rapidly raise or lower the barrel, or tube, of the instrument. The *fine adjustment*, or *micrometer-screw* (*m*), serves the purpose of very gradually and

Fig. A.—Speucer's stand No. 0, showing objectives, eye-piece, and nose-piece (one-half full size).

15

more accurately raising and lowering the tube, so as to bring the object into accurate focus. The *nose-piece* (*R*) consists of a collar attached to the lower, or distal, end of the tube. It permits of several objectives being attached to the microscope in such a manner that by simply rotating the nose piece the various objectives attached thereto may be brought into use successively. The most useful form for general work is the triple nose-piece designed for three objectives, as shown in the cut. The *substage condenser* [*C* (Abbe's is most employed)] consists of a system of lenses arranged immediately beneath the central opening of the stage. The object of this arrangement is to condense the light reflected from the mirror so that it becomes focused upon the object, thus furnishing brilliant illumination. The iris-diaphragm is arranged immediately below the condenser, for the purpose of modifying the light passing from the reflector through the condenser. Some instruments are provided with a second iris diaphragm, situated above the condenser and immediately beneath the stage. This diaphragm is employed when the condenser is not in use, and *vice versâ*. The *draw-tube* (*A*) is a very important part of all high-grade microscopes, because, by its skillful manipulation, the errors due to refraction occasioned by the varying thickness of the cover-glasses can be largely corrected. Thus, by varying the length of the draw-tube an effect may be produced upon the image similar to that of making the back lens approach or recede from the front lens of the objective by means of the correction-collar. Those beginning microscopical work will do well to familiarize themselves with the influence exerted by varying the length of the draw-tube, and this is best accomplished by the study of some delicate test-object of the diatom order, such as *Pleurosigma angulatum*, keeping in mind the rule that, the thinner the cover-slip, the longer should be the tube, and *vice versâ*. In order to facilitate adjustment, the draw-tubes of the best microscopes are furnished with a scale in millimetres on the side of the tube.

For urinary work—as, indeed, for sputum and blood examinations—an instrument of medium size is more convenient than the larger stands with heavy, mechanical stage. The substage should possess all the essentials for arrangement and modifica-

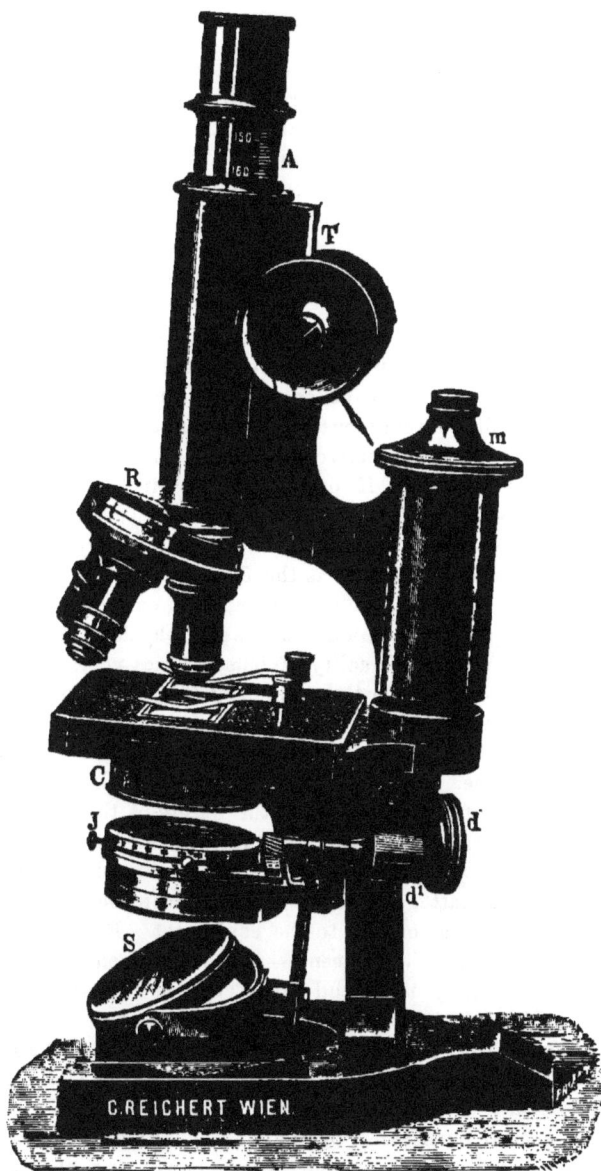

Fig. B.

tion of light, such as an Abbe condenser, iris-diaphragm, and preferably a lateral swinging mirror, or reflector. The mechanical stage is not a necessity, although a great convenience. Nearly every maker now supplies a portable mechanical stage that can be fitted to his particular instrument when desired. Special care should be exercised in choosing a microscope that the mechanical adjustments (notably the rack-and-pinion movement and micrometer-screw) work with smoothness, ease, and precision.

The most important accessories of the microscope are the *objectives*, and the quality of all microscopical work must depend very largely upon their excellence. In urinary work a $\frac{1}{4}$- or $\frac{1}{5}$-inch (5 or 6 millimetres) focus objective, and a $\frac{1}{7}$- or $\frac{1}{8}$-inch (3 to $3\frac{1}{2}$ millimetres) focus objective are an absolute necessity. To these should be added, if practicable, a $\frac{1}{12}$-inch oil-immersion objective, especially if the work is to include bacterial search. With the above-named equipment of lenses all urinary work may be readily compassed as well as the fields of blood and sputum examinations; in short, such equipment will cover the ordinary range of purposes for which the general physician requires a microscope. While the quality of lenses varies very much, one can scarcely go astray in this particular if one patronizes only the leading manufacturers, among whom may be mentioned Zeiss, Reichert, Leitz, and Hartnack, abroad, and especially Spencer at home. It is of prime importance to keep in mind the suitability of the objectives to the nature of the work to be done rather than to purchase a certain set, consisting of a low and a medium power, because of a certain saving in the price of the outfit. The latter course is sure to result in the necessity of adding another lens or two to the equipment. For the coarse examination of urinary sediments—as casts, epithelium, etc.—no lens is so generally useful and satisfactory as a $\frac{1}{4}$- or $\frac{1}{5}$-inch objective. If left to the maker or dealer, as a rule a $\frac{1}{2}$- or $\frac{1}{3}$-inch is usually furnished; but this is too low a power. Reichert's No. 5 will be found very suitable. Zeiss, unfortunately, does not make an appropriate power: his *C* is too coarse and his *D* is too fine for this special purpose. For the more minute examination of urinary sediments,—*i.e.*, for studying the morphological features

ot casts, blood- and pus- corpuscles, epithelium, and granular
sediments,—a comparatively high-power objective is necessary :
about a ½- or ¼-inch focus. Reichert's 7a and Zeiss's D are suitable for this special purpose. The slightly-higher magnifying
power of the former has an advantage in searching for tubercle
bacilli and other micro-organisms if one has no immersion objective in his equipment. As before mentioned, the Spencer Lens
Company turns out objectives that are in every way first class.
The objectives just considered belong to the class of dry lenses;
i.e., lenses that are used without the interposition of any medium
between the lens and the object examined, save the atmosphere.
These objectives answer excellently for coarse and medium
powers of magnification. When, however, very high power is
required, a dry lens, from its short focus, must approach very
near the object, and in so doing much of the light bearing the
image of the object is lost through refraction. *Refraction* is the
term expressing the optical fact that light passing from one
medium to another whose densities differ (as air and water) becomes bent in its course. Thus, the rays composing a pencil of
light, in passing from the object through the cover-glass, many
of them, especially near the periphery of the pencil, become so
bent as to fall outside of the arc of the front lens of the objective,
thereby reducing its power of definition. With coarse- or medium-
power objectives—the focus being much longer, the lens much
larger, as is also the pencil of light itself—comparatively few of
the rays are lost, and consequently the resolving power of the
lens is not materially impaired through refraction.

The *oil-immersion objective*, also termed the *homogeneous
system*, is so constructed that when in use the pencil of light
passing from the object through the objective traverses only media
of the same refractive power. This is accomplished by placing
between the lens and the cover-glass a medium which refracts
light to the same degree as do the cover-slip and the lenses composing the objective. For this purpose a drop of oil possessing
the identical power of refraction as the glass is placed on the
front of the lens or on the cover-slip, and the observation is made
through the oil (usually cedar-oil). It will be perceived that
thus there is no loss whatever of light through deflection, or re-

fraction, as in the dry system of objectives. Immersion lenses are usually constructed for use with oil or water as the intervening medium, and they cannot be used as dry lenses, neither can dry lenses be used as immersion objectives. Low- and medium-power dry objectives are usually made in fixed mountings. Many high-power dry objectives have a screw, or *correction-collar*, attachment. By means of the correction-collar the back lens can be brought near or removed farther from the anterior lens. The object of this device is to permit of a limited amount of correction, which is found necessary on account of the varying thickness of the cover-slips, for it is found that an objective of medium or high power corrected for a thin cover will not give its best results with a thick one, unless provided with a correction-collar adjustment. With the immersion system correction is unnecessary, as it has already been shown that the refractive power of the immersion fluid is identical with that of the cover-slip.

In using medium- and, above all, high- power dry objectives which have no correction collars too little attention has hitherto been given to the proper selection of the cover-slips. From that which has already been said regarding refraction, it will be gleaned that upon its proper regulation depends much of the defining power of the objective. Cover-glasses for use with medium- and high- power dry objectives should be selected with scrupulous care, and only those turned out as No. 1 by reputable makers or dealers should be purchased. Such cover-slips are carefully selected as to thickness, and, as rule, vary in this respect within a comparatively narrow range,—say, from ten to twenty micromillimetres (0.1 to 0.2 mm.). The objectives turned out by most prominent makers in dry fixed mounts without correction-collars are adjusted for a thickness of cover-glass varying only from one and one-half tenths to two-tenths millimetre (0.15 to 0.2 millimetre). Where very fine work is to be done, it is better to measure the exact thickness of cover-slips by means of a cover-gauge, and, having ascertained their exact individual thickness, put them away in boxes accurately labeled for future use. This is of special importance in cases in which objectives with correction-collars are to be employed. In such cases the exact thickness of the individual cover-glass should be recorded on the glass slide or mount, and the collar can then be adjusted accurately for the thickness of the cover-glass when the object is under examination. The author never makes permanent mounts without recording the exact thickness of the cover-glass upon each mount.

The tube, or barrel, of the microscope is made in two forms—a long and large tube (8 to 10 inches long), known as the English tube, and a short tube (6½ inches : 160 millimetres), known as the Continental tube. The latter is the more desirable form, because most of the best makers of objectives adjust them for use with the Continental tube.

Apochromatic objective is the term introduced by Professor Abbe to designate a form of lens made by combining new kinds of glass with natural mineral (calcium fluoride, or fluoride, or fluor-spar). Apochromatic objectives were introduced in 1886, with the object of attaining the higher kind of achromatism, in which rays of three spectral colors are combined in one focus, in place of rays of two colors, as in the case of the ordinary achromatic objectives. The special features of these objectives when used with compensating eye-pieces [1] are as follow : (*a*) Three rays of different colors of the spectrum are focused at the same point, leaving a minute tertiary spectrum only, while with achromatic objectives—made from crown and flint glass—only two different colors are brought to the same focus. (*b*) With apochromatic objectives the correction of spherical aberration is obtained for two different colors in the brightest part of the spectrum and the objective shows the same chromatic correction for the marginal, as for the central, part of the aperture. In the achromatic objective correction of spherical aberration is limited to one color, the correction only covering the central part of the aperture, the objective remaining undercorrected spherically for red rays and overcorrected for the blue rays. (*c*) Apochromatic objectives admit the use of very high eye-pieces ; unfortunately their high price excludes them from general use. Fortunately,

---

[1] The compensation eye-pieces are especially constructed for use with achromatic objectives, in place of the old form of Huggian eye-pieces used with achromatic objectives.

they are not a necessity, save in photography, where it is important to abolish chromatic aberration.

Illumination constitutes a most important feature in the use
of the microscope, and upon its proper management very much
of the efficiency of the work depends. Direct unmodified sun-
light is unsuitable and should not be employed for ordinary
work. North light is most uniform, and is preferable, especially
that reflected from white clouds, rather than from a clear sky.
It is best when working to face the light, in order to avoid the
shadows produced by the hands in adjusting the mirror or other
parts of the instrument. A shade should be used to protect the
eyes and one may also be employed to shade the stage. The use
of lamp-light should, as far as possible, be avoided; when necessary, however, it is best to insert a slip or two of blue glass between the reflector and the object to soften the light. The beginner should practice with the eye as near the eye-piece as possible,
also to use each eye alternately, keeping the eye open which is
not engaged in observation. Too much light is to be avoided,
as it injures the eyes and detracts from the quality of the work.
In urinary work *oblique* and *axial light* are most suitable, but
especially the former. By the term *axial* light is understood
light reaching the object with the rays parallel to each other
and to the optic axis of the microscope, or a diverging or converging cone of light whose axial, or central, ray is parallel with
the optic axis of the microscope. In each case the object is uniformly illuminated.

By the term *oblique* light is meant light in which parallel
rays from a plane mirror form an angle with the optic axis of
the microscope; or, if a concave mirror is used, the light is
oblique when the axial ray of the cone of light forms an angle
with the optic axis. Since oblique light is by far the most efficient for the examination of most urinary sediments, it is well to
bear in mind the following methods of employing it: (*a*) The
first and best method is by placing the reflector to one side of
the stage, since by this means almost any angle of obliquity may
be secured without limiting—cutting off—any material part of
the pencil or cone of light from the reflector. For these reasons,
other things being equal, an instrument is preferable that is ar-

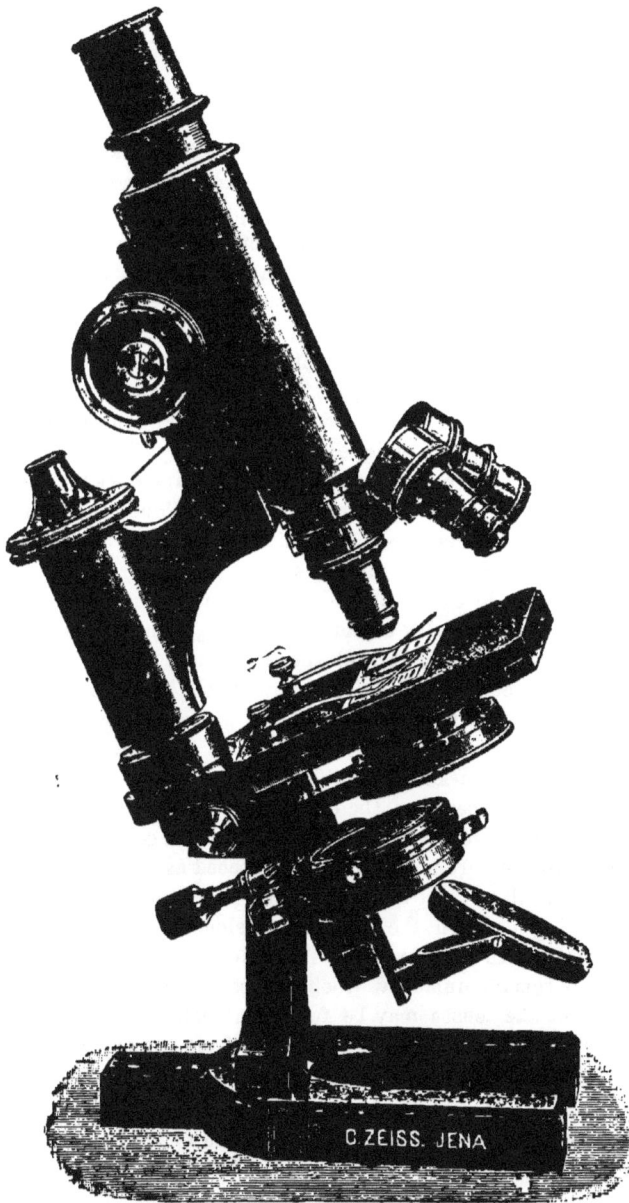

Fig. C.—Zeiss's stand IVa, showing objectives, eye-piece, and nose-piece, without swing-out condenser (one-half full size). The above form of Zeiss stand is that most used in America for general medical purposes, and is recommended as the most suitable form of this make for urinary work.

(225)

ranged with a swinging reflector rather than one with the reflector fixed; *i.e.*, only movable perpendicularly. (*b*) Microscopes provided with a fixed reflector—*i.e.*, only movable perpendicularly and on a plain with the axis—usually have a sliding stop which, by moving aside, cuts off much of the direct rays from the object, admitting only or largely the peripheral rays, which must fall somewhat obliquely upon the object. (*c*) A third method of obtaining oblique light is that with the Abbe condenser as follows: First, focus the light from the condenser upon the object, then rack the condenser down until the focus is considerably below the object; the rays coming from the condenser come to a point (focus) below the object, and continuing past the focus they decussate and diverge so that, when they fall upon the object, all but the axial rays are oblique. As a summary of the make-up of the microscopical equipment suitable for urinary work it should consist as follows: A medium-sized stand with Continental tube; two eye-pieces, or oculars; substage, fitted with Abbe condenser; an iris-diaphragm; a triple nose-piece, and preferably a swinging reflector; three objectives should complete the equipment, viz.: a $\frac{1}{4}$-inch and a $\frac{1}{8}$-inch dry objective, and a $\frac{1}{12}$-inch oil-immersion objective. The workmanship of the instrument should be first class in all particulars. While undoubtedly the best microscopes are those at present turned out by Zeiss, unfortunately they are also the most expensive. The author knows of few microscopes at present in the market that fulfill all of the above conditions at the moderate expense of the one shown in the cut[1] (Fig. *D*).

Care of the Microscope.—Should the lenses or oculars become blurred or spotted from dirt or dust, proceed as follows: If on the ocular, it will be discovered by rotating the ocular in the tube while looking through the instrument, since the dust will be seen to rotate with the eye-piece; if the dust be upon the objective, it will remain unmoved when the ocular is rotated. The condition of the lenses may be further ascertained by holding them toward the light and at a distance from the eye, when, if the lenses are clean, a clear picture of the reduced image of the

---

[1] The equipment shown in the cut (Fig. *D*) is sold by Richards & Co., 12 East 18t} }treet, New York, and 108 Lake Street, Chicago, for $100.

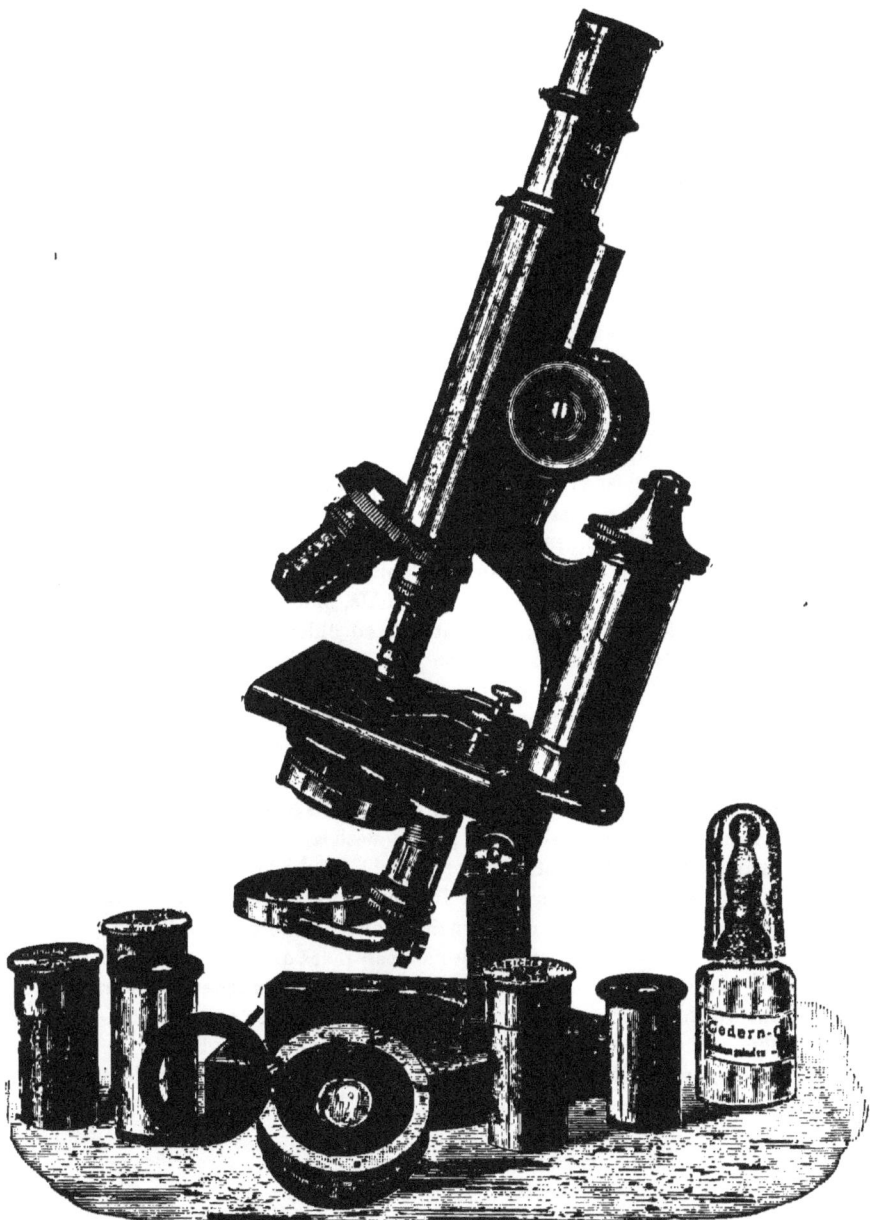

Fig. D.—Reichert's stand No. IIIb, showing objectives, eye-pieces, iris-diaphragm, Abbe condenser, and round, dustless nose-piece.

(227)

window will be seen. It is recommended to remove dust from the optical parts by means of lens-paper or a soft-linen cloth. The author prefers silk, however, in all cases, as it leaves no fluff on the glass. The lens may first be breathed upon; if this does not succeed, the silk may be moistened with a little distilled water; should this prove ineffectual, the silk may be slightly moistened with alcohol, but care must be taken that there be no excess of alcohol on the silk, otherwise it may get between the lenses and dissolve the cement and thereby render the lens useless. After using the oil-immersion objective it should be cleaned as follows : Wipe off the homogeneous liquid with a silk pocket-handkerchief; then, in the case of cedar-oil, wet one corner in benzin and wipe the front lens with it; immediately after wipe with a dry part of the silk. The cover-glass of the preparation may be cleaned similarly.

Glass surfaces should never be touched with the fingers. The oculars and objectives should never be allowed to fall upon the table, much less so upon the floor. All parts of the instrument should be kept free from liquids, more especially acids, alkalies, alcohol, benzin, turpentine, and chloroform. When an objective is left in position on the instrument, an eye-piece should also be left in position in the upper end of the tube to prevent dust from accumulating on the back lens of the objective. To clean the mechanical parts of the microscope Gage advises the use of some " fine oil " (olive-oil or liquid vaselin and benzin, equal parts) on a piece of chamois-leather or on lens-paper, rubbing the parts well; then with a clean, dry piece of chamois-leather wipe off most of the oil. If the mechanical parts are kept clean in this way, a lubricator is rarely needed. In cleaning lacquered parts benzin alone answers well, but it should be quickly wiped off with a clean piece of lens-paper. Do not use alcohol or ammonia, as they dissolve the lacquer and mar the finish of the instrument. The special features of the microscopes turned out by different makers should be studied in the catalogues, which are very complete and, for the most part, admirably illustrated.

EXAMINATION OF URINARY SEDIMENTS.

Preparation of the Sediment.—The sediment having first been thoroughly concentrated by means of the centrifugal apparatus, as detailed on page 149, the next point of importance is to so arrange it that it can be examined in detail under the microscope. For the coarser examination the most satisfactory method is to make a temporary mount of two or three drops of the sediment in a shallow cell on a glass slide. For this purpose slides are sold containing in the centre a hollow, ground-out depression, or cell, for the sediment. A much better arrangement, however, is to prepare a few slides as follows : By means of a turn-table and a fine sable-brush a ring of gold-size should be made, three quarters of an inch in diameter, in the centre of an ordinary glass slide ; after thoroughly drying, two or three more layers of gold-size should be added to the ring, when the cell will be found of the requisite depth. After drying four or five days, the slide is ready for use. A number should be made up and kept on hand ready for use as required. Such cells have the great advantage of affording a perfectly flat and even field ; so that one does not require to refocus the microscope each time the slide is moved, as in the case of the ground-out cell. In mounting the sediment the author is in the habit of proceeding as follows : First having thoroughly cleaned and dried a slide containing a cell, and a slide perfectly plain with no ring, as well as two or three circular cover-slips ($\frac{3}{4}$-inch circles), these are all arranged standing slantingly against the edge of some object about half an inch thick, such as a tile, on the table. The best way to clean the slides and covers is to hold them under the warm-water faucet ; then dry them with an old, but clean, soft-silk handkerchief. The slides should be tipped against the tile with the ring surface downward in order to avoid dust from settling in the cell. Next take up a slender and rather narrow-pointed nipple pipette and carry the point down the sediment-tube to within an inch or so of the point of the tube ; then stop, and by gentle, but steady, pressure upon the rubber nipple expel from 3 to 6 or 7 bubbles of air ; now carry the point of the pipette to the bottom of the sediment-tube, release the pressure from the rubber nipple, and the sediment will quickly flow into the point of the pipette. Next, with one hand withdraw the pipette from the sediment-tube, and with the other hand take up the slide with the cell, turn it over, and deposit within the ring from 2 to 4 drops of the sediment. Next, while holding the slide steadily in a horizontal position in one hand, lay down the pipette from the other hand, and take up a pair of fine-pointed forceps, and with the latter seize a cover-glass by its edge, turn it over, and bring its edge to the outer edge of the ring composing the cell; lastly, gently lower the cover-glass slowly and evenly over the cell so that the sediment spreads out evenly, leaving no air-bubbles beneath the cover-glass. More or less excess of sediment usually escapes from beneath the cover on the slide. With a small slip of filter-paper take up this excess, and the slide is ready for examination. Thus prepared, the sediment can be examined in any position, and any obliquity of the instrument or of light may be employed, without disturbing the sediment, and the latter remains without changes from evaporation for hours. The author in special cases varies the above method of manipulation somewhat, according to the quantity and character of the sediment. Thus, it is not uncommon to meet with urine so heavily charged with pus or other cellular elements that when concentrated with the centrifuge it is unsuitable for examination undiluted. In such case the pus, or it may be blood-cells, so completely fill the field as to hide certain more delicate and less highly-refracting structures—as hyaline casts—from view, and the latter are likely to escape detection. In such cases the following method will usually succeed better: After taking up the sediment in the nipple pipette, discharge but 1 drop or so of it in the cell on the glass slide; then cleanse the pipette and take up a few drops of distilled or filtered water, or better of the clear urine above the sediment, and discharge from 1 to 3 drops of the latter into the cell with the sediment; agitate the whole till evenly mingled, and lastly lay on the cover-glass as already described. Thus diluted, not only is there greater chance of detecting the presence of casts, but the morphological features of the cellular elements are also more plainly made out. Having mounted one or two slides with cells as just described, the plain slide remains. Upon the centre of this plain glass slide is next deposited 1 or 2 drops of the sediment, and then a thin clean cover-glass is gently lowered over it. The excess of sediment escaping from beneath the cover-slip is now taken up by means of filter-paper, and the slide is ready for the microscope. This last mount secures a very thin film or layer of concentrated sediment between the slide and a thin cover-glass, very suitable for examination under a higher power. This form of mount should be examined at once, since evaporation is more rapid than with the cell-mount.

Microscopical Search.—The slides having been prepared, as already described, the microscope is placed in a convenient light on rather a low table, and arranged for use as follows : The ¼-inch objective, if on the nose-piece, is swung into position ; the slide with cell-mount is placed in position upon the upper surface of the stage, the tube of the microscope is racked down until the front lens of the objective is very near the cover-glass ; the concave surface of the reflector is next turned toward the central opening in the stage, the condenser is swung aside ; the iris-diaphragm is brought up under the stage very close to the mount, and the diaphragm is closed down to a minute opening. Lastly, the reflector is swung well to one side, and while looking through the microscope the reflector is turned about between the fingers until it throws the light upon the sediment to be examined. Next, the tube is very slowly racked upward, until some of the outlines of the sediment begin to appear, then by turning the fine-adjustment screw the instrument is focused. It will be found somewhat difficult in the case of beginners to readily find the exact focus at all times, and a few hints upon this subject may be of assistance. Great care should be exercised in lowering the tube of the microscope toward the mount by means of the rack and pinion, while looking through the instrument, because the eye, being engaged in looking through the microscope, cannot judge distance with any degree of accuracy; consequently the objective is exceedingly liable to be forced violently against the mount by this practice, to the ruin of both. It is a much safer practice to rack down the tube while looking sidewise across the stage, until the objective is very close to the stage and beyond the plane of focus ; then, while looking through the tube very slowly reverse the course movement, raising the tube until the outlines of the sediment appear, finally effecting accurate adjustment by means of the micrometer-screw. Having focused the instrument upon the sediment and regulated the light, the slide may be searched as follows: Slowly move the slide forward on the stage until the edge of the ring appears in the lower right-hand corner of the field of vision; next move the slide to the right in a direct line until the opposite side of the ring appears in the lower left side of the field; next move the slide toward the body of the instrument a little more than the width of the field; next move the slide in a direct line to the left until the edge of the ring again appears in the right side of the field. Continue these movements in the order named until the upper edge of the ring appears in the field, when the whole contents of the cell will have passed under observation. In going over the slide, if the fingers are employed to move the mount instead of the mechanical stage, the slide may be moved by two methods, as follows: (a) Mark with the eye the extreme lateral border of the field by fixing the vision on some structure, as a pus-corpuscle or epithelial cell, then quickly move the slide sidewise till the structure appears at the opposite border of the field, scan the field deliberately, and then again move the slide the width of another field, and so on until the slide is exhausted. (b) Another method is: by a steady, but continuous, movement carry the slide along, affording time for the eye to take note of all the details of the moving picture, only pausing as something of special note appears in the field. In either case from time to time move the fine-adjustment screw from right to left, and reverse, to be sure that the instrument is in focus.

Casts.—In searching for casts note any elongated structures appearing in the field, and study carefully their outlines under different ranges of focus; note their morphological features. Note carefully the character of the body of the supposed casts, whether granular or clear; if any cellular elements are attached, such as epithelia, blood, or pus; and, if any of these be found attached, observe if they be well preserved, as indicated by sharply-defined outlines, or if they be granular and ragged, and partly disintegrated or broken down. Note if there be any small, round, highly-refracting globules attached to the cast,—in other words, fat-globules. Observe carefully the relative sizes of the casts, as the predominance of certain sizes in a manner indicates the seat of the most prominent pathological changes in the kidneys. Thus, if the casts found are mostly of the small, narrow order, we may assume that the more prominent changes in progress are situated in the outer border of the cortex near the capsule; if the medium-size casts predominate, it indicates more deeply-seated changes,—viz., in the middle zone; while if the large, broad casts are numerous, we infer that the morbid changes involve the straight collecting tubes in the areas of the pyramids. There will be little difficulty experienced in detecting granular and cellular casts, as they are, comparatively speaking, highly refracting and stand out prominently in the field of vision. It is altogether different with hyaline casts, which are often exceedingly difficult to find. This is because, in the first place, they are very transparent, and have very feeble refractive powers; and, in the second place, they are usually sparse. The following hints may aid the beginner in his search: Bear in mind the facts that cellular elements, notably pus- and blood-corpuscles,—as, indeed, also epithelia,—are comparatively heavy and usually settle to the bottom of the cell, while by their comparative lightness hyaline casts often partly float, and do not settle so low. The instrument should first be brought to a focus upon some epithelial or other cell; then the focus should be changed to a higher field, indicated by the cell becoming fainter or more hazy in outline. If now the field be darkened by closing down the diaphragm to a small point, and the reflector kept well to one side by slowly moving the slide about, an occasional film or thread of mucus will appear. This is very near the proper range of focus and illumination, and, by carefully going over the slide under such conditions, if hyaline casts are present they will be found. In thus going over a slide, if any elongated shadow appears the focus should be changed upon it, and one will often be rewarded by observing the well-defined outline of an undoubted hyaline cast emerge from the shadow.

Diagnosis of Casts.—The distinguishing features of renal casts under the microscope are: (a) Uniformity of marginal out-

16

lines throughout the greater part or whole of the structures. The margins of casts, it is true, may be in places indented, and for considerable distance in places even ragged and broken look, ing; but, for the most part, the outline is even and uniform, and stands out sharply cut as the edge of a rule. It does not look accidental in appearance, but rather, as it really is, molded. (*b*) For the most part, renal casts are uniform in their individual diameters,—that is to say, they do not suddenly appear bulged in one place and narrowed in others. It is true that in cases of very long casts—which are comparatively uncommon—a con-siderable difference may be observed in the diameter of such casts at their distal portions; but this difference is gradually— never suddenly—reached. The body of a renal cast never ap-pears split. It may be found curved and even twisted, though the latter is rarely the case. Lastly, renal casts, as found under the microscope, for the most part are either short or of medium length, not unlike a finger; very rarely they attain considerable length. The reason of this is because they are fragile and easily fractured, and they are subject to many accidents before they can be brought to view under the microscope. (*c*) The ends of renal casts are either rounded like the end of the finger or they are abrupt and ragged, plainly showing evidences of fractures. The ends of renal casts never appear split or bifurcated.

False casts, or pseudocasts, are structures very commonly met with in the urine, and are a source of more or less confusion to the beginner. For the most part, false casts consist of mucous threads, and are nearly always to be found in mild grades of irri-tation of the bladder and urinary tract, though sometimes they come from some of the minute ducts along the urinary passages. False casts, for the most part, possess distinct features which are in striking contrast with true casts, as they appear to the experienced observer. Indeed, nearly all the essential features of false casts are directly opposite to those of true renal casts. Thus, the outlines of false casts are indistinct or non-linear, and are more or less irregular. Their diameters, consequently, are exceedingly variable; in one place small or narrow, and im-mediately after they swell out in spindle forms. Their ends are often tapered out in fine, slender points, or they may be split, bifurcated, or branched; but they never possess the smooth, rounded ends or abruptly-fractured terminals of true casts. They are rarely, if ever, cellular or granular, but under a high power they appear marked with longitudinal striations, or markings, as though largely composed of fibres. They are often wavy and ribbon-like in form, and they sometimes appear twisted and bent into grotesque forms. In short, false casts lack the regularity and design, as it were, both in outline and detail, that true casts possess in their essential composition.

Epithelia.—Various forms of epithelia are met with in nearly every specimen of urine, normal and abnormal, and careful note should be made of their form, size, and the condition of their protoplasm or contents. They are highly refracting, and, consequently, very plainly visible under the microscope, their outlines standing out sharply and distinctly. The general character of epithelia found in the urine corresponds to the three varieties found throughout the body, viz. : (1) *flat*, or *squamous;* (2) *cuboidal;* and (3) *columnar,* or *cylindrical.* All epithelia are granular, and possess one or more nuclei, though the latter may not always be visible, but may, through maceration in the urine, drop out, leaving vacuoles. The character of the granulation varies, sometimes being coarse, at other times fine. Epithelia are subject to certain changes in the urine ; by the absorption of water they swell up and become more regular in outline; indeed, the smaller forms may thus become spherical. The small-sized cuboidal and columnar epithelia have always attracted the most interest in urinary sediments, since such cells come from the convoluted tubes and straight collecting tubes of the kidneys, though not exclusively from these sources. By careful study of the description and illustration of the various forms of epithelia, on pages 186 and 187, no difficulty should be experienced in recognizing these structures under the microscope.

Crystals.—Little difficulty will be experienced in recognizing, much less in finding, the various forms of crystals met with in the urinary sediment. For the most part, they are comparatively large and highly refracting, and are therefore best viewed by low powers and moderate illumination. A few exceptions to this rule occur, notably in the case of the small envelope-shaped crystals of calcium oxalate which are often very minute. As a rule, the brown-colored crystals met with in the urine are uric acid or ammonium urate, and their various forms should be studied so that they are familiar pictures. As a rule, the large, clear prisms are triple phosphate, while the smaller forms of highly-refracting stars, envelope shapes, and dumb-bell forms are oxalate of calcium. These comprise the more common forms met with. Special and unusual forms should be studied. (See pages 156–177.)

Pus- and Blood- Corpuscles.—In order to make a satisfactory study of the morphological features of pus- and blood- cells in the urinary sediment, resort must be had to a higher power than in the case of casts and epithelia. The student should proceed as follows : A $\frac{1}{4}$-inch to $\frac{1}{8}$-inch objective is placed in proper position and the plain-mounted slide (without cell) is then placed upon the stage in position for examination. The concave reflector is adjusted, as already described, to secure oblique illumination, and the diaphragm is opened so as to admit more light, since greater illumination is necessary. The form and characteristics of pus- and blood- cells have been so fully described in another section (Section VII) that but little requires to be added here. It may, however, be stated that confusion is sometimes experienced by beginners in distinguishing pus- from blood- cells,—nay, even the small, round epithelia have been mistaken for one or the other, but especially for pus-corpuscles. This is not likely to occur when all three are present at once in sufficient numbers to admit of careful study; but where only scattering cells of one order are present a mistake is more likely to occur. It is well for the student to keep in mind the relative sizes of these structures as a general guide. Pus-corpuscles are nearly always to be found in the urinary sediment, and these should first be sought and identified as the small granular disks with multiple nuclei, two or more. They constitute the medium-sized round cells met with in urinary sediments. Somewhat smaller in size—about one-third smaller—will be noted the pale, non-granular, non-nucleated hemispherical disks, or red blood-cells ; while at least one-third larger than pus-cells will be noted the granular, very large, nucleated round cells, the so-called renal epithelia.

After the outlines and general features of casts, epithelia, pus- and blood- corpuscles have been recognized, a closer study should next be made of their minute features under the higher-power lens. Careful search should be made for evidences of pathological changes in all these structures,—more particularly for evidences of fatty changes, as evidenced by the presence of small, spherical, highly-refracting globules of fat in the proto-plasm. The state of preservation of the cells should be noted,— i.e., whether they retain their uniform unbroken outlines or if they are partly broken down and the protoplasm essentially altered. Finally, note should be taken of the quantity and so far as possible the character of the granular sediment, which may consist of albuminous granules or amorphous inorganic elements.

MICRO-ORGANISMS.

The presence of micro-organisms, their motility, and even some of their morphological features may often be made out among other elements of the urinary sediment when prepared as already described. A satisfactory examination in detail of these bodies, however, is not practicable without some special preparation and staining. As very little has been written, com-paratively speaking, of the preparation of urinary bacteria for examination under the microscope, the author will here outline the method found convenient and useful in his laboratory, since it is exceedingly simple.

The urine having first been thoroughly sedimented in the centrifugal machine,—much better with the special attachment provided for this purpose,—a few cover-glasses of known thick-ness are cleaned and arranged conveniently for use. One of the clean cover-slips is placed upon a clean card, and upon the centre of the cover-slip is placed 1 or 2 drops of the highly-concen-trated sediment. Next, the sharp corner of a freshly-torn slip of filter-paper is allowed to touch the extreme margin of the sediment. The filter-paper should be held steady till saturated with the liquid elements of the sediment; then a fresh piece of filter-paper should be applied to another point at the margin of the sediment. This should be continued until the aqueous elements of the sediment have been as far as possible removed, leaving a glistening, gelatinous-looking residue. Another clean cover-slip is next seized in a sharp-pointed forceps and gently lowered over the sediment, exactly covering the slip that holds the sediment. The two cover-glasses should next be seized between the previously-cleansed thumb and forefinger of one hand and gentle pressure made, when the remaining aqueous elements of the sediment will ooze from between the cover-slips. Next, with a soft clean silk pocket-handkerchief wipe the edges of the covers by a circular movement, holding the thumb and finger more or less tightly as the axis around which the covers revolve while being wiped. Gradually increase the pressure, both in holding the cover-slips and wiping, until the organic elements of the sediment are spread out in a thin film between the cover-slips and all the aqueous elements are wiped away. Finally, slide the cover-slips quickly apart, dry the films, and, lastly, pass them two or three times through a Bunsen or spirit- flame.

It will be perceived that this simple process frees the sediment from its aqueous elements, and the latter carries with it the dissolved urinary salts which would, if left, greatly interfere with the subsequent staining, or, by crystallization on the cover-slip in drying, mar the preparation.

With regard to staining, nearly every work on bacteriology is replete with instructions and directions how to proceed, and the subject is, indeed, so extensive that special works upon the subject should be consulted in order to attain proficiency. Certain general stains are very useful for urinary sediments, among which may be mentioned carbol-fuchsin, gentian-violet, methyl-blue, etc. The first named may be made use of for staining both gonococci and tubercle bacilli, and therefore it should always form a part of the laboratory equipment. As a general stain for micro-organisms, found in the urine, the author prefers thionin in the form of a 5- to 7-per-cent. aqueous solution.

Mounting.—The prepared and stained cover-slips containing the micro-organisms are mounted very simply, as follows : A bottle of good Canada balsam thinned with xylol should be kept on hand, as well as some perfectly clean thin glass slides of good quality. The cover-slips having been thoroughly dried, a clean glass slide is placed upon a white card, and by means of a glass rod a small drop of balsam is let fall upon the centre of the slide, taking care that no air-bubbles are attached thereto. Next, with a fine forceps take up the cover-slip and gently lower it, film-side down, over the drop of balsam on the slide. If the drop of balsam is not too large the balsam will flow evenly under the cover-slip, little or none running out upon the free surface of the slide. If the drop be large the extra balsam may be wiped away with a piece of filter-paper. The preparation, after proper labeling, is permanent and ready for examination at any time.

Bacterial Examination.—The microscope should be arranged for high-power work as follows : The diaphragm should be thrown wide open and the condenser brought up close to the stage, so that the light is focused upon the film. The plane surface of the reflector is turned toward the stage of the microscope, as parallel rays are preferable with the condenser. The draw-tube should be so arranged that the distance from the upper edge of the objective to the top of the tube where the eye-piece fits in is just 160 millimetres ($6\frac{1}{2}$ inches). The top of the stage should be shaded from direct light from any source. The instrument should be placed in nearly a perpendicular position,—*i.e.*, with the tube perpendicular if the immersion objective is to be employed, otherwise the immersion fluid is apt to drain away from the field of work. With the immersion objective in position, and the light arranged, the tube is lowered slowly, after first having placed a large drop of immersion fluid on the centre of the cover-slip. While lowering the tube with the rack-and-pinion movement, the observer should look across the stage with the eye nearly on a level with the latter, and while so doing bring the tube slowly down until the front lens of the objective touches the immersion fluid. The observer should next look through the instrument, adjust the light properly, and move the slide about until some distinct tint appears in the field. The stain having been recognized, the micrometer adjustment will readily be made to bring the microscope in focus with the micro-organisms, which may now be examined in detail.

SECTION IX.

GRAVEL AND CALCULUS.

CONCRETIONS of a more or less hard and dense character are liable to form in the urinary passages. These bodies are variously termed, according to their size, location, etc., *sand, gravel, stone,* and *calculi.* These formations are called *primary* when they are deposited from urine which has undergone no decompositional changes, and are the result of some original defect of, or foreign addition to, the composition of the urine. The *secondary* formations, on the other hand, are due to decomposition of the urine with resulting precipitation of its elements, and, for the most part, comprising those ammoniacal changes of the urine resulting from inflammatory disorders of the lower urinary passages. Concretions of small size—not too large to make their way spontaneously through the urethra—have been somewhat arbitrarily termed gravel; while, on the other hand, the larger concretions have received the name of stones, or calculi. Concretions vary greatly in size, some of them being so minute as to require the microscope for their recognition, while others attain the enormous size of an orange or even larger. The smaller concretions mostly emanate from the kidneys or renal pelvis, while those of large size come from the bladder.

The most practical classification of urinary concretions is that which corresponds to the chief constituents of which they are composed. This comprises the following divisions: (1) *uric acid;* (2) *urates;* (3) *calcium oxalate;* (4) *cystin;* (5) *xanthin;* (6) *urostealith;* (7) *basic calcium phosphate;* (8) *calcium carbonate;* (9) *calcium phosphate with ammonio-magnesium phosphate.* These are all strictly of urinary origin, and composed of urinary ingredients. In addition to these, at least two other classes of concretions are met with, whose origin is extra-urinary, viz., *prostatic calculi* and *fibrin* or *blood concretions.*

Concretions may consist exclusively of one ingredient, uric acid and calcium oxalate being the most frequent examples.

(241)

Far more frequently, however, two or more primary deposits occur in separate and alternate layers, constituting the so-called "*alternating*" calculus, the most common of which are those composed of uric acid and calcium oxalate. The number and thickness of these layers in alternating calculi vary very greatly according to the age of the calculi and the frequency of the changes giving rise to the alternating deposits. The thickness of the layers also varies to some extent inversely with their number.

If a calculus consist of but one primary substance, its arrangement is usually stratified, exhibiting a larger or smaller number of concentric layers. Such are usually uric-acid, calcium-oxalate, and phosphatic formations. The great majority of concretions consist of a central division or *nucleus*, and an excentric division or *body*. There may be in addition, especially in old or large calculi, an external envelope or *crust*, which is nearly always. phosphatic. The nucleus varies much in size and composition. It may consist of the same material as the body of the calculus, especially when the latter is made up of such primary deposits as uric acid, calcium oxalate, etc. Frequently, however, the nucleus consists of some organic product, as blood-clots, coagulated mucus, or epithelium. Exceptionally foreign bodies introduced from without into the bladder become the nuclei for subsequent calculous formations.

The causes of urinary concretions comprise (*a*) the conditions favorable for their origination; (*b*) the conditions favoring their formation and growth. The conditions favoring the origin of calculi have received some light through the researches of Carter, who found that the actual nucleus nearly always consists of globular forms of urates and calcium oxalate, rather than crystals of these substances; and, furthermore, that a colloid matrix was always an essential element of the nuclear formation. Rainey and Ord have furthermore shown that the globular forms of urates and oxalates referred to are only produced when precipitation occurs slowly in the presence of a colloid medium. From this it would appear that morbid conditions of the urinary passages accompanied by exudations of colloid matter, such as mucus or albuminoids, constitute the initial step in the formation of calculi. In the presence of such colloids precipitation of urates, oxalates, etc., occurs, which, combining with the colloids, form globular aggregations that constitute the basis for subsequent development of stone. The conditions favoring the growth and development of concretions embrace chiefly the states of the system or of the urine itself which favor precipitation of the urinary solids. Under the first set of causes may be mentioned digestive disorders, organic diseases of the kidneys, and diseases of the urinary passages, most notably cystitis. The conditions of the urine favoring the growth of calculi are variable, and in some cases directly opposite. Among these causes may be mentioned (a) excess in the urine of the precipitated substance; (b) overacidity of the urine. This disposes more especially to the growth of uratic concretions, as it diminishes the solvent powers of the urine over uric acid and the urates, thus leading to their precipitation. (c) An alkaline condition of the urine. This may lead to precipitation of the phosphates or carbonates of calcium and magnesium. If the urine be alkaline from fixed alkali, the earthy phosphates or carbonates of calcium may be precipitated and favor calculous growth of that order,—a rare condition, however. If the urine be alkaline from volatile alkali—ammonia—the result is precipitation of triple-phosphate of ammonium and magnesium,—an exceedingly common occurrence. (d) Deficiency of urinary salines and of sodium and alkaline phosphates in the urine weaken its solvent powers over the urates and cause precipitation. (e) Lastly, deficiency of the normal urinary coloring matters greatly weakens its solvent powers and favors precipitation of various constituents, especially uric acid and the urates.

URIC-ACID CONCRETIONS.

Calculi composed entirely or chiefly of uric acid comprise the great majority of stones met with in human urine. Statistics give the proportion to all other forms at from 75 to 90 per cent. Indeed, the great frequency of this form of gravel led early observers to the conclusion that all stones were composed of uric acid in whole or in part; hence the name lithuria, for a long time, was applied to gravel in general, and even at the present time is sometimes erroneously applied.

Uric acid may be passed in the form of crystalline clusters, or as smooth, spherical bodies, ranging in size from a pin-point to that of a grain of wheat, or, again, in roughened concretions as large as a pea. These are derived from the kidney or renal pelvis, and may be washed out with the urine singly or in numbers at different intervals. The passage of uric-acid gravel from the kidneys is usually attended by more or less pain, sometimes, indeed, so acute that it has received the name of "*renal colic.*" All these uric-acid concretions present a yellowish-brown or reddish appearance. If retained in the bladder, these small concretions grow, more or less rapidly, into round or oval or elongated and flattened stones. Their surfaces are tuberculated and irregular, and they vary in weight from a few grains to four or five ounces. They are hard and brittle, and, upon section, they may be seen to be marked with concentric laminæ.

Uric acid may be recognized by the murexid test (see page 33). It is soluble in solutions of lithium, potassium, and sodium hydrate and also in piperazin, while in solutions of sodium and potassium bicarbonate it is insoluble, as well as in water and dilute acids. Since the combination of uric acid and lithium is more soluble in water than its combination with sodium or potassium, it has become popular to treat the so-called uric-acid diathesis with mineral waters containing a few grains of lithium carbonate to the gallon. As Bunge has shown,[1] however, "this *naïve* idea simply implies ignorance of Berthollet's law in reference to the diffusion of bases in the economy, as no such solvent action of uric acid in the economy is obtained thereby."

The formation of uric-acid calculi is most frequent in early and late life; that is to say, in children and old people. A highly-acid state of the urine is the most essential condition for deposition of this form of concretion. Uric acid frequently forms the basis of calculi which, subsequently, become the nuclei for further formation with other urinary constituents, constituting mixed calculi. The most frequent of these secondary formations is calcium phosphate; less frequently, calcium oxalate. Should the urine become alkaline at any time from

[1] Text-book of Physiological and Pathological Chemistry, London, 1890, pp. 356 and 357.

volatile alkali, triple phosphates will be deposited and add to the growth of the calculus. This, indeed, is almost an essential in the late stages of calculous life in the bladder, since alkaline urine is one of the features of vesical inflammation which attends nearly all calculi of any size, especially during their late sojourn in the bladder. The conditions favoring the deposition of uric acid have been more fully considered in another section of this volume (page 32).

URATE CONCRETIONS.

Urates rarely constitute the sole constituent of stone in adults, but with uric acid they are frequently met with in the calculi occurring in children. On the other hand, in connection with calcium oxalate, urates form a large percentage of calculi found. Urate concretions have been regarded as consisting of ammonium urate, but calcium and sodium are to be found in them quite generally upon analysis. They do not attain a large size, as in the case of uric-acid stones, rarely being found larger than a small marble. They are usually of a light-grayish color and multiple in number, two or more being found together. They are not so hard and dense as the uric-acid calculi. Like uric acid, urate stones are nearly always deposited from acid urines; the exceptions to this rule consist chiefly of the mixed stones of triple phosphate and ammonium urate, which are the outgrowths of ammoniacal urine.

The urate deposits possess a great tendency to form infarcts in the renal tubules during early infancy. These consist of ammonium and sodium urate, which form yellowish-brown lines, often reaching from the papillæ to the bases of the pyramids, following the lines of and blocking the interiors of the large, straight, uriniferous tubes. They occur most frequently from the second to the tenth day after birth, although they may occur as late as ten or twelve weeks after birth. It is significant that these infarcts are not found in the kidneys of stillborn children. The conditions leading to these deposits in infants is physiological rather than pathological,—due to the highly-concentrated state of the urine at birth; the deficiency of aqueous elements in the urine does not permit either of solution or washing out

of the uratic components of the urine, which, upon the establishment of respiration, become greatly in excess.

In febrile conditions in children deposits of sodium urate
are common, and no doubt frequently form the nuclei and primary deposit for the subsequent development of gravel. The
frequent occurrence of calculi in children may be largely attributed to this source.

CALCIUM-OXALATE CONCRETIONS.

Oxalate-of-calcium concretions are met with most often as
large, rough, dark, tuberculated masses commonly called "*mul-
berry calculus.*" Less often they occur as small, rounded, smooth,
dark-grayish bodies, called "*hemp-seed*" calculi. Calcium-oxalate concretions are extremely hard and brittle, and when crushed
present sharp, angular lines of fracture. The nucleus often consists of uric acid or urates, or it may be colloid. Pure calcium-
oxalate calculi are often met with, but much more frequently
mixed calculi of calcium oxalate and uric acid occur in alternating
strata around a mixed nucleus. Less often calculi are met with
consisting of oxalate of calcium as the basis, surrounded by a
more or less deep incrustation of triple phosphate.

The urine associated with calcium-oxalate calculi is always
acid, unless in cases of long standing, in which the stone has set
up cystitis, when it may be found ammoniacal. In short, the
condition of the urine in these cases is very similar to that with
uric acid, as might be expected from the fact that calculi are so
often encountered composed of alternating layers of these two
substances.

CYSTIN CALCULUS.

This form of concretion is comparatively rarely met with in
practice. Although not usually attaining the large size of uric-
acid calculi, they may be found in the bladder exceptionally of
considerable dimensions. As a rule they are of medium size,
oval or cylindrical in form, with finely-granular surfaces, over
which may be seen small crystals of a decidedly yellow color.
Although these concretions are rather soft and compressible,
they break with a crystalline fracture. Upon section they present a radiated appearance of yellow color not unlike bees-wax,

but they turn gray upon exposure to light. The causes of cystin deposits in the urine have already been discussed in the preceding section. The most notable feature in this connection is its tendency to occur repeatedly in members of the same family. Cystin is readily recognized by its ready solubility in ammonia, depositing, upon evaporation, its beautiful and characteristic six-sided crystals. It is also soluble in mineral acids as well as in the fixed alkalies and their carbonates; while it is precipitated by vegetable acids and ammonium carbonate.

XANTHIN CONCRETIONS.

Xanthin calculi are exceedingly infrequent; in fact, they are perhaps the rarest of all concretions met with in the urine.[1] They seem to be entirely confined to young subjects, none as yet having been met with either in adult or in advanced life.

Xanthin concretions are of a whitish, yellowish-brown, or cinnamon-brown color, of medium hardness, with amorphous fracture, and on rubbing appear like wax. They vary in size from a pea to that of a hen's egg. They burn completely when heated on platinum-foil. They give the xanthin reaction with nitric acid and alkali, which should not be mistaken, however, for the murexid reaction.

CALCIUM-PHOSPHATE CONCRETIONS.

Phosphate-of-calcium calculi are among the rarer forms of unmixed concretions. They possess a chalk-like appearance and break with an amorphous fracture. They may be dense in structure or loose and spongy. Two forms of calcium-phosphate calculi occur: (a) Round or oval form, varying in size from a small bean to a hen's egg. These are of a white, chalky appearance, of friable surface, and break with an amorphous fracture. These are usually vesical calculi of elderly people, especially dyspeptic people with alkaline urine. (b) The second form is irregular, sometimes branched in shape, of a grayish-white color, compact in texture, brittle, and of porcelain-like fracture. These are usually found in cysts and pockets of the urinary channels, and appear to be of local origin.

---

[1] Less than a dozen cases are at present on record.

The earthy phosphates are often abundantly deposited in the urine, but, owing to their amorphous form, they possess little tendency to concrete; otherwise this form of calculus would doubtless rank among the most frequent of urinary concretions. It is well known that patients may void alkaline urine, turbid from undissolved earthy phosphates, for months without stone formation.

AMMONIO-MAGNESIUM-PHOSPHATE CONCRETIONS.

Calculi composed exclusively of this salt are uncommon; but, on the other hand, it very frequently forms the exterior of other forms of calculi, such as uric acid or calcium oxalate. Ammonio-magnesium-phosphate concretions are always significant of ammoniacal urine, and herein lies the explanation of its ·frequency as a secondary deposit upon other primary concretions. The mechanical irritation set up by any primary calculus of much size invariably leads to ammoniacal decomposition of the urine, usually in the bladder. The result of these changes is prompt precipitation of triple phosphates of highly-crystalline nature, and therefore readily tending to concrete upon any nucleus in the bladder. It has been stated that triple phosphate calculi are rare; this applies to concretions of marked size.

MIXED PHOSPHATE CONCRETIONS.

The most common variety of mixed phosphatic calculi are those composed of calcium phosphate with triple phosphate of ammonium and magnesium. This has been termed the " fusible calculus," because under the blow-pipe it fuses into a black, enamel-like mass. These calculi often attain a large size. They are of grayish-white color, often covered on the surface with bright, glistening points,—triple phosphatic crystals. In texture they are friable and somewhat spongy, often composed of concentric strata, easily fractured into thin laminæ, the fractured surfaces often presenting bright deposits of crystalline triple phosphates. This calculus is very soluble in mineral acids, but is insoluble in water and alkalies ; its chief characteristic is its fusible property under the blow-pipe.

These calculi nearly always form upon some other primary

nucleus, such as uric acid or calcium oxalate, and they often incrust fungous or other growths of the bladder. They are always the result of ammoniacal urine. and originate chiefly within the bladder.

CALCIUM-CARBONATE CONCRETIONS.

Concretions of carbonate of calcium belong to the rarer forms of calculi met with in human urine, very few authenticated cases being on record. They are small in size, of smooth surface, gray or bronze-like in color, and often very hard in texture. They are mostly spherical in shape, often translucent, and on section present numerous concentric lines. The nucleus is usually multiple, and upon crushing these calculi they break into sharp fragments.

OTHER FORMS OF CONCRETIONS.

A few additional varieties of concretions are occasionally met with in the urine.

Fatty Concretions.—These consist of fatty matters saponified by the alkalies of the urine. This substance has received the name of *urostealith*. These concretions are soft and friable, usually of brownish or yellowish color, and often incrusted with phosphates. They burn with a flame when heated on platinum-foil, and give off an odor similar to resin or shellac. These calculi have only been met with a few times.

Indigo Concretions.—Indigo has been met with in the urine in the form of calculus. It is doubtless derived from the indoxyl-potassium sulphate of the urine, which may be changed by highly-acid urine into indigo. Such cases are exceedingly rare, however, but one or two having been recorded.

Fibrin and *blood concretions* are mentioned as having been met with in the urine, but these must be looked upon as anomalous occurrences, except as forming the nucleus of other calculous growths.

Prostatic Concretions.—Prostatic calculi may be mentioned as sometimes encountered in the urine, although not strictly speaking urinary products. Sir Henry Thompson found these calculi invariably in the adult prostate at the autopsy. They

17

are found in the follicles of the prostate; at first consisting
entirely of albuminoid matter, they later become impregnated
with earthy matter, and ultimately become as hard as other
forms of calculi. It is exceptional that they give rise to special
symptoms; only so when they attain an unusual size and en-
croach on the gland-tissue or project into the urethra. It is
only exceptionally that these concretions are spontaneously discharged with the urine.

Clinical Differentiation.—It would add greater precision to
the treatment of calculus if the nature of the stone could be
made out. The accuracy with which this can be compassed will
vary considerably in different cases. The most trustworthy information is obtainable from examination of calculi already
voided with the urine. In a very considerable percentage of
cases the habit of spontaneous expulsion of small concretions
has been established. If the discovery of a retained calculus be
not too remote from the spontaneous passage of a concretion
which has been secured and examined, and, furthermore, if the
characters of the urine correspond, it furnishes the most certain
evidence of the nature of the retained calculus. It is of the
utmost importance, therefore, to preserve all concretions that
may be spontaneously voided with the urine, however minute
they may be, and to make an accurate note of their composition,
as well as of the characters of the urine at the time they were
voided.

In the absence of the above information it is still possible,
in a large majority of cases, to make out with very great probability the composition of the retained concretion. The evidences
here are to be drawn from the characters of the urine—notably
its chemical reaction and the character of the deposits. In
addition to these, certain inferences are to be drawn from the
constitutional peculiarities of the patient, as well as the age,
together with the known frequency of the different forms of
concretions considered with regard to their location.

If the urine be frankly acid *at the time voided*, the calculus is
nearly certain to be either uric acid or calcium oxalate. Uric-
acid calculi being by far the more frequent, the chances are in
favor of the presence of that form of concretion. If, in addition

to the *sharply-acid* reaction of the urine, *directly* upon cooling it precipitates crystals of uric acid, and the symptoms appear in vigorous subjects addicted to the liberal use of meat diet, it becomes reasonably certain that the calculus is of the uric-acid form.

If, on the other hand, the *freshly-voided* acid urine precipitate habitually the calcium-oxalate crystals, the patient being of sedentary habits and prone to indulgence in the use of both saccharine and proteid foods, it may be pretty confidently concluded that a calcium-oxalate calculus is retained. As already observed, calculi are very frequently met with composed of alternating layers of uric acid and calcium oxalate, because they are both the outgrowth of acid urine, as well as largely of similar habits of living. But the predominance of either variety may usually be made out by repeated observation of the features just considered. Cystin and xanthin concretions are also met with in acid urine, but these formations are so rare that they may, for practical purposes, be ignored, unless the presence of these substances be discovered in the urinary sediment, in which cases the probability of the presence of these rarer forms of calculi may become a proper subject for consideration.

The forms of calculi met with in *alkaline urine* next demand attention. For the most part these are vesical calculi. As a rule, they are more complex in composition, because the conditions of the urine during the formation and growth of these concretions are subject to greater change. Renal calculus is frequently composed of a single constituent, because its origin and growth are practically in the same location and subject to the same conditions of the urine throughout. With vesical calculus, on the other hand, the nucleus may originate in the kidney when the urine is acid, or even alkaline from fixed alkali, while its subsequent development occurs in the bladder, where the urine may be ammoniacal, as a consequence of its residence there or otherwise, and its growth will be influenced by the elements consequent to ammoniacal urine.

The most frequent calculus met with in alkaline urine is that composed of mixed phosphates, or the calculus with uric-acid or urate nucleus covered with mixed phosphates. If the urine

be alkaline from fixed alkali, the calculus is pretty sure to consist of calcium carbonate or phosphate. In such cases the urine
has usually long been alkaline, and the calculus is likely, accordingly, to consist uniformly of the same substance throughout.

If the urine be ammoniacal the nucleus and body of the
calculus is very likely to be composed of different substances.
The nucleus may be uric acid, urates, or oxalate of calcium, but
the crust is pretty sure to consist of mixed phosphates. Re-
membering the rapidity with which urine undergoes ammoniacal
changes in vesical disturbances, special care should be observed to see if the urine be ammoniacal in these cases *at the
time it was voided,* otherwise the observer may be greatly misled
in his conclusions. The intensity of ammoniacal reaction of
the urine, the deposit of phosphatic fragments, and the amount
of deposit of triple-phosphate crystals, together with the grade
of the accompanying cystitis, will furnish some idea of the age
and magnitude of the phosphatic calculus.

ANALYSIS OF CALCULI.

In conducting the analysis of urinary calculi the size, color,
form, and density of the concretion should first be noted, as
these often indicate, with considerable probability, their composition, or at least the direction in which the chemical examination
should be pursued. Since many calculi are composed of more
than one deposit, in order to ascertain with greater precision the
several components, section should first be made of the calculus
by means of a fine saw. If the calculus be brittle it will often
answer the purpose to break it into as large pieces as possible.
Upon section or fracture, portions should be scraped from the
different-appearing strata for separate examination. A portion
of the calculus should first be subjected for some time to a red
heat upon platinum-foil, either over a spirit-lamp or by means of
a blow-pipe. In the latter case the best method is to lay the
platinum-foil on a plaster-of-Paris cast, when the powdered calculus and foil may be raised to any desired degree of heat without danger of burning the fingers. If upon ignition little or
no fixed residue be left, the calculus is composed of some of
the organic deposits—as uric acid, ammonium urate, xanthin,

cystin, proteid substances, or urostealith. If, on the other hand, the fused calculus leave a considerable residue, it consists of some of the inorganic bases, either alone or in combination, such as urates of sodium, potassium, or ammonium, calcium oxalate, calcium carbonate, calcium phosphate, or ammonio-magnesium phosphate.

If the concretion burn up and leave little or no residue, it is necessary next to proceed by chemical methods to determine which of the organic deposits it be composed :—

Uric Acid.—A portion of the powdered concretion is submitted to the murexid test (see page 33), and if the characteristic color-reaction be obtained with nitric acid and ammonia, the calculus consists of uric acid or ammonium urate. A portion of the finely-powdered calculus is subjected to boiling water, when, if complete solution be effected, the calculus is ammonium urate; but, if slightly or not at all solvent, it may be concluded that it is a uric-acid calculus. If further confirmation be necessary, let the solution stand until cool: ammonia will be evolved when treated with potassium hydroxid, if ammonium urate be present; and red litmus will be turned blue if suspended over the solution. If the result be negative—no ammonia present—the calculus is uric acid.

Xanthin does not give murexid reaction, but its solution in nitric acid upon evaporation leaves a bright citron-yellow residue, insoluble in potassium carbonate, but soluble in potassium hydroxid, with resulting deep reddish-yellow color.

Cystin does not give murexid reaction. Owing to its contained sulphur, if dissolved in potassium hydroxid and a little lead acetate be added, upon boiling a black precipitate of lead sulphide forms and imparts to the solution an inky appearance. Cystin also dissolves in ammonia, and upon evaporation crystallizes in regular hexagonal plates. If dissolved in hydrochloric acid and slowly evaporated, it forms diverging crystals arranged in sheaf-like form.

Protein concretions do not give murexid reaction ; upon heating they evolve the odor of burnt horn or feathers. They are insoluble in water, alcohol, and ether, but are soluble in potassium hydroxid.

Urostealith gives no murexid reaction, but dissolves in ether, and yields fatty acids upon boiling with baryta-water. It dissolves in potassium hydroxid when heated and becomes saponaceous.

If the concretion be incombustible and leave, after ignition, a relatively large residue, it is necessary next to proceed to determine its composition, as follows :—

Urates of the Fixed Alkalies (Sodium and Potassium Urates).—In order to isolate these fixed bases the concretion is finely powdered, and after boiling in distilled water is filtered. The urates pass through the filter in solution, while the less soluble uric acid remains on the filter. The solution is next evaporated and then ignited, and the residue consists of the fixed bases. If the residue turn moistened turmuric paper brown, it is either potassium or sodium ; if it be the latter, it imparts to the flame of the blow-pipe a yellow color.

Magnesium and calcium, if present in the residue, may be dissolved in dilute acids, and, upon addition of sodium phosphate and ammonia, the calcium and magnesium are precipitated as ammonio-magnesium and calcium phosphates.

Calcium Oxalate.—These concretions first blacken upon heating, but upon further ignition they finally leave considerable white ash, which dissolves in hydrochloric acid *with* effervescence. If this solution be neutralized with ammonia, and oxalic acid be added, characteristic envelope-shaped crystals of calcium oxalate are precipitated, and may be recognized readily by the microscope.

Calcium carbonate, like calcium oxalate, at first blackens upon ignition, but ultimately burns white, leaving considerable infusible ash. Calcium carbonate, however, is distinguished from calcic oxalate by its highly-characteristic property of dissolving in hydrochloric acid *with* effervescence. It will be remembered that the fused ash of calcium oxalate—not the calculus—dissolves in hydrochloric acid with effervescence.

Ammonio-magnesium phosphate with more or less *calcium phosphate* usually occur together, and as such constitute the mixed phosphatic or fusible calculus. Upon ignition this calculus melts into an enamel-like mass. Upon prolonged ignition

ON HEATING THE POWDER ON PLATINUM-FOIL, IT

Does not burn		Does burn	
The powder when treated with HCl		**With flame**	**Without flame**
Does not effervesce			The powder gives the murexid test
The gently-heated powder with HCl			*The powder when treated with KHO gives*
The powder when moistened with a little KHO			

Test / description	Substance
No noticeable ammonia reaction	Uric acid.
Strong ammonia reaction	Ammonium urate.
Does not give murexid test. The powder dissolves in HNO_3, without effervescence. The dried yellow residue becomes orange with alkali, beautiful red on warming	Xanthin.
Flame pale blue, burns a short time. Peculiar sharp odor. The powder dissolves in ammonia, and six-sided plates separate on the spontaneous evaporation of the ammonia	Cystin.
Flame yellow, pale, continuous. Odor of resin or shellac on burning. Powder soluble in alcohol and ether	Urostealith.
Flame yellow, continuous. Odor of burnt feathers. Insoluble in alcohol and ether. Soluble in KHO with heat. Precipitated herefrom by acetic acid and generation of H_2S	Fibrin.
Effervesces	Calcium carbonate.
Effervesces	Calcium oxalate.
No NH_3, or, at least, only traces of NH_3. Powder dissolves in acetic acid or HCl. This solution is precipitated by ammonia (amorphous)	Bone-earth (magnesium and calcium phosphate).
Abundant ammonia. The powder dissolves in acetic acid or HCl. This solution gives a crystalline precipitate with ammonia.	Triple phosphate (mixed with unknown amount of earthy phosphate).

they do not show an alkaline reaction like calcium oxalate and carbonate. The fused ash of this calculus dissolves in hydrochloric acid *without* effervescence.

The excellent table on preceding page, from Heller, shows at a glance the chief features of the analysis of the various calculi, and will be found to greatly facilitate the analysis of calculus by the student.

PART II.

URINARY DIAGNOSIS.

DISEASES OF THE URINARY ORGANS AND URINARY
DISORDERS.

THE URINE IN OTHER DISEASES

.

(257)

SECTION X.

DISEASES OF THE URINARY ORGANS, AND URINARY DISORDERS.

URINARY diagnosis, as considered in the subsequent pages of this work, will include : *First*, diagnostic data derivable from the urine which relate directly to pathological conditions of the urinary organs themselves. *Second*, diagnostic data derivable from the urine which relate to pathological conditions, either local or general, but the prominent feature of which is some marked and characteristic departure from the normal condition of the urine itself. *Third*, diagnostic data derivable from the urine which relate to pathological conditions primarily independent of the urinary organs, in which the latter may or may not become involved.

In order to bring the range of urinary diagnosis more fully within the field of practical clinical work, the plan will be followed of first describing the changes effected in the urine by the various forms of disease, followed by a brief epitome of the leading clinical symptoms, and, where necessary, also the differential features in each case.

A perusal of the nine preceding sections will familiarize the student with the process of secretion and excretion of the urine ; the chemical and microscopical characters of the latter, both normal and abnormal ; and the clinical significance of the various morbid products met with in the urine in the course of disease. In order now to compass the entire groundwork of knowledge essential for practical urinary diagnosis, it only remains to consider the most approved methods of physical examination of the urinary organs themselves. This requires a practical knowledge of the regional anatomy of the urinary organs, and, while doubtless most readers of this volume are already familiar with the subject through the numerous and excellent text-books on anatomy now in general use, it is yet believed that a brief and practical survey of the subject here will facilitate the study of the subjects shortly to be considered.

ANATOMICAL CONSIDERATIONS.

The Kidneys.—These are two large, glandular organs situated in the upper and posterior part of the abdominal cavity, on either side of the spinal column. Each kidney is about four inches in length, two inches in its transverse diameter, and rather more than one inch in thickness. These dimensions vary somewhat in individual cases. The left kidney is ordinarily slightly longer and narrower than the right kidney. The weight of each kidney is from 4 to 5 ounces, the male kidney being ordinarily 2 or 3 drachms heavier than that of the female. The left kidney in both sexes weighs about 100 grains more than the right one. The combined weight of both organs, in proportion to the body-weight, is about 1 to 240.

In form the kidney resembles a "*haricot,*" or "*kidney-bean.*" It is compressed from either side, presenting an anterior and posterior surface, both of which are slightly convex, the anterior surface most so. The outer border presents an elongated convex line, while the inner border is concave, with a deep notch in the centre,—"*the hilum.*" The upper and lower extremities of the kidney are slightly wider than the middle of the organ, the upper being somewhat the wider of the two. The anterior surfaces of the kidneys look obliquely outward and forward from either side of the bodies of the vertebræ. The posterior surfaces of the kidneys—rather more flattened than the anterior—look obliquely backward and inward toward the spines of the vertebræ. The upper end of the kidney—somewhat knobbed and larger than the lower—is nearer the spinal column, and has a slightly more posterior position than the lower end. The inner border of the kidney, at its upper part, is about one inch from the middle line of the body; the outer border, at its lower part, is three and three-fourths inches from the middle line of the body. The outer or convex border of the kidney looks obliquely upward, while the concave or inner border looks obliquely downward and forward.

The kidneys are situated deep in the loins, on either side of the vertebral column. The upper border of the kidney corresponds with the space between the eleventh and twelfth ribs, while the lower border corresponds with the middle of the third

lumbar vertebra. The pelvis of the kidney is about on a level
with the spine of the first lumbar vertebra. During deep inspi-
ration both kidneys are usually depressed by the diaphragm
about half an inch, *though not always so.*

" An horizontal line passing through the umbilicus would lie
just below the lower borders of both kidneys; while a vertical line

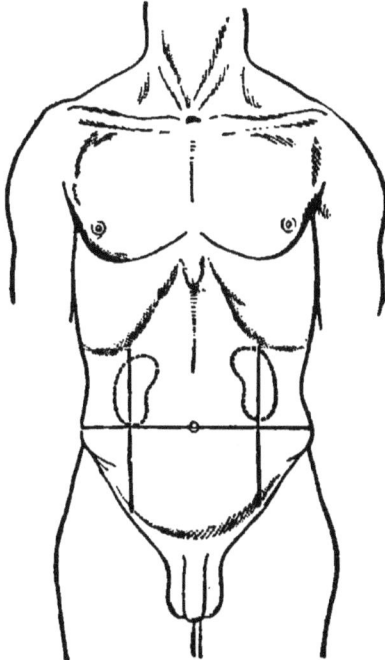

FIG. 36.—TOPOGRAPHICAL RELATIONS OF KIDNEYS, ANTERIORLY.
(After Morris.)

extending perpendicularly upward from the middle of Poupart's
ligament to the costal arch would pass directly over the kidney,
slightly external to its median line. Posteriorly, a line parallel
with and one inch from the vertebral column, extending from
the lower edge of the tip of the spinous process of the eleventh
dorsal vertebra to the lower edge of the spinous process of the
third lumbar vertebra, would fall just inside of the inner border

of the kidney.   If now two lines be drawn, from the ends of the line just described, horizontally outward for two and three-fourths inches, and if the outer ends of these two lines be joined by a perpendicular line, the whole kidney would normally lie within the four lines described " (Morris).

The kidneys rest on the crura of the diaphragm, on the anterior lamella of the posterior aponeurosis of the transversalis

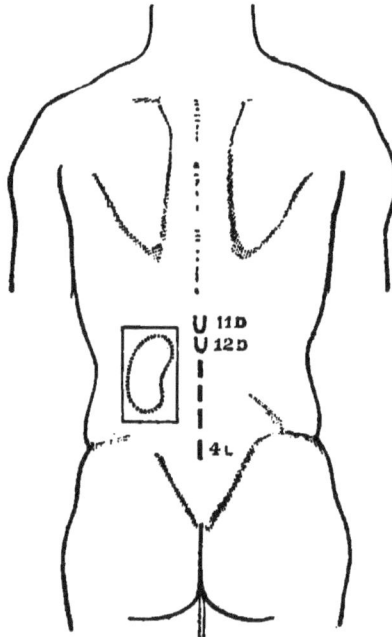

FIG. 87.—TOPOGRAPHICAL RELATIONS OF KIDNEYS, POSTERIORLY.
(After Morris.)

muscle.   To a slight extent they also rest upon the psoas muscle. The *right kidney* is somewhat lower than the left, owing to the position of the liver, which it touches by its suprarenal capsule at its upper end; then the peritoneum passes over its anterior surface near the upper end, and the duodenum and commencement of the transverse colon are in contact with it where they are uncovered by peritoneum.   The *left kidney*, rather higher than the right, is covered in front by the great end of the stomach, the spleen, and the descending colon. The front of the organ touches the fundus of the stomach, and then comes in contact with the pancreas and, lower down, with the commencement of the descending colon. The external border of the left kidney, in the upper two-thirds of its extent, is in contact with the spleen. (See Fig. 38.)

The kidneys are surrounded by a thick layer of fat contained

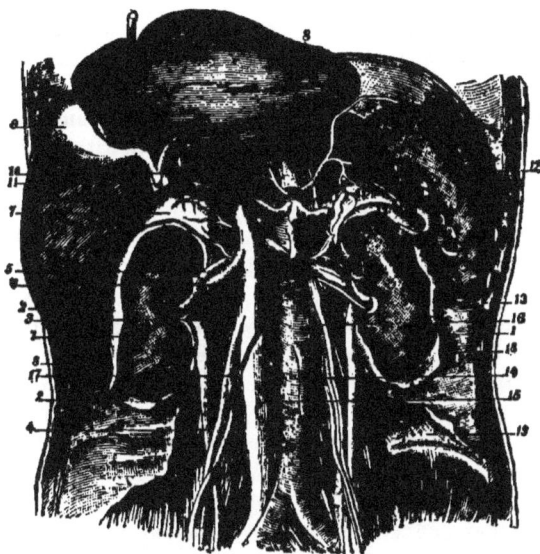

FIG. 38.—RELATIONS OF THE KIDNEYS. (After Sappey.)

1-1, the two kidneys; 2-2, fibrous capsules; 3, pelvis of the kidney; 4, ureter; 5, renal artery; 6, renal vein; 7, suprarenal body; 8-8, liver, raised to show relations of its lower surface to right kidney; 9, gall-bladder; 10, terminus of portal vein; 11, origin of common bile-duct; 12, spleen, turned outward to show relations with left kidney; 13, semicircular pouch on which the lower end of the spleen rests; 14, abdominal aorta; 15, vena cava inferior; 16, left spermatic vein and artery; 17, right spermatic vein, opening into vena cava inferior; 18, subperitoneal fibrous layer or fascia propria, dividing to form renal sheaths; 19, lower end of quadratus lumborum muscle.

in the meshes of a loose areolar tissue and constitute the "*tunica adiposa.*" It is thicker and more abundant posteriorly than anteriorly, but everywhere it completely invests the fibrous capsule of the organs. The amount of fat contained in the tunica adiposa is subject to great variation in different subjects. This fact should not be forgotten, since in stout persons it may be so pronounced as to mislead one as to the size of the kidney itself. On the other hand, in spare subjects, the fatty elements of the tunica adiposa may become so far absorbed that this tunic becomes loose, and its connections with the kidney and surrounding parts are relaxed so that the kidneys are capable of a very considerable degree of mobility.

The *capsule* of the kidney is a thin, smooth, firm, and closely-fitting envelope. Composed of numerous firm, elastic fibres, it possesses considerable power of stretching and contracting, regulated by the degree of vascular tension of the kidney. The capsule adheres, by minute fibres of connective tissue and capillary vessels, to the surface of the kidney, from which, however, it can be readily separated in the healthy organ without dragging any of the glandular structure of the organ proper with it. The capsule, following the notch or hilum in the renal substance, passes into the sinus of the kidney and becomes continuous, around the bases of the papillæ of the pyramids, with the stronger external fibres and elastic tissues of the calyces and pelvis. The *pedicle* of the kidney is composed of the dilated upper end of the ureter, the renal artery and vein, a quantity of connective tissue, and a large number of lymphatics and nerves. The relations of the vessels and ureter to each other in the pedicle are as follow: From above downward, *artery, vein, and ureter;* from before backward, *vein, artery, and ureter.* This arrangement occasionally varies.

The kidney is liberally supplied with blood; indeed, out of all proportion so, according to its relative size. The *renal artery* is of large size and arises from the aorta a little below the origin of the superior mesenteric artery, the right usually arising a little lower than the left. As the aorta lies to the left of the median line, the right renal artery is longer than the left and crosses behind the vena cava inferior. Before reaching the notch or hilum of the kidney each artery divides into four or five chief branches, which sink into the sinus behind the corresponding branches of the renal vein and in front of the pelvis. Deep in the notch of the kidney these branches break up into a number of smaller branches, which leave the veins between the calyces and enter the substance of the kidney between the papillæ.

The *renal vein* is a short, wide vessel, and, like the artery, takes an almost horizontal course. Its primary branches—four or five in number—issue from the hilum in front of the arterial branches, and then the vein continues in front of the artery until it joins the vena cava. The left renal vein is joined by the spermatic vein, both right and left renal veins receiving branches from the suprarenal capsule of their respective sides.

The *nerves* of the kidney consist of filaments from both the sympathetic and cerebro-spinal systems. They accompany the renal artery, and are derived from the renal plexus and the lesser splanchnic nerve.

The kidneys are surmounted by two small, yellowish, flattened bodies,—the *suprarenal capsules,*—which dip slightly downward over the upper borders. The right one is somewhat triangular-shaped, the left one semilunar. They are connected with the kidneys by the common investing areolar tissue, and each capsule is marked on its anterior surface by a fissure which appears to divide it into two lobes. The right suprarenal body is closely adherent to the posterior and under surface of the liver; the left lies in contact with the pancreas and spleen. Both capsules rest against the crura of the diaphragm, on a level with the tenth dorsal vertebra, and by their inner borders are in relation with the great splanchnic nerve and semilunar ganglion.

The Renal Pelvis.—As the ureter passes upward it loses its cylindrical form on a level with the lower end of the kidney, and it there begins to expand into a large, funnel-shaped dilatation, which is known as the "*pelvis*" of the kidney. After entering the hilum or notch the pelvis divides into two or three primary tubular branches, which in turn end in several short truncated but wide pouches, named *calyces* or *infundibula*, the mouths of which receive the papillæ as "does a glove the fingers." A single calyx often surrounds two or three papillæ, so that the calyces are fewer in number than the pyramids of the kidney.

The Ureter begins at the lower, pointed end of the funnel-shaped renal pelvis, at a point about the level of the lower border of the kidney, and extends, in length from fourteen to

sixteen inches, to the base of the bladder, into which it opens by
a constricted, slit-like opening, after having passed obliquely for
nearly an inch between its muscular and mucous coats.

The ureter is a cylindrical, membranous tube, about the
diameter of a goose-quill; but the lumen of the tube is not
uniform. The ureter, in passing downward and inward to the
brim of the pelvis, lies directly behind the peritoneum, resting on
the psoas muscle, and is crossed by the spermatic vessels. In
the pelvis it enters the peritoneal fold constituting the posterior
false ligament of the bladder, and runs downward and forward
by the side of the bladder, entering the wall of the latter about
two inches from the ureter of the opposite side. In the female
the ureters pass by the neck of the uterus, about an inch from
the latter.

The Bladder.—This is a hollow, musculo-membranous organ
situated behind the pubis within the pelvis, in front of the rectum
in the male, the uterus and vagina intervening between it and
the rectum in the female. The shape of the bladder varies with
the age, sex, and degree of distension of the organ. In infancy
it is conical in form and projects above the pubis. In the adult,
when empty, it is small and triangular in form, situated deeply
in the pelvis, flattened from before backward, and rises on a
level with the upper border of the pubic symphysis. When
slightly distended, the bladder is rounded in form and partly
fills the pelvis; when *greatly* distended it is oval in shape and
rises into the abdominal cavity, sometimes extending as far as
the umbilicus. It is largest in its vertical diameter, and its long
axis is directed obliquely downward and backward. When moderately distended (containing one pint) it measures about five
inches in length by about three inches in width. The bladder is
divided into summit, body, base, and neck. The summit constitutes the upper, rounded border of the organ, which is covered
by peritoneum. The body of the bladder, posteriorly, is also
covered by peritoneum, but anteriorly it is uncovered by that
membrane. The base of the bladder is directed downward and
backward, resting, in the male, upon the rectum; in the female,
lying in contact with the lower part of the cervix uteri, and adherent to the anterior vaginal wall. The neck of the bladder is

the constricted portion continuous with the urethra; in the male, surrounded by the prostate gland.

PHYSICAL EXAMINATION.

For examination of the *kidney* by means of percussion the patient should lie upon the abdomen, across a rather hard pillow. In this position there will be found, in the lumbar region of the normal subject with normal kidneys, a space between the last rib and the pelvic brim, rather less than two inches broad,—five centimetres,—which elicits a dull note upon sharp percussion. Anteriorly, this dullness is abruptly exchanged for tympanitic resonance as the intestines are approached. The dull note is continuous upward and outward beyond the limits of the kidney: on the right side continuous with that due to the liver; on the left side continuous with that due to the spleen; while below, the pelvic brim (within which the lower border of the kidney lies) prevents the lower border of the kidney from being defined. It will, therefore, be seen that the *normal kidney* elicits but little information upon percussion, owing to its unfavorable position for that purpose, for even moderately-enlarged kidneys cannot, with percussion, be thus outlined. "Obscured by the thickness of the abdominal walls, covered in part by the lower ribs, liver, and spleen, in part arched over by the vertebral processes, covered by the body of the sacro-spinatus muscle, the lateral border of which closely corresponds with the convex border of the kidney, it will be readily seen that these organs present the greatest obstacles to percussion." In abnormal conditions of the kidney, however, notably those of large tumor, percussion becomes of very decided utility, but for such purposes the patient should be placed upon the back, in the position for palpation.

Palpation of the kidney is best conducted by placing the patient upon the back, with the thighs slightly flexed and somewhat separated from each other. The examiner should approach the side of the patient which he desires to palpate, and with one hand upon the anterior wall of the abdomen he should pass the other hand behind the patient, pressing deeply with his fingers from behind forward in the renal region (between the lower border of the ribs and the iliac crest), pushing firmly forward any tumor against the opposite hand.

In pathological conditions of the kidneys, notably if the organs be very much enlarged or displaced, physical examination often elicits valuable diagnostic information. Thus, in the case of morbid growths, some knowledge of their nature is obtainable by the sense of touch and manipulation. Thus, the organs may feel smooth, uneven, globular, lobulated, fluctuant, soft, or dense. They move but slightly with respiration. If tumor be present, the kidney may leave its normal bed beneath the diaphragm and be seen in front; but a *normal* movable kidney is not visible upon anterior inspection.

A circular, symmetrical swelling between the borders of the ribs and the pelvic brim, extending posteriorly toward the spine, with œdematous condition of the skin and tissues beneath, may point to perinephritis with perinephritic abscess. Tenderness upon pressure is obtainable in acute, but rarely, if ever, in chronic, diffuse nephritis. It is present in renal stone, especially if the latter has excited inflammation. In hydronephrosis it is usually present, and in perinephritis it is especially prominent. Large formations, as carcinoma, sarcoma, hydronephrosis, pyonephrosis, perinephritis, and echinococcus, are plainly palpable ; the latter may show, by quick, short, bimanual percussion-strokes, a peculiar whiz,—the "*hydatid vibration.*" Since the kidney lies behind the peritoneum, when it becomes enlarged by growths so as to extend forward it usually pushes before it the ascending or descending colon against the anterior abdominal wall ; in such case the colon may be made to furnish valuable differential knowledge, because other abdominal growths, being intra-peritoneal, push the colon aside, and therefore furnish no tympanitic note from this source. Since the colon is best distinguished when it contains air, it is often advisable to inflate it for diagnostic purposes. Movable kidney is known by its form, mobility, size, often its capability of replacement, and occasionally pulsation of the renal artery may be felt.

Although palpation is usually sufficient to reveal the presence of renal tumors of any considerable size, *anterior* or *lateral* percussion is usefully employed for confirmative purposes. In the author's experience, one of the most valuable methods is that of anterior auscultatory percussion for this special purpose. By placing the stethoscope over the centre of a viscus or tumor, and by the finger-tips *very gently* tapping the abdominal wall in a radiating direction from the instrument, the outline of the body upon which the stethoscope rests can be made out with great precision by the impulse and pitch of the note conveyed to the ear.

The method of diagnosticating unilateral dislocation of the kidney by bilateral percussion, upon the theory of differential bilateral resonance, although formerly much relied upon, has, upon wider experience, proved untrustworthy.

The differential features between renal tumors and those of adjacent organs often require most careful consideration. Thus, the differential features between a moderately displaced right kidney downward and a distended gall-bladder, an echinococcus cyst or other growth upon the lower border of the liver is not readily made out by palpation or percussion. Respiratory mobility, if pronounced, may, with considerable degree of certainty, exclude the kidney. Capability of replacement, on the other hand, so that the tumor disappears, proves the tumor to be renal. Movable left kidney is differentiated from movable spleen by palpation and percussion. Palpation may reveal characteristic notches in case of the spleen; while in movable kidney the pulsations of the renal artery may sometimes be felt by deep pressure at the hilum of the organ. The course and relations of the colon, as ascertained by percussion, are also here valuable guides. Respiratory mobility may or may not accompany splenic enlargement ; if present it argues against the renal nature of the tumor.

The ureters are so inaccessible that they furnish but little information upon palpation or percussion. A few surgeons, notably Simon, have repeatedly felt the ureters by anæsthetizing the patient and introducing the hand into the rectum. Recently palpation of the ureters *per vaginam* has come to be practiced. This offers no special difficulty, as the ureters can be plainly felt for nearly three inches of their lower extremities, and growths or stones in this portion of the canal can be clearly made out. To a less extent palpation of the ureter may be practiced *per rectum* with the finger, but only about the last inch or so of the tube can thus be ordinarily defined, as it lies within the bladder-wall.

Abdominal palpation is rarely successful in ureteral examinations, and only in cases of very spare subjects, when the ureters are greatly distended or are occupied by large growths. The vesical orifice of the ureter may be inspected by means of the cystoscope, and morbid conditions of that part of the canal can be determined satisfactorily. Catheterization of the ureters, as yet, is applicable only to limited cases—in the female. This immensely valuable means of diagnosis, especially desirable for renal as well as ureteral purposes, must very shortly be perfected in connection with the cystoscope. The author's instrument, devised by Leiter for this purpose, has never proved satisfactory; nor can any of the instruments in present use be depended upon to meet the purpose.

The bladder is only noticeable on external inspection in cases of extreme distension, when it rises into the abdominal cavity. Palpation is applicable in moderate distension of the bladder above the symphysis pubis. It may also be practiced *per vaginam* and *per rectum* either by one hand or, often with advantage, bimanually. Sir Henry Thompson proposed and practiced digital exploration of the bladder by opening the urethra at or about the membranous portion and making a passage sufficient to admit the index finger into the bladder. While the operation is in itself usually a harmless one, more recent measures (the cystoscope) for the most part render it unnecessary. Percussion over the region of the bladder reveals an area of more or less extended dullness, according to the degree of distension of the organ. But percussion of the bladder for the purpose of distinguishing tumors of the vesical region is scarcely necessary, since catheterization will usually quickly determine if they be of vesical origin or not.

A very decided advance in our methods of inspecting the bladder has taken place since 1887, when Nitze first published his methods of examination of the bladder by means of electric illumination.[1] Since then the complete work of Fenwick on "Electric Illumination of the Bladder and Urethra " (1888) and Nitze's " Text-Book of Cystoscopy " (1889) have furnished in detail all the technique of this method of examination of the bladder. By means of the cystoscope we are now able to obtain most complete knowledge of a large number of pathological conditions of the bladder. " Its use plainly discloses ulcerations, their character and extent. It enables us to see diverticula, to find and locate foreign bodies, to not only plainly see stones, but also to ascertain their size, number, shape, character, and even to percuss them with the instrument, the encysted stones no longer escaping detection. Above all, the diagnosis of morbid growths of the bladder by this means is rendered comparatively easy and sufficiently early to render far more efficient their treatment." Certainly, no longer can a diagnosis of obscure disease of the bladder be considered complete without the use of the cystoscope.

Renal Hyperæmia.

Hyperæmia of the kidneys is met with in two forms : (a) *active or acute hyperæmia*, consisting of active determination of arterial blood to the kidneys; (b) *passive hyperæmia*, or venous stasis, consisting of a retention of venous blood in the kidneys, usually the result of some obstruction to the venous circulation, either local or general.

ACUTE RENAL HYPERÆMIA.

This condition may be said to mark the initial stage of nearly all forms of acute nephritis. It may be brought about by numerous causes. In the course of eruptive fevers and inflammatory diseases, such as diphtheria, erysipelas, pneumonia, and acute rheumatism, the kidneys, in common with other internal organs, become more or less pronouncedly hyperæmic. Toxic influences and certain irritants are very common causes of renal hyperæmia. If the toxin or irritant be quickly removed, the normal condition of the renal circulation is rapidly established

---

[1] "Contribution to Endoscopy of the Male Bladder," Archiv f. klin. Chir., vol. xxxvi, p. 661.

again; but if protracted sufficiently long, the prolonged hyper-æmia is very apt to result in active nephritis.

A number of substances when swallowed are capable of inducing active renal hyperæmia, the best known of which are cantharides, turpentine, sulphuric and other mineral acids, phosphorus, cubebs, and potassium chlorate. Acute renal hyperæmia may also be induced by local auto-irritation, as in the case of lithuria and oxaluria when long continued. Exposure to cold and moisture constitutes a frequent cause of the condition under consideration. In the majority of such cases the hyperæmia subsides without, perhaps, having attracted special attention, but sometimes it passes on into acute nephritis. In late stages of diabetes the kidneys usually become hyperæmic, and even albuminuria and mild grades of nephritis may also result.

The Urine.—In acute hyperæmia of the kidneys the urine contains more or less blood, the quantity depending upon the degree of congestion present. In mild cases only a few scattering corpuscles are to be seen, while in active congestion the urine may be very bloody. The color of the urine will depend mostly upon the degree of hæmorrhage. Albumin is always present, but usually in small amount, rarely exceeding 10 or 15 per cent. bulk measure. The urine usually contains a few renal casts. mostly of the hyaline order, and of small size; and free epithelium from the renal tubules may often, though not invariably, be observed. The quantity of urine at first is increased, and, corresponding to this, the specific gravity is somewhat reduced, and the proportion of solids is reduced, though the absolute solids may be normal. The urine retains its normal acidity.

If the congestion continue long, most of the physical characters of the urine just named are apt to become reversed. The specific gravity becomes increased, as do the solids, while the quantity becomes diminished. In some cases of acute renal hyperæmia, especially when induced by cantharidis, fibrinous coagula may appear in the urine.

Leading Clinical Features.—When the cause is toxic, the symptoms are to be looked for in the resultant effects on other organs of the substance ingested. Locally, more or less frequency of micturition is present, sometimes with pain, urgency, and perhaps vesical tenesmus. Some pain is apt to be present in the region of the kidneys, but, in the absence of this, some tenderness may usually be elicited upon deep pressure in the same location. In febrile forms the usual features of pyrexia are to be observed, more or less marked, according to the grade and character of the fever present.

PASSIVE RENAL HYPERÆMIA.

Venous stasis or passive hyperæmia of the kidneys is not at any time a primary renal disease, but is always secondary to

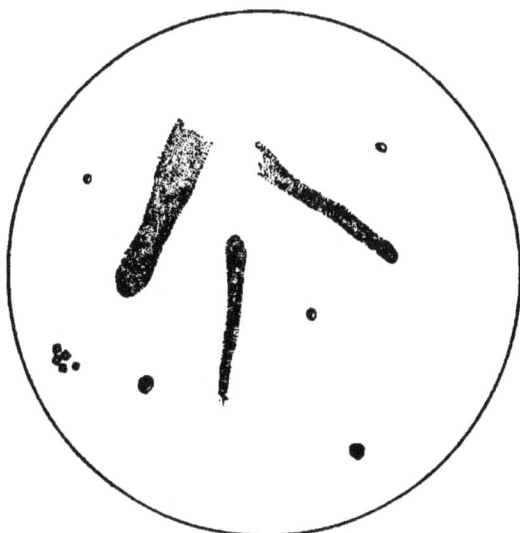

FIG. 89.—URINARY SEDIMENT IN PASSIVE HYPERÆMIA OF THE KIDNEYS.
(After Peyer.)

some obstructive disease of the heart or circulatory organs. Mitral disease of the heart and valvular insufficiency, or stenosis, are the active causes of most of these cases.

The Urine.—The quantity of the urine in uncomplicated cases is always diminished, and the specific gravity is increased, usually ranging from 1.025 to 1.030. The color of the urine is dark-brownish red, and the chemical reaction is frankly acid. The transparency of the urine is somewhat diminished, owing to the presence of undissolved urates and increase of mucus. The quantity of uric acid is relatively, and sometimes absolutely, increased, and free uric-acid crystals are usually to be seen in the sediment, as are the amorphous urates. There is usually no reduction, either in the relative or absolute quantity of urea present. The urine usually contains a small and variable amount of albumin; sometimes merely traces are present; occasionally, though rarely, it is absent, while sometimes it may reach one or two grammes per litre; the degree of stasis, seemingly, does not correspond with the degree of albuminuria. The sediment usually contains a few hyaline casts of small size, and, occasionally, scattering blood-discs or nuclei may be seen attached to these casts (Fig. 39). A few scattering blood-corpuscles are usually to be seen in the sediment; occasionally this is considerable, although such is more rarely the case than the appearance of the urine would indicate, owing to the concentration of the urine.

Leading Clinical Symptoms.—These are very characteristic, and comprise dropsy, mostly of the feet and lower extremities; general cyanosis, dyspnœa, hacking cough; prominence of the large veins, notably those of the abdomen; weak, thready pulse, and cardiac lesions.

Differentiation.—Venous stasis of the kidney is to be distinguished from interstitial nephritis by the increased volume of the urine in the latter disease, its low specific gravity, pale color, normal transparency, spare deposit; absence, as a rule, of blood from the deposit, and constant deficiency of urea, both relatively and absolutely. To these may be added, as the most prominent clinical features of interstitial nephritis, a full, hard pulse, *always* showing increased tension; cardiac enlargement (left ventricle hypertrophied), visual disorders in late stages, absence of dropsy till very late, chronic uræmic disturbances, and the habit of nocturnal micturition,—all of which are absent in passive renal hyperæmia.

Acute Diffuse Nephritis.

This is the so-called acute Bright's disease, and is marked by very pronounced and characteristic features, clinically as well as urinary. The causes include those already considered as provocative of acute renal hyperæmia, any of which, if of sufficient intensity or duration, are capable of bringing on acute diffuse nephritis. By the use of toxic drugs, as a matter of actual experience, acute nephritis is rarely induced, probably because the exciting cause is rarely prolonged sufficiently to bring about high grades of nephritis. Acute diffuse nephritis is most often met with in practice as a result of the acute infectious fevers, notably scarlatina, pneumonia, typhoid fever, diphtheria, relapsing fever, and epidemic influenza. Less often violent exposure to cold and certain local affections of the skin, including extensive burns, erysipelas, carbuncles, etc., give rise to this disease. Lastly, pregnancy must be recognized as the causative factor of a very considerable number of these cases.

The Urine.—In acute diffuse nephritis the urine possesses the following typical characteristics : The quantity is invariably diminished, and sometimes extremely so. At the height of the disease but a few ounces of urine may be voided during the whole twenty-four hours ; but later on, if improvement occur, the quantity gradually increases until the normal volume, or even more, may be reached. If the quantity of urine rise above the normal it may be taken as an evidence that the acute character of the disease is modified, and the tendency is toward resolution. On the other hand the urine, at the very height of acute nephritis, may become practically suppressed, and, if this continue, death may be invariably predicted within a very few days.

The specific gravity of the urine depends upon the quantity voided, and, as already shown, this varies with the course of the disease; so will this feature of the urine. In the early and very acute stage of the disease the specific gravity of the urine usually rises above normal, often reaching 1.025 to 1.030 or even higher. With continuance of the disease the tendency is toward a lowered specific gravity of the urine, corresponding with the increased volume.

The color of the urine varies considerably at different periods

of the disease, the variation depending in part upon the quantity
of urine voided, and in part upon the quantity of contained
blood. As a rule, the color of the urine is dark, more or less
approaching chocolate color. The transparency of the urine is
diminished, the urine presenting a smoky, opaque appearance,
in which the normal lustre is completely lost. A diminution of
this character of the urine denotes changes tending toward resolution. The chemical reaction of the urine, uninfluenced by
medication, is always sharply acid; but upon the use of alkaline
salts, so much employed in treatment, the urine is often found
to be alkaline.

The gross quantity of the urinary solids is diminished in
acute nephritis, the urea suffering the most pronounced reduction. The relative amount of solids varies with the volume of
urine excreted; so that in the early stage, marked by great reduction in the volume of urine, the relative amount of solids
may be normal or above. It is important to make the distinction
here, however, that the gross solids for twenty-four hours are
always reduced. Upon convalescence the gross solids are increased, especially the urea and chlorides, which were merely
held back. At the height of the disease the urea is often reduced to 100 grains or even less for twenty-four hours. The
urine contains albumin in variable, but always large amount in
acute nephritis. It may reach as high as 2 per cent., or even
more by actual weight; so that upon coagulation it nearly fills
the test-tube. More frequently, however, the range is in the
vicinity of ½ to 1 per cent.,—5 to 10 grammes per litre (Esbach's
method). A few cases of acute nephritis following scarlatina
are recorded, in which the urine was free from albumin. It is,
however, rare in such cases that albuminuria is absent throughout the whole course of the disease; more often it is of sudden
onset at some stage, and occasionally it has been observed in
intermittent form.

The degree of albuminuria is considered to mark the degree
and course of acute nephritis toward a favorable or unfavorable
termination; and while the degree of albuminuria can rarely be
taken as a safe guide in this direction in general, it may be more
depended upon as such in this special form of nephritis than

perhaps in any other form of renal disease. On the whole, a continuous diminution in the quantity of albumin in the urine in acute nephritis may be accepted as evidence of progress toward resolution.

The presence of blood in the urine may be regarded as one of the essential features of this disease, though varying greatly in amount in different cases as well as at different periods in the same case. Fluctuations are frequent during the course of the disease, and, indeed, the blood may alternately appear and disappear. Hæmaturia is developed early, being in most cases among the first symptoms noticed, while it usually subsides much earlier than does albuminuria. The quantity of blood lost in the average case of acute nephritis is very considerable if the disease continue long; this is partly evident by the increasing pallor of these subjects. Hæmaturia may be regarded as a valuable prognostic indication in this disease, being rarely absent in severe cases; its appearance marks, with early and great certainty, relapses of the acute process, when previous progress was favorable.

The urinary sediment in acute diffuse nephritis is large in quantity, usually brownish in color from admixture with blood, urates, and coloring matters. Microscopical investigation of the sediment discloses the presence of red blood-corpuscles in larger or smaller numbers. These are somewhat altered, and appear "*washed out*" and ragged, unless in cases of marked hæmorrhage, when they present more nearly their normal appearance. Some pus-corpuscles are to be noted in the field, but rarely in any considerable number. Cellular forms are characteristic of this deposit; mostly small, round, uninuclear cells from the renal tubules, which may be present in great numbers; while less numerous are the narrow-pointed, small-tailed cells from the renal pelvis. The epithelium is well preserved, and affords characteristic pictures of these structures under the microscope. Renal casts are present in large numbers, and may have attached to them (a) blood-corpuscles, (b) leucocytes, (c) renal epithelium (Fig. 40). The above varieties of casts are characteristic of the beginning of acute nephritis, or the disease at its height; but they are subject to alterations in character as the disease con-

tinues some time.   With advancing changes consequent to
the disease, disorganization of the epithelium occurs, and we
find the metamorphosed casts, such as the dark, granular,
and broad, hyaline ones, with more or less organic molecular
débris.   With advance toward resolution the quantity of sediment diminishes and the casts become less and less numerous.
As already noted, the uric acid of the urine is increased in acute
nephritis, the chlorides are diminished, and in very acute cases

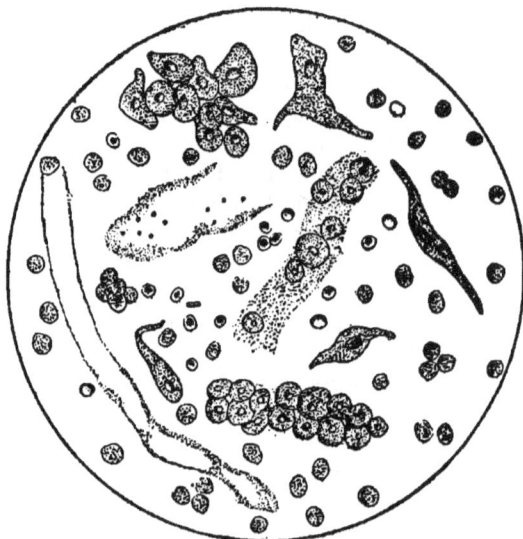

FIG. 40.—URINARY SEDIMENT IN ACUTE NEPHRITIS.   (After Peyer.)

the latter may disappear altogether.   When advancing toward
recovery the volume of urine increases, and with this diuresis
the chlorides and urea become markedly increased, having been
held back by defective eliminative power of the kidney during
the height of the disease.

   Leading Clinical Features.—The most prominent clinical
features of typical acute diffuse nephritis are as follow : Dropsy,
which is always present, and of a general character, involving
the face, hands, feet, and cellular tissues in general.   A very

noticeable pallor steals over the patient, though less prominent than in the chronic form of the disease. The temperature rises to 100° or 102° F., and the pulse is over 100, full, resisting, and marked by increased tension. Headache is present, often severe, persistent, and most often frontal. Nausea is frequent and often attended by vomiting. Acute uræmia is common, often manifested by acute visual disorders, stupor, temporary paralysis, and sometimes convulsions. The appetite is abolished, thirst is prominent, and there is dull, aching pain or stiffness felt in the loins and tenderness upon deep pressure in the renal region.

CHRONIC DIFFUSE NEPHRITIS.

This disease may be a sequel of the acute nephritis just considered, but more often it develops· insidiously from the beginning. It is more apt to result from the acute form when the latter is the outgrowth of scarlatina, pneumonia, diphtheria, or some of the acute infectious fevers. The student is advised to make a most careful differential study of this disease, more especially with regard to amyloid disease of the kidney, with which it has since the days of Bright himself been frequently confounded, with disastrous results to the treatment, since they are almost diametrically opposite in character.

The Urine.—The quantity of urine, as a rule, is diminished in progressive chronic diffuse nephritis. Although no such marked reduction occurs as in acute nephritis, yet a reduction of 40 or 50 per cent. of the normal volume is not uncommon. The fluctuations of the daily quantity are marked, more so than in the acute disease. In the late stages the volume of urine increases, and in chronic cases tending toward secondary contraction of the kidney the quantity of urine often exceeds the normal amount. This is because interstitial changes have been set up and the symptoms tend to conform to the interstitial variety of nephritis. The specific gravity of the urine is below normal. In cases marked by unusual reduction in the quantity of urine, the specific gravity may rise above normal. This, however, is unusual, and results from concentration of the urine. With a normal volume of urine, and often much less, the specific gravity rules below 1.020, and in late stages tending toward renal contraction the specific gravity of the urine often sinks to 1010 or below.

The color of the urine varies from pale lemon to dark brown, more often approaching the former than the latter color. The urine is always cloudy, the more so as the quantity is diminished. When the volume of urine is nearly normal the color is often very light, but the transparency is always diminished more or less, the appearance of the urine being of a hazy, dirty character. This depends upon the invariable presence in the urine of a large amount of sediment, consisting of epithelium, renal casts, pus-corpuscles, and molecular matter.

The urine always contains albumin, and usually in large quantities. In fact, albuminuria may be said to reach its maximum as a symptom in this form of disease, often reaching as high as 3 or even 4 per cent. by actual weight. In such cases coagulation becomes so pronounced with reagents that an accurate estimate of the quantity by bulk measurement can only be made by largely diluting the urine previous to testing. While the average quantity of albumin in the urine in this disease, therefore, ranges very high, it fluctuates markedly in different cases as well as in the same case from time to time. It would seem to maintain a fairly constant ratio to the specific gravity of the urine in most individual cases ; more especially so when the changes in the specific gravity are sudden, or are observed over short periods of time. Thus, if the specific gravity of the urine be 1.014 one day, while the next day it rises to 1.018, a decided increase in the amount of the albumin is sure to be noted. These changes, however, are only relative; the absolute loss of albumin for twenty-four hours remains pretty uniform over short periods of—say—a few days. Any decided increase in the absolute quantity of albumin in these cases indicates an extension or aggravation of the disease. In cases characterized by great chronicity, more especially in those cases tending toward contraction of the kidneys, the quantity of albumin in the urine often becomes reduced both relatively and absolutely.

The solids of the urine suffer more or less reduction in this disease, urea and the chlorides most notably so. Occasionally, when dropsy is subsiding under diaphoretic measures, there may be a temporary increase of the solids of the urine, especially that of urea, which may even exceed the normal. This, however, is of brief duration, and, after a time, falls back again below the normal standard.

The urinary sediment furnishes the key to the diagnosis of this disease. As already stated, the sediment is relatively large in quantity and consists of casts, white blood-corpuscles, epithelium, and cellular remnants. The casts are numerous and of

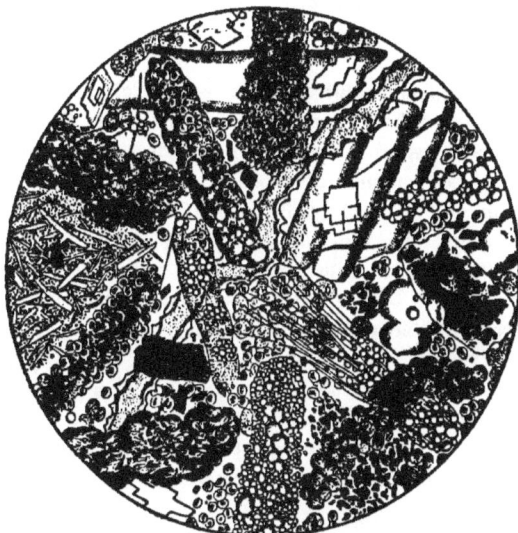

FIG. 41.—URINARY SEDIMENT IN CHRONIC DIFFUSE NEPHRITIS, SHOWING RESULTS OF FATTY CHANGES IN PROGRESS. (After Peyer.)

nearly all known varieties, but the most distinctive ones are the *dark granular*, *broad hyaline*, and more especially the so-called *fatty casts* (Fig. 41). In the more recent cases the casts may be less numerous, and, as a rule, the hyaline, slightly dotted, or faintly granular ones, as well as those dotted with cell-fragments, predominate. The longer the disease continues, the more numerous the casts become, and, moreover, the more predominant become the dark granular casts, the broad casts from the large, straight tubes, and the casts with fat-droplets attached to them.

19

The fatty casts may almost be said to be characteristic of this condition. Red blood-corpuscles are rarely met with in this lesion, perhaps only in cases which have recently sprung from the acute form of nephritis. On the other hand, leucocytes are always to be found in larger or smaller numbers. A very marked sediment of granular *débris* is observed in this lesion, consisting of broken-down cellular elements.

Epithelial cells from the renal tubules are to be found, sometimes in numbers. They are less perfectly preserved than in the acute lesion, disorganization of structure being everywhere apparent.

Leading Clinical Features.—The leading clinical features of chronic diffuse nephritis are briefly and concisely the following : First and most prominently dropsy, which is progressive, obstinate, general, and sooner or later extreme,—involving the cellular tissues and ultimately the serous cavities. Anæmia is no less marked and striking, palpable in the pallid, puffy face and dough-like extremities and body, and pale mucous surfaces wherever visible. Debility is prominent and progressive ; these patients being feeble and helpless, often bedridden. Emaciation is progressive, but masked by the dropsy. The appetite and digestion fail, owing to the charged condition of the blood with effete products which the kidneys fail to eliminate. Uræmia, when present, is of the less active or chronic order, coma and convulsions being rare, except at the close or the result of acute complications.

CHRONIC INTERSTITIAL NEPHRITIS.

Under the above head will now be considered the diagnostic features of those usually slowly-advancing chronic processes, which ultimately terminate in granular contraction or atrophy of the kidneys, known as renal cirrhosis or chronic Bright's disease. Many of the early writers seem to have confounded this lesion with diffuse nephritis, at least so far as to consider it the outgrowth of that primary lesion. We now know that while renal contraction is sometimes the result of long-continued diffuse nephritis, yet the overwhelming majority of cases begin not only independently of that lesion, but are essentially interstitial and atrophic processes from the beginning, and, moreover, are the outgrowth of totally-opposite conditions of the system and habits of life from those in chronic diffuse nephritis. The previously robust, hearty, and overnourished are almost invariably the subjects of the interstitial lesion; while, for the most part, the opposite class of people are more commonly the subjects of the other-named lesion.

It may be premised that primary interstitial nephritis is one of the most stealthy and insidious of all diseases in its manner of approach, giving rise to few, if any, noticeable symptoms until in progress for a number of years,—often ten to fifteen. The lesions, though wide-spread, including the heart and arterial system, are yet almost imperceptible in their manifestations in the early stages; at the same time they are slowly progressive and permanent in character. Notwithstanding all this, with due care and minute scrutiny of all the surrounding features of the case, interstitial nephritis may always be diagnosticated, however early and slight the lesion, if only attention be called to the matter; and the method of compassing this will now be considered.

In the study of interstitial contracting kidney it should be borne in mind that, as a rule, it is accompanied by a progressive hypertrophy of the left ventricle of the heart in at least 80 per cent. of the cases. While the cardiac hypertrophy is in progress, the symptoms, both urinary and general, are pretty uniform and invariable. If the patient survive sufficiently long, however, the hypertrophied heart undergoes degenerative changes, and with the consequent heart-failure many of the characters of the urine, as well as the general symptoms, change completely. If this fact be kept in mind it will serve to prevent the confusion so apt to arise in consequence of the variability of the symptoms in different cases, as well as in the same case at different periods of the disease.

The Urine.—In typical interstitial nephritis the urine is increased in quantity, is slightly paler than normal in color, perfectly transparent, rather sharply acid in reaction, and the specific gravity somewhat below the normal range. Albumin is usually present in small quantity; only a few scattering casts are present, and these are of the narrow, perfectly hyaline order; renal epithelium and cellular elements are rarely observable, but uric-acid and calcium-oxalate crystals are often to be seen under the microscope. The chlorides of the urine are nearly normal in quantity, the urea more or less deficient, and the phosphates are reduced considerably.

A more minute analysis of these features shows the following characters: The quantity of urine is usually increased from the beginning. This polyuria is maintained uniformly and progressively until a comparatively late period of the disease, when heart-failure sets in and the volume of urine often then sinks below normal; nor can it in such cases again be maintained regularly up to the normal standard during the remainder of the patient's life. The specific gravity of the urine becomes progressively lowered; in the beginning a falling off of but two or three points is usual; later on the reduction is more marked, though it never descends as low as in chronic diffuse nephritis or amyloid disease, but in pronounced cases it ranges between 1.010 and 1.016. With heart-failure and consequent diminution of the volume of urine the specific gravity rises somewhat, and may even approach again the normal standard, after having remained for years constantly reduced.

While albuminuria is the rule in this lesion, many exceptions have been noted. The exceptions are often apparent rather than real, because albuminuria of interstitial nephritis is notoriously intermittent in character, sometimes disappearing for days and weeks, to return again and again, regardless of the stage of the disease. It is probable that if these so-called nonalbuminuric cases were kept under constant observation albumin would be found in the urine in many of them some time during the course of the disease. This has been the experience of the author, although he has met with a few cases in which the urine was absolutely free from albumin throughout. So long as interstitial nephritis remains uncomplicated the quantity of albumin in the urine is invariably small, usually ranging below 10 per cent. volumetric measurement by the author's centrifugal method. General or local disturbances, such as "catching cold," mild febrile attacks, etc., quickly increase the albuminuria. In late stages of this lesion associated with cardiac failure albuminuria becomes augmented considerably, the quantity of albumin may reach from 30 to 40 per cent. bulk measure. In no case, however, does albuminuria approach the extreme grade in this lesion that it does either in acute or chronic diffuse nephritis.

Both the relative and absolute amount of urea in the urine begin to suffer reduction from the beginning. At first it is slight, but as the disease advances it becomes a constant and, in many cases, a marked feature. It is not at all uncommon, in advanced interstitial nephritis, to note a reduction of the absolute amount of urea of from 50 to 75 per cent. More or less reduction is also to be noted of the quantity of all the urinary solids, the chlorides suffering the least reduction and the phosphates most. With regard to the phosphates in particular, a diminution in quantity of the phosphates in the urine may be regarded almost as constant a feature of this lesion as the presence of albumin.

Casts from the renal tubules are probably always present in the urine in this lesion, but they are rarely numerous, sometimes extremely sparse and difficult to find. This is due to the fact that they are of such delicate, hyaline, non-refracting character that the most careful search is necessary to detect them; besides, their small numbers, rendered still more sparse by the accompanying polyuria, it often becomes necessary to concentrate the sediment in order to find them. As the disease advances the casts become more numerous, and often they show fine granulations; they are largely of the narrow hyaline and granular orders.

The crystalline deposit in the urine in this lesion consists chiefly of uric acid and calcium oxalate, both of which are often to be noted together. For the most part these deposits are noticeable in the early stages of the disease. The uric acid is precipitated chiefly in consequence of the diminished pigmentation of the urine in this lesion, rather than in consequence of the excess of the former. The oxalic deposit occurs most often in gouty subjects.

On the whole, the urinary sediment in this lesion is remarkably small in quantity and practically free from cellular elements, save those that are common to normal urine. It is not unusual to find, even upon standing twenty-four hours, or centrifugation of the urine, little or no sediment noticeable to the naked eye. Toward the termination of the disease, however, a sediment is usually noticeable, in consequence partly of the more concentrated state of the urine, as well as the wider extension of the lesion.

Interstitial nephritis renders the kidneys extremely prone to take on subacute or even acute attacks of nephritis upon exposure to certain causes, especially that of cold and febrile or inflammatory diseases. In such cases the quantity of albumin in the urine becomes markedly increased, the volume of urine diminished, and the sediment becomes more pronounced and approaches, in its special features, those of acute nephritis already described. In such cases it is necessary, in addition to the urinary examination, to carefully regard the history and general clinical features of the case, in order to diagnosticate the true conditions present.

Leading Clinical Features.—In typical cases of chronic interstitial nephritis we may look for the following clinical features: The patient habitually rises at night once, twice, or oftener to void urine which, to the eye, appears normal in its transparency and nearly so in color. The pulse is *always* full, hard, and resisting to the finger, and marked by decided tension as measured by the sphygmograph. The second cardiac sound, as heard best in the second right intercostal space, within an inch and a half of the sternum, is *always* distinctly accented,—sharper and louder than normal. In most cases—at least 80 per cent.—the normal area of cardiac dullness is more or less extended below and to the left, and, in many cases, notably if the lesion be advanced, this feature is very prominent. Disorders of vision are common some time during the course of this lesion, not very frequently early, but almost certain in late stages. Uræmic disorders are encountered during the course of the disease in some of the following forms: Mild post-cervical neuralgia is very common, almost characteristic; diarrhœal attacks, which mark eliminative efforts of the system vicariously; dyspnœa, which often appears of an asthmatic type; drowsiness, coma, and sometimes convulsions.

Attacks of bronchitis are common and difficult to get rid of; winter cough of the aged frequently owes its origin to this cause. Acute inflammations of the pleura, lungs, or peritoneum are prone to be suddenly kindled and run a fatal course. Dropsy is absent, save in advanced cases, and then it is due rather to the cardiac failure than to the renal lesion.

The early diagnosis rests upon the following points: A previous condition of robust health is usual; age, over 40 years; patient rises habitually at night to void urine of normal appearance; the pulse is full and hard (never weak); the second sound of the heart is abnormally loud; the urine is deficient in urea; small quantities of albumin are usually present, and hyaline casts are to be observed under the microscope if the sediment be concentrated.

The diagnosis of the advanced lesion can scarcely be overlooked by the most superficial observer. The plainly-observable hypertrophy of the heart; the presence of uræmic disorders, as headache, dyspnœa, visual defects, diarrhœal attacks, and perhaps drowsiness at times; together with certain changes in the urine, notably albuminuria, deficiency of urea, the presence of casts of the hyaline and granular order almost exclusively, serve both to call attention to the disease and mark its special character.

Interstitial contracting kidney in many of its features is not unlike passive hyperæmia of the organs. The differential features of the two have already been noted in connection with the description of the latter. (See page 273.)

AMYLOID DISEASE OF THE KIDNEY.

Amyloid change in the kidney is a local manifestation of a general constitutional defect; moreover, this lesion is seldom confined to the kidneys, but nearly always involves the liver, spleen, and gastro-intestinal tract. Amyloid, or, as it is sometimes termed, waxy degeneration, is essentially the outgrowth of a cachectic condition of the system, and most often follows in the wake of syphilis, chronic suppurative processes, such as abscesses, extensive ulcerations, or necrosis. Tuberculosis is frequently traceable in the family history of these cases.

The changes in the kidneys in amyloid disease are very marked, and give rise to a very pronounced train of symptoms; the latter, upon superficial examination, might be mistaken for those of chronic diffuse nephritis. Both the lesions and symptoms, however, are essentially and widely different in character. The author desires to emphasize in the strongest possible manner the importance, therefore, of carefully distinguishing these two renal lesions, since over and over again he has been a witness to the melancholy results of such error. Repeatedly have these cachectic, ill-nourished subjects come under his observation in the last stages of the disease, who had long been consigned to the starvation process of a "milk diet," under the impression that their albuminuria was the result of nephritis.

The Urine.—The characteristic features of the urine in typical amyloid lesions of the kidneys are concisely as follow: The volume of urine is above normal, the color lighter than usual; the transparency is unchanged, the specific gravity is low, albumin is present in marked quantity, and the sediment is very slight in quantity, containing little or no cellular elements ; and but a moderate number of casts are present, most of which are of medium size and broad, hyaline orders.

Before considering more minutely these features of the urine it may be premised that the characters of the urine in this lesion of the kidney are exceedingly apt to fluctuate rather widely in different cases; but, notwithstanding this fact, well-marked diagnostic characteristics are not difficult to trace throughout the progress of the disease in most if not, indeed, in all cases.

The quantity of urine in amyloid lesions of the kidneys rules above normal from the beginning, and in most cases the increase is decided. The volume of urine is subject to temporary periods of falling off, and at such periods it may fall below the normal standard. These periods of temporary reduction in the volume of urine may often be accompanied by corresponding attacks of diarrhœa. The specific gravity of the urine is pretty uniformly reduced in marked cases, ranging from 1.008 to 1.014. Sometimes, however, even in cases attended by a marked degree of albuminuria, the specific gravity ranges as high as 1.016 to 1.018, and in such cases the prognosis is more favorable. In fact, the few cases of ultimate recovery from this disease observed by the author have, without exception, been cases attended by a comparatively high range of specific gravity of the urine. Cases, on the other hand, are not infrequently met with in which the specific gravity of the urine sinks remarkably low, perhaps lower than in any other form of renal lesion,—1006 and even 1004. Such

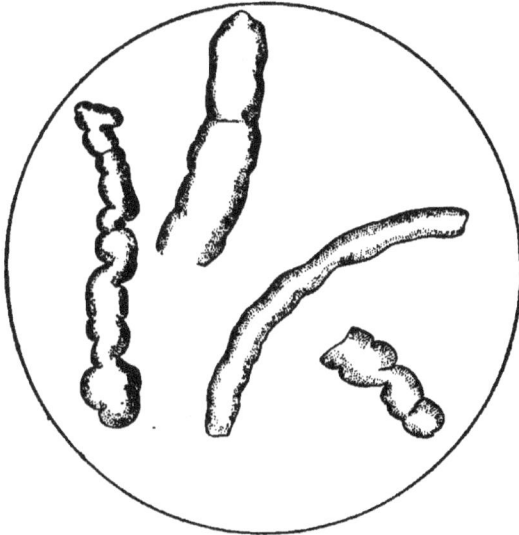

Fig. 42.—Waxy Casts in Urine of Amyloid Disease of the Kidney.
(After Peyer.)

features are, however, only met with in very late stages, and are usually associated with marked polyuria.

The presence of albumin in the urine may be regarded as an essential feature of amyloid lesions of the kidneys. The urine not only always contains albumin in this lesion, but the albumin is present in considerable, often in large, amount. The usual range is about 4 to 7 grammes per litre (Esbach's method), though it is not uncommon for it to rise to double that amount. The course of albuminuria in amyloid lesions of the kidney is quite variable as to quantity; in the early stages it may be slight,

though subject to sudden increase, and with the advent of polyuria may again fall off. The urine always contains globulin in
this lesion, often in larger quantity than that of serum-albumin.

Some reduction of the urinary solids is usually to be noted
in this lesion. The urea is slightly below the normal standard,
probably due chiefly to the lowered state of general nutrition,
rather than to the influence of the lesions over the function of
the kidneys, since uræmia is rare in uncomplicated amyloid
kidneys.

The casts are subject to some variation in number and variety.
With the polyuria the casts are often very scarce and almost
exclusively of the hyaline order. On the other hand, when casts
are comparatively numerous, so-called waxy, yellowish, refracting
casts may be present, and occasionally dark, granular ones. The
chief distinctive feature about most of the renal casts met with
in this lesion is their comparatively large size and hyaline character.

The urinary sediment is comparatively small in quantity in
this lesion, in fact unnoticeable, as a rule, to the naked eye, and
it is practically devoid of cellular elements throughout if the
disease remain uncomplicated.

Leading Clinical Features.—Amyloid lesions of the kidneys
appear in the wake of the so-called wasting diseases, or are often
preceded by syphilis or some exhausting suppurative process.
These patients appear unhealthy and plainly cachectic, except
in a few syphilitic cases. The skin assumes a sallow or bronze-
like tint, the tongue is nearly always heavily coated, dyspepsia
is prominent, and diarrhœal attacks are common. Dropsy is
present in most cases, but exceptionally it may be absent until
late. Uræmia is exceedingly rare. The liver and spleen become
enlarged some time during the course of the disease in the majority of cases. These patients are weakly, anæmic, and enfeebled, with small, thready pulse and cold extremities; but
the predominant features throughout are the disorders of the
stomach and bowels; dyspepsia or diarrhœa or both demand
almost constant attention as the disease becomes advanced.

The distinction between amyloid disease of the kidneys and
chronic diffuse nephritis hinges upon the following points: In

nephritis the urinary sediment is large in quantity, and contains a large number of casts, including epithelial, dark granular, and fatty casts, as well as those with fragmentary cellular elements attached. Leucocytes, cellular elements, and granular *débris* are prominent features of the sediment. Dyspepsia and diarrhœa are not especially prominent features, nor is cachexia a common accompaniment. The liver and spleen are not enlarged, but anæmia is very pronounced. In amyloid disease the reverse of the above features prevail.

CYSTIC DISEASE OF THE KIDNEY.

This disease is met with in two forms: (*a*) As a congenital obstructive disease, usually associated with absence of the ureter, or other malformation interfering with escape of urine. (*b*) As a disease of adult life, and independent of the congenital form; and in many of its features allied to chronic interstitial nephritis, —indeed, by some authors considered a form of the latter.

Practically, the latter form only possesses a clinical interest to physicians, and to this form the following considerations apply :—

It is not uncommon to find cysts of considerable size in chronic contracting kidney as the result of distal constriction of the uriniferous tubes, which result in proximate dilatation by the urine. In the disease under consideration, however, the cystic formation, though undoubtedly the same in origin (tubular dilatations), yet it so greatly exceeds all other changes in the kidney that the organ increases in bulk sufficiently to entitle it to rank among abdominal tumors. The shape of the kidney is, in the main, retained, and the weight of the organ may reach from 2 to 16 pounds. The disease is almost uniformly bilateral. Dickinson noted but 1 case out of 26 in which the disease was confined to one kidney. Both the medulla and cortex of the organ are replaced by cysts varying in size from a pin-point to the size of grapes or walnuts, the larger ones being usually in the centre of the organ. They contain fluids which vary in color, some being pale straw-colored, deep-yellowish, purple, or bloody. In consistence the contents of the cysts may be serous, viscid, syrupy, caseous or almost solid, consisting of fat mole-

cnles, epithelinm, crystals of cholesterin, uric acid, and triple
phosphates. The cysts do not intercommunicate or terminate
with the large conducting tubes, calyces, or renal pelvis, but are
essentially closed cavities.

The Urine.—This somewhat resembles the urine in chronic
interstitial nephritis. In typical cases the urine is albnminous,
the quantity of albumin varying from 5 to 30 per cent. bulk
measure. The urine is pale in color when free from blood; of
low specific gravity, varying from 1.010 to 1.015, though it has
been noted as low as 1.005. Renal casts are usually found, nearly
always of the granular order and large size. The urine contains
blood at intervals, and sometimes in large quantities. In one of
the author's cases the hæmaturia was so severe and persistent for
months that the patient became blanched and anæmic to an extreme degree, notwithstanding absolute rest in bed and the use
of styptics.

Hæmaturia largely contributes toward exhaustion in many
of these cases. In most cases pus is present in the urine in
moderate amount. The quantity of urea is markedly reduced,
both relatively and absolutely. The phosphates are sometimes
increased in quantity, which is rarely the case in interstitial
nephritis. The chlorides and sulphates suffer but little change.
Triple-phosphate crystals are frequently noted in the sediment,
notably in late stages of the lesion, when, as is usual, more or
less cystitis is present.

Clinical Features.—The most prominent clinical features of
the disease, aside from the urine, are enlargement of the left
ventricle of the heart, without valvular disease, and increased
arterial tension, as shown by the sphygmograph. The skin is
pale and sallow, and cachexia is apparent. The patients are
always adults, mostly between the ages of 45 and 60 years.
Hæmaturia is prominent, recurrent, obstinate in character, and
often profuse. There is usually tumor in renal region ; bilateral,
though often unequal in size; soft, but non-fluctuant, and preserving the shape of the kidney. In late stages of the lesion
there are nausea, vomiting, headaches, suppression of the urine,
coma or convulsions, the latter being the most frequent cause
of death. Less frequently death results from exhaustion (through

renal hæmorrhage), bronchitis, pneumonia, or pulmonary œdema attended by severe dyspnœa.

Cystic disease of the kidney is distinguished from chronic interstitial nephritis by the non-fluctuant swelling in the sides, the recurrent and often severe hæmaturia, and the sallow, cachectic appearance of the patient. From cancer it is distinguished by the absence of pain and slower progress in cystic disease. In cancer there is rapid growth of the tumor, which is of nodular outline and unequal resistance. The age, in cancer, is usually either under 5 years or over 50, while in cystic kidney the most common age is from 40 to 55. Finally, the aspirating needle will determine if the tumor be cystic or of solid growth.

SECTION XI.

RENAL TUBERCULOSIS.

IN the light of recent facts and investigations, tuberculosis of the genito-urinary tract, through hetero-infection, must be considered rare, if indeed possible. Certainly, so far as the kidneys are concerned, it seems out of the question. While careless or unclean catheterization may cause infection of the prostate, involving the vesicula and extending to the epididymis, the anatomy of the urethra and the fact that tubercle bacilli do not multiply in the urine or possess in themselves any degree of motor power negative the view heretofore held by some, that infection may occur through coitus, or the infection reach the kidneys or upper urinary tract through the urine.

Tuberculosis of the kidney occurs in two distinct forms: (*a*) *acute miliary tuberculosis* and (*b*) *local caseating tuberculosis* or "*scrofulous kidney.*" The miliary form is mostly met with in children under 10 years of age. It is pretty uniformly bilateral, although the organs often differ in degree of infiltration.

The caseating or scrofulous kidney is most frequent in young and middle adult life, although it may be met with late in life, and it is rare under 10 years of age. Scrofulous kidney is nearly as often unilateral as it is bilateral. Of the two forms of renal tuberculosis the miliary form is about twice as frequently met with as is the caseating or scrofulous kidney. Miliary tuberculosis is always associated with tuberculosis in other parts of the organism, most often phthisis pulmonalis, tubercular meningitis, and tabes mesenterica. This form rarely gives rise to distinctive symptoms, being merged for the most part into general tuberculosis, and, therefore, scrofulous kidney only deserves special consideration in this connection.

Chronic localized tuberculosis of the kidney, renal phthisis, tubercular pyelitis, tuberculous pyelonephritis, or scrofulous

(294)

kidney, as it is severally known, begins usually at the papillary apices, in the calyces, or renal pelvis, and from thence by the blood- and lymph- channels it extends to the kidney proper. Deposits of cheesy matter infiltrate the renal papillæ, and in the course of a few weeks these form irregular, softened areas, which by progressive infiltration spread deeply inward, involving the parenchyma of the kidney. The organ becomes, in consequence, enlarged and lobulated. The renal pelvis and ureter, on the other hand, become contracted in consequence of thickening of the mucosa, and later on the ureter often becomes choked or blocked by softened and caseous masses detached from ulcerated caseating surfaces above. At the same time the tubercular nodules within the kidney, after reaching considerable size, undergo necrotic changes and break down, forming irregular, rudely-globular cavities. These, as they enlarge, become pyriform in shape and extend until they coalesce, and at length the whole medulla and most of the cortex become involved. The destructive process continuing, the contents of the renal cavities at length burst into the renal pelvis and the whole organ becomes practically an abscess-cavity. If the ureter be pervious the urine washes down the *débris*, which presents characteristics soon to be considered in detail. Should the ureter, however, become permanently blocked, dilatation and sacculation of the kidney result,—practically a tubercular pyonephrosis. In case one kidney remain uninvolved the diseased organ either becomes an hydronephrotic cyst or a shrunken, " putty-like " mass ; in both cases little, if any, of the secreting tissue proper can ultimately be found. The disease is usually a chronic one in character, and during its course neighboring organs are often involved by direct extension through the capsule of the kidney, more especially the liver and spleen.

The Urine.—Polyuria from tubercular irritation is probably the earliest urinary change. The quantity of urine is increased and the calls to micturate are more frequent than normal. Traces of albumin and a few blood-corpuscles are usually to be found also before destructive ulceration sets in. The urine is usually pale and murky in appearance, of somewhat lowered specific gravity, and of acid reaction.

When ulcerative changes set in the urine presents the following changes more or less marked, according to the degree of degeneration in progress. The urine is of a pale, milky color, the transparency is diminished, the specific gravity is below normal, and the reaction is, as a rule, alkaline. The urine contains pus in gradually-increasing quantity. The pus is less variable in quantity from day to day (unless the ureter becomes blocked) than in most other forms of pyuria. The pus imparts to the urine a more or less pronounced milky appearance, which does not completely subside as in most other conditions of pyuria, but much of the pus remains in suspension even after long standing; this is very significant of renal tuberculosis. The urine contains blood in rather more than 25 per cent. of the cases. If the blood come from the renal pelvis it is usually small in quantity, often unappreciable to the naked eye. Some, times very marked hæmaturia occurs, usually at intervals; and this denotes ulcerative changes within the renal parenchyma. As the disease becomes advanced the urine usually becomes ammoniacal and markedly offensive, contains ropy mucus, and deposits triple phosphates, small caseous masses, and renal *débris*. The urine is albuminous, sometimes highly so; always in excess of the ratio, due to contained pus and blood.

The bacillus tuberculosis of Koch is present in the urine in most cases after necrotic changes set in, and its discovery is diagnostic. The tubercle bacillus can often be demonstrated in the urine by the methods already considered, though this is by no means so easy as in the sputum, owing to the fact that they are relatively few and scattered in the average sample of urine. It is safer, therefore, when suspected, but not found by direct examination, to resort to cultures in gelatin after the usual manner and the inoculation of animals.

Leading Clinical Features.—The leading clinical features of renal tuberculosis are concisely as follow : Polyuria and dysuria, the latter prominent and progressive. The bladder will not tolerate the urine, and is only free from pain or distress when empty. Pain begins about the middle of the act of micturition and continues to the close, but, as a rule, not after. Some pain is usual in the renal region, accompanied by tenderness upon deep pressure. The evening temperature is usually from 2 to 4 degrees higher than normal, or sometimes periods of fever occur, lasting for several days, followed by periods of remission. These patients suffer from profuse night-sweats, loss of appetite, debility, and emaciation in the course of the lesion; while cough and diarrhœa are scarcely less frequent accompanying features of the disease. Uræmic complications are rare.

From calculus the distinction hinges on the slow development and irregular degree of pyuria, which is usually preceded by slight hæmaturia in calculus. The general nutrition is better preserved than in tuberculosis. The pain on micturition is relieved at the close in tuberculosis, while in stone it is increased. Absence of temperature and constitutional symptoms characterize the progress of calculous disease, while they are prominent in tuberculosis.

RENAL CANCER.

Primary cancer of the kidney may appear in the form of carcinoma, sarcoma, and rarely as lymphadenoma. Sarcoma is most frequently met with in childhood, while carcinoma is mostly met with after 40 years of age, the encephaloid being the most frequent variety of the latter, though occasionally the melanotic growth is met with. Primary cancer usually attacks but one kidney; only exceptionally is it bilateral. Of 59 cases collected by Ebstein, 31 involved the right kidney, 23 the left, and 5 both kidneys. Encephaloid cancer of the kidney sometimes attains an enormous size,—14 to 56 pounds.

Renal cancer is more frequent in males than in females. Though slow of development, often remaining quiescent for years, when once it has begun it rapidly progresses toward a fatal termination,—usually within a year or two. The disease begins in the fibrous stroma of the cortex or in the tubular epithelium; sometimes, however, primary invasion begins in the lymphatics about the hilum. Wherever the primary lesion begins, the whole organ eventually becomes infiltrated. There is usually appreciable tumor after the disease becomes thorougly established.

The Urine.—The most prominent feature of the urine in cancer is hæmaturia. In carcinoma this is especially pronounced, often

20

recurring and frequently uncontrollable. In the other forms of cancer hæmaturia may sometimes be absent throughout. Hæmaturia is variable, occurring sometimes early, and subsiding as the disease progresses, when it may disappear and not return. Again, the disease may become advanced before blood appears in the urine. As a rule, when palpable tumor is present there is hæmaturia. The hæmaturia is somewhat characteristic in its irregular intermittency, appearing and disappearing at intervals without apparent cause; while often profuse, it is yet rarely so excessive as to rapidly produce anæmia or exhaustion. Albumin is usually present in the urine in small quantity; always present if there be blood. The quantity of urine is usually increased unless the ureters become blocked by blood-clots. Pus is present but in small quantity, save in advanced cases, attended by decided destructive changes in the kidney; and even then the quantity of pus is remarkably small in amount, considering the extent of necrotic changes in progress.

In carcinoma the urine frequently contains acetone, even before there is advanced emaciation. Frequent micturition is the rule, and it may be so pronounced as to call attention chiefly to the bladder when only the kidney is involved. The presence of organized substances in the urine, such as epithelium, casts, etc., are of little diagnostic value in renal cancer. Cancer-cells are not recognizable in the urine in this disease, and those alleged to have been found were doubtless transitional epithelium, which is often present in considerable quantity if malignant disease invade the renal pelvis. The only significant feature in this connection would be the discovery of particles of the morbid growth with distinct alveolar structure in the urine, but in malignant disease limited to the parenchyma of the kidney this is practically unknown.

Leading Clinical Features.—Increasing tumor is the almost invariable rule, which is to be looked for in the anterior lumbar region, between the costal arch and the crest of the ilium. If large the tumor approaches the umbilicus, extending upward and downward into hypochondrium and iliac, and even to the inguinal regions. The tumor is usually lobulated or presents obtuse margins; the lobulations often possess unequal degrees of hardness.

The tumor is nearly always fixed. Pain occurs early, usually persistent in character, though sometimes intermittent. It is most severe in the affected kidney, but may be reflected to neighboring parts. The character of the pain is dull and aching, often paroxysmally increased, but not greatly aggravated by body movements. At first the pain is usually slight or even vague, and at times absent, becoming again severe and prolonged. From dull in the beginning it may later on become lancinating, either spontaneously or evoked by pressure, but not by movements. When tumor becomes large and presses upon the larger trunk-nerves, pain often extends to the chest, across median line, and downward to the hips and limbs, simulating sciatica. The pressure exerted by tumor when the latter is large often causes œdema of the feet and legs, ascites, and prominence of the superficial abdominal veins, as well as constipation, disturbances of the stomach, and icterus. The constitutional symptoms include emaciation, anæmia, cachectic appearance,—browning or sallowing of the skin,—failure of strength and vitality. Uræmia is rarely, if ever, present, unless nephritis co-exist. Accidental or complicating features are sometimes added; such as paraplegia from spinal pressure, vesical paralysis with retention of urine, asthmatic or laryngeal dyspnœa, and spasmodic cough from extension of the pathological process.

Differential Features.—From hepatic tumors cancer of the kidney differs as follows: The former have no intestine in front of them; the dullness upon percussion is uniform throughout. Renal tumors lie in part behind ascending colon, which passes obliquely up and to the left, giving clear note of percussion on lower and inner margin. *Splenic enlargement* presents rigid, thin borders instead of round and lobulated ones, and, moreover, as a rule, splenic tumor has more mobility, and deep percussion often elicits intestinal resonance through its substance. There is usually antecedent history of ague or intermittent fever, leucocythæmia, etc., in splenic enlargement.

RENAL CALCULUS.

Calculi may originate in the secreting structure of the kidney —usually in the tubules—or in the renal calyx, but their development is most common within the renal pelvis; although this sometimes takes place in the infundibula, calyces, or even in the dilated tubules, which form cavities for their location in the parenchyma of the organ.

Renal calculus is usually unilateral, though many exceptions occur to this general rule. The calculus when large is usually single, the smaller ones being more apt to be multiple.

Renal calculus occurs at all ages, including intra-uterine life. It is, however, most common before 15 and after 50 years of age. In young people and children calculus is most frequent among the poor, while calculus in advancing life is most common in people of comfortable circumstances and luxurious habits. As a rule, the calculus in infancy is of the ammonium-urate variety; that in young adults, uric acid; that after 40 years of age is usually calcium oxalate.

The Urine.—Blood appears in the urine in the vast majority of cases of renal calculus, and presents the following features: Hæmaturia not profuse, but appears in repeated paroxysms; *increased by exercise.* The blood is intimately mingled with the urine; is not bright in color, but smoky-brownish or porter-colored. The volume of the urine is not increased in uncomplicated renal calculus, but is rather diminished. The urine is usually sharply acid and of high color. The above are usually the early features of the disease, before pyelitis begins. After pyelitis is established the urine undergoes the following changes: Pus and mucus appear in the urine in greater or less quantities. More or less frequency of micturition is present, and the act is accompanied by uneasiness,—sometimes amounting to pain. This may be so pronounced as to lead to an impression that cystitis exists. The deposits in the urine are significant, but care should be taken to secure only primary deposits,—not those due to decomposition changes. Centrifugal sedimentation of the urine is the only trustworthy method of securing this. Urates and oxalates are often observed in the sediment, the former frequently in quantity. With the advent of pyelitis, more or less phosphatic deposit is to be found in the urine. Epithelium in greater or less quantity is found in the sediment in advanced renal calculus, most often the angular and spindle form. Lastly, the appearance of small-sized concretions in the urine often furnish diagnostic data of the highest value.

Leading Clinical Features.—These consist of dull aching pain situated deeply in the loin, usually unilateral, and often radiating along the ureter toward the testicle, down the thigh, and sometimes extending as far as the foot. The pain may be sharp and lancinating at times; intensely severe paroxysms are occasional features (renal colic), lasting a few hours and then suddenly subsiding. The ordinary pain of renal calculus is invariably increased by exercise, either walking or riding; it is, therefore, more marked in the evening than in the morning. Tenderness upon deep pressure anteriorly is to be found, especially if the calculus has excited inflammation. Gastric disturbances are common, including nausea, vomiting, and periods of more or less disorder of digestion,—acidity, flatulence, etc.

Differential Features.—The early stages of renal tuberculosis are most likely to be confounded with renal calculus. In renal tuberculosis there is usually tuberculous history in the family, and often tuberculosis may be found elsewhere in the patient, as in the joints and glands, and the age is usually from 20 to 40. Polyuria is not prominent, and renal colic is rare save late, and then less severe, usually not causing retraction of the testicle. Hæmaturia is more persistent, but not so much influenced by movements or exercise. The urine is cloudy from the beginning, of low specific gravity, depositing more pus, and albuminuria is early and more pronounced. Tubercle bacilli usually are present in the urine, and inoculation of animals with urine deposit induces tuberculosis. General symptoms are prominent, such as anæmia, emaciation, weakness, rapid pulse, and evening temperature, with night-sweats. These features are, for the most part, absent in stone.

From malignant disease the distinction rests upon the more decided hæmaturia of the former, which often results in anæmia. Hæmorrhage is uninfluenced by movement, and therefore it occurs at night as well as by day.

RENAL EMBOLISM.

Renal embolism consists of an impacted thrombus which has formed in some part of the circulatory system,—usually in the heart,—and is carried by the blood-current to the kidney, where it blocks one of the renal vessels. Recent endocarditis furnishes the most frequent source of renal embolism. As the fibrinous clots accumulate upon the cardiac valves, sooner or later they become detached in whole or in part, and those from the left side of the heart are liable to be carried by the blood-current into either kidney. The anatomical results of embolism are very constant and striking, and in few organs are they more often noticed at the autopsy than in the kidneys, though they are not so frequently recognized during life as in some other locations, notably in the lungs or brain.

The Urine.—Changes in the urine are very striking in renal embolism, although they are not wide-spread. Sudden and pronounced albuminuria is almost invariable. Albumin may appear in the urine in a few hours after the attack, but sometimes not for twenty-four hours or so after impaction. The albuminuria is marked in degree for from two to five days; it then gradually diminishes and completely disappears, or leaves only mere traces after two, three, or four weeks. The specific gravity of the urine is decidedly increased in the beginning, often reaching 1030 to 1035; it gradually lowers from day to day, and the normal range is reached after a few days or weeks. The quantity of urine is decidedly diminished at first, the color is dark brown, and the reaction is sharply acid. Blood is usually present from the beginning in variable, but rarely excessive quantity. Some degree of pyuria is to be noted, but this is rarely pronounced.

The urine contains casts in this lesion, often in considerable numbers. At first they are mostly hyaline; later on epithelial casts appear, as well as those with pus-corpuscles attached to them. After the first five to seven days the casts become less numerous, they are mostly hyaline, and at length they disappear from the urine. It will be seen from the above features that the urinary changes begin abruptly, but that the urine progressively and steadily resumes its normal characteristics, and, in from a few days to three or four weeks, all manifestations of urinary disturbance pass away. The block, if aseptic, remains in the kidney, but ceases to be a source of renal irritation.

Leading Clinical Features.—A previous history of endocarditis is the rule. The impaction is followed at once by sudden pain in the renal region, often accompanied by chills, which latter may be repeated, and some irregular action of the heart, with sense of præcordial oppression, or clogging, and frequently dyspnœa. Some slight elevation of temperature is usual. If the pain be very severe vomiting is usual, and even some degree of collapse may follow.

URÆMIA.

The intimate toxic character of uræmia as yet remains an unsolved problem, only known to us by its results. Our knowledge of this matter, therefore, is inexact as yet, and remains for the chemico-physiologist to unravel. In the light of our present knowledge of this subject, the only conclusions that seem justifiable are as follow: (a) All theories attempting to explain the cause of uræmia through the action of any single product or toxin must be abandoned as fallacious, as the toxin is undoubtedly multiple. (b) The general cause of uræmia lies in the failure of the kidneys to excrete the urine in part or in whole, and that the urine or its primary elements, as retention products, act as direct toxins upon the organism, evoking the symptoms termed uræmic. (c) That the most successful attempt at isolation of these products of the urine to date we owe to the investigations of Bouchard. (See Section V, page 143.)

Uræmia may appear as an acute and rapidly overwhelming toxicosis, causing coma, convulsions, and death within a few hours; or it may linger for weeks or for months as a milder form of toxic disturbance, with symptoms such as somnolence, restlessness, headaches, nausea, attacks of diarrhœa, dyspnœa, visual disorders, and general disturbance of nutrition.

The Urine.—This furnishes the key to the diagnosis of uræmia with great uniformity. The essential feature of the urine in uræmia is diminution of the absolute amount of solids, but more especially of urea. The quantity of urea excreted, instead of being 500 grains for an average body-weight, becomes reduced to 200 or 100, and even less than 50 grains in some cases. As a rule the activity of the symptoms bear an inverse ratio to the quantity of urea excreted, and therefore, in those cases attended by extreme diminution of the excretion of urea, the symptoms are sure to be pronounced and threatening so long as this continues. The uric acid, chlorides, phosphates, and sulphates of the urine also suffer marked reduction in uræmia, but, with the exception of uric acid, these are probably of no special significance. As a rule the volume of the urine is diminished, and the degree of diminution varies through all degrees up to complete suppression. The specific gravity of the urine, notwithstanding the decreased volume, is also decreased, and sometimes markedly so, descending frequently to 1.008 or below. Exceptions to this rule are noted sometimes in acute diffuse nephritis, when the volume of urine is reduced to a few ounces, the urea being still decreased both relatively and absolutely; but the febrile condition accompanying the acute nephritis causes some increase in the other solids, which proportionally become excessive, and thus raise the specific gravity of the urine sometimes even to the normal range.

With regard to the morbid constituents of the urine in uræmia, albumin is the most constant feature, and is present in all grades, from mere traces up to 2 or 3 per cent. by actual weight, depending upon the nature of the associated lesion. It should be borne in mind, however, that while albumin is usually present in the urine in uræmia, exceptional cases occur in which it is said to be absent; though this is rare. Even in those cases of chronic interstitial nephritis characterized by absence of albumin in the urine, the exciting cause of a uræmic attack, especially if acute, is apt to be of sufficient congestive character to cause, at least, mild albuminuria. It may, therefore, be repeated that active uræmia is *extremely rare* without accompanying albuminuria.

The urinary sediment in uræmia includes a very wide range of morbid products with no very constant associated features. We may have casts, epithelium, pus, blood, bacteria, together with crystalline or amorphous deposits of urates, phosphates, oxalates, etc. The only products that may be considered at all constant are renal casts, which are rarely—perhaps never—absent in

uræmia. Sometimes they are extremely sparse and may be overlooked without due care; notably so in chronic interstital lesions
of the kidney, in which their form is often limited to the small,
perfectly-clear, non-refracting variety of casts, which are con-
fessedly difficult to find; but failure to find them does not, without every precaution, prove their absence. With our improved
methods casts should be found when present. The nature and
number of renal casts will depend upon, and correspond to, the
character of the renal lesion present, which need not be repeated
here. For full consideration of this subject consult Section
VII, page 189.

Corresponding Clinical Features.—Among the milder symptoms of uræmia may be mentioned dyspepsia, flatulence, nausea,
occasional diarrhœa, neuralgia or headaches, vertigo, dyspnœa of
an asthmatic type, bronchial catarrh, and various nervous disturbances, such as insomnia, restlessness, mental depression,
numbness of certain parts of the body, drowsiness, and certain
visual disorders.

The more pronounced symptoms include severe headache,
usually frontal; vomiting, extreme nervousness, twitching of the
muscles, drowsiness; pulse increased to 100 or over, usually hard
and tense; temperature lowered unless some inflammatory action
be associated; tongue coated with dry, brown fur; breath foul
(uræmic), and often more or less profound stupor or coma, or
convulsions, or both.

Differential Features.—Uræmic coma may be mistaken for a
variety of conditions, notably apoplexy, epilepsy, alcoholic or
opium narcosis.

Uræmic coma may be known by the following features: The
subjects are usually young or middle-aged; previous attacks are
unlikely; renal lesions are present in some form; appearance of
the patient is pallid, sometimes cachectic; pulse increased to
100 or over; the pupils tend to dilate; the respirations may or
may not be hastened; breathing is stertorous and *labial;* unconsciousness is not complete, the patient may be partly aroused
by efforts; a peculiar odor of breath is present (uræmic); the
convulsions are of recurrent order, and, as a rule, albuminuria is
present.

Apoplexy is differentiated as follows: The age of the patient is nearly always past medium life, often advanced; previous attacks unlikely; heredity marked; granular kidney often associated; appearance of patient normal; pulse slow, full, 60 per minute or under; pupils unequal; respirations slow, stertorous, and *deeply guttural;* insensibility complete and profound; patient cannot be aroused; hemiplegia is present.

Epilepsy is most common under 30 years of age; previous attacks are the rule; the appearance of the patient is dusky, purple, gradually becoming paler; pulse slightly accelerated, small, feeble, and dicrotic; temperature normal, or a degree or so above; pupils normal; respirations stertorous, guttural, unsteady; unconsciousness is not complete, coma of brief duration; great muscular relaxation present.

Alcoholism is common to all adult ages; previous attacks are the rule. The features are suffused and bloated, the lips livid, and the expression vacant; pulse frequent, small, and feeble; temperature slightly lowered, pupils dilated; respirations are deep and slow; stertor is intermittent; breath is alcoholic; vomiting is common; the conjunctivæ are injected; the features are swollen, and the subject can usually be partly aroused.

Opium coma is most frequent in the young, and may be habitual or accidental. The features are shrunken, pallid, cyanotic; expression is ghastly; pulse usually slow and feeble; temperature not increased, rather lowered; pupils contracted; respiration slow, shallow, and feeble, and opium may be detected in the breath.

HÆMOGLOBINURIA.

Hæmoglobinuria constitutes a condition characterized by the escape of the blood coloring elements by way of the urine, very little, if any, of the corpuscular elements of the blood accompanying the pigmentary elements. From whatever general source it may originate, it is primarily due to dissolution of the red corpuscles of the blood, which permits the coloring matters to escape in solution. As an occasional phenomenon it may be met with in the course of certain infectious diseases, extensive burns, and in various forms of poisoning. In addition to this it occurs as an idiopathic disease of intermittent character, and to such the following considerations are intended especially to apply :—

This disease is most common in males,—three or four to one. It occurs at all ages, from 3 to 52 years, but most often between 20 and 50. Malaria seems to be the most prominent historical feature of these patients, while cold is undoubtedly the most frequent exciting cause.

The Urine.—The appearance of the urine in the intervals between the attacks is perfectly normal; but with the attack its appearance becomes at once strikingly altered, apparently bloody. The color assumes a dark-red, port-wine, or porter color, and is somewhat turbid or smoky in appearance, and deposits, upon standing, an abundant chocolate-like sediment. The specific gravity of the urine varies from 1.015 to 1.030, the average range being slightly above normal,—1.023 to 1.025. The reaction of the urine may be acid or faintly alkaline, and the volume is somewhat above normal. In most cases the quantity of urea is increased. The urine gives a highly albuminous reaction, and further testing shows the presence of globulin.

The urinary sediment is chiefly made up of amorphous granular matter,—doubtless disorganized blood-corpuscles,—in which are often to be seen minute crystals of hæmatin. Casts are usually present, chiefly dark, granulated ones, though often, also, hyaline casts may be found. Many of the casts are made up of hæmoglobin. Calcium-oxalate crystals are usually present, and less frequently are uric-acid crystals found. Blood-corpuscles are either absent or only a few scattering ones are to be found. The urine gives the characteristic blood reaction with guaiacum and ozonic ether ; even the interparoxysmal urine often shows the blue reaction. The spectroscopical examination of the urine shows the two absorption bands between Frauenhofer's *D* and *E* lines characteristic of oxyhæmoglobin. Renal epithelium is often seen in the sediment, sometimes deeply stained by the blood-pigment. Amorphous urates are usually present in abundance, especially as the attack is subsiding. The chlorides of the urine are usually deficient, the phosphates and sulphates in excess, and so-called indican is not infrequently present in considerable excess.

Prominent Clinical Features.—The symptoms of idiopathic hæmoglobinuria are distinctly paroxysmal, beginning with chill, —sometimes continued rigors for an hour or more,—which are usually due to previous exposure to cold. The exposure, however, does not cause the disease, but merely provokes the paroxysm, as is proved by the fact that so long as the patient remains warm he continues free from symptoms. The chill is accompanied or followed by retching, and often vomiting, as well as pain in the back and limbs; often with retraction of the testes. General malaise succeeds with yawning and stretching. Sometimes tenderness in the renal region is to be elicited upon deep pressure. Thirst, headache, and drowsiness are frequent features, and the skin sometimes becomes jaundiced. In from half an hour to two hours the patient voids more or less port-wine-colored urine, The urine retains this abnormal color for two or three passages; the whole attack usually being completed in twenty-four hours or less time; more rarely it may continue for several days. The attack is often succeeded by griping pain in the umbilical region. and more or less pallor and weakness succeed the attack for a day or two. Urticaria is an occasional accompaniment of the disease.

The temperature is lowered during the cold stage (96° F.), but often rises above normal when the chills subside. After the attack the patient remains apparently well for a longer or shorter time,—it may be for months,—until again exposed to cold. Nephritis is not an infrequent result; protracted cases are characterized by repeated paroxysms.

CHYLURIA.

This disease usually arises in consequence of some lesion of the lymphatic system, whereby the chyle is diverted from the natural channels into some part of the urinary tract. As an idiopathic disease chyluria has heretofore been almost exclusively confined to the tropics, or to those who have spent much of their lives there. As such it depends upon the invasion of the blood and urinary tract by a parasite,—the *Filaria sanguinis hominis,*—as first pointed out by Dr. Lewis, of Bengal, and already described and illustrated (page 210). Besides the endemic form, the disease is occasionally met with in people who have never lived in the tropics, and, when thus occurring, it may be considered an accidental condition, brought about by traumatisms or diseases which have established communications between the lymph-channels and the urinary passages. The accidental form of the disease is comparatively rare; at least nine-tenths of the cases met with, even in temperate climates, are the result of infection in the tropics of people who previously there resided and contracted the disease.

The endemic form prevails widely in the tropics, including especially India, China, the West Indies,—notably in Barbados, Trinidad, and Demarara,—also Cuba, Bermuda, Brazil, Mauritius, the Isle of Bourbon, and South Australia. The disease attacks indifferently both natives and foreigners, males and females, and shows but little preferences as to age of the subject. The peculiar and interesting nature and habits of the parasite which causes this disease have already been fully described (p. 211).

The Urine.—The peculiar condition of the urine in chyluria furnishes the key to the recognition of the disease. The appearance of the urine is characteristically milky, and it so remains upon standing for days without settling, in consequence of the finely molecular division of the contained fat, thus permitting it to remain in suspension. It is unusual for oil-globules of any size to be found in the urine in this lesion; indeed, the emulsion is so complete that only minute granular matter is seen. Sometimes, upon standing, the fat rises to the surface of the urine and collects in cream-like flakes. The quantity of fat found in the urine in chyluria varies greatly, depending largely upon the quantity and quality of the food taken; the urine of digestion (after food) is richest in fat, while that of fasting contains the least.

If the urine be shaken with ether the fat is dissolved and the urine assumes its normal color and appearance. In addition to fat, chylous urine usually contains blood in sufficient quantity to impart a very noticeable pink color to the fluid. The pink tint is fainter than would be expected in proportion to the quantity of blood actually present, the opacity caused by the fat greatly obscuring the coloration due to the blood-corpuscles. Upon standing, however, the contrast becomes striking; the bright sediment of precipitated blood is then plainly visible. This precipitated blood-clot becomes more pronouncedly pink upon exposure to the atmosphere, as first pointed out by Dr. Vandyke Carter.

A notable characteristic of chylous urine is its tendency to spontaneous coagulation upon standing. If the urine be at all rich in fibrin, shortly after it is voided it will coagulate into a firm, vibrating, jelly-like mass resembling corn-starch *blancmange.* Unfortunately, sometimes coagulation takes place in the urinary channels, notably the bladder, and may give rise to most distressing symptoms until it be dissolved or broken up and removed. The clots which form after the urine is voided often become very firm, and long retain the form of the vessel in which the urine stood; if in bottles they may even have to be broken in order to remove the coagulum. The coagulation of chylous urine depends directly upon the fact of the almost constant presence of fibrin in the urine, although the quantity varies considerably. At times it is insufficient to cause coagulation. The quantity of fibrin present is usually in inverse ratio to the amount of contained molecular fat.

The uniform presence of albumin in the urine is attested by the constant coagulation of the urine by heat or other albumin precipitants. Corpuscular elements are sometimes present in the urine, besides red blood-cells resembling lymph-cells, as well as large oval and rounded cells which microscopically and chemically evince the characteristics of epithelium. The urine is usually devoid of renal casts unless nephritis be excited by the disease; and since the urine always contains fibrin, this would indicate that the lesions are not situated in the kidneys, but rather in the conducting channels of the urine. Filaria are sometimes found in the urine, especially of the tropical form of the disease, if sought for in the night urine. In urine excreted during the day they are rarely to be found, owing to the fact that the parasite is quiescent during the day. Pus-cells are more or less numerous in the sediment.

The solids of the urine suffer some reduction owing to the

drain upon the elements which go to furnish nutrition. The specific gravity is lowered to a moderate, rarely to an extreme, degree,—1.016 to 1.010. The urea, chlorides, and sulphates are usually deficient, especially the two former, while the phosphates are often considerably increased. It is common to find a considerable deposit of uric-acid crystals in the freshly-voided urine upon cooling.

Leading Clinical Features.—The clinical symptoms are rather negative; dropsy, uræmia, and frequent micturition being absent. There is usually an indefinite dragging pain in the back and loins, especially preceding the attacks. Anæmia becomes more or less marked according to the extent and continuance of the drain. Loss of strength and depression are prominent features during the escape of chyle; these, however, are at once relieved if this cease. Tuberculosis often becomes a complicating feature of very chronic cases. The disease pursues an intermittent course,—especially so the tropical endemic,—due to successive ruptures of lymphatics; the accidental form is more uniform in its course. The duration of the disease is indefinite, but always chronic, lasting from ten to forty-seven years.

DIABETES INSIPIDUS.

This disease has been variously designated under the terms *diuresis, polyuria, polydipsia,* and *hydruria*. Little or nothing definite is known as to the pathology of the disease; it is not improbable that it is caused by a number of different morbid conditions. The disease is much more frequent in males than in females. It may appear at any age, but in the majority of cases the disease appears between 5 and 30 years of age. From the number of alleged causes of the disease by various authors, it is very evident that nothing definite is known of the etiology, save in those cases that can be distinctly traced to traumatisms, intracranial growths, or other lesions of the nervous system.

The Urine.—The chief features of the urine are : enormous increase of volume, lowered specific gravity, and absence of both sugar and albumin. The daily volume of urine not infrequently reaches from 15 to 40 pints. The urine is pale in color, almost, in fact, watery in appearance, and the specific gravity ranges from 1.002 or 1.003 to 1.007. The reaction is feebly acid or neutral. Upon standing, the urine soon becomes ammoniacal and turbid from precipitation of earthy phosphates, and gives a rather offensive, fish-like odor. The urea, while proportionately reduced, is, in fact, absolutely increased considerably above the normal range. Uric acid is apparently greatly deficient, and it is even claimed to be often absent. The increase of urea and deficiency of uric acid favor the presumption that the latter undergoes conversion into the former. The chlorides, phosphates, and sulphates are more or less increased, more especially the phosphates, which sometimes become greatly excessive. Albumin is usually absent from the urine, although in protracted cases it is often present in small quantities. Inosite is frequently present in the urine, as Strauss claims, merely as the result of irrigation of the tissues, since he succeeded in producing the same condition, experimentally, upon subjects by administering copious draughts of water.

Prominent Clinical Features.—The most prominent symptoms of this disease are as follow: Inordinate, constantly-recurring thirst, which is only briefly quenched by copious draughts of water. Less constantly the appetite is increased. These patients are sensitive to cold and are easily chilled, the temperature being somewhat lowered. The tongue is dry, and more or less discomfort is experienced in the stomach; pain and diarrhœa are often present. The skin is dry, pinched, and dusky. The patient becomes spare and weak, though, exceptionally, fair strength and health is maintained for years. In late stages of the disease, œdema of the lower extremities sometimes appears.

DIABETES MELLITUS.

Saccharine diabetes constitutes a perverted state of the elaborative functions in which certain elements which go to make up nutrition—notably starches and sugars—fail to reach their normal destinations in the economy. The symptoms evoked are partly due to lack of nutrition and partly to the damaging effects of the waste products (chiefly sugar) upon the tissues. The direct cause of the disease is an impaired functional capacity of the liver in its glycogenic relations. This may, however, be induced through impaired nervous influences, which may be central or reflected. In addition, some complemental relationship exists between the functions of the liver and pancreas, which often permits lesions of the latter organ to evoke the phenomena of diabetes mellitus. The precise nature of this relationship is unknown.

Something over 30 per cent. of the cases can be traced hereditarily. The disease is notably frequent among Hebrews. It attacks males twice more frequently than females. It occurs most often between the ages of 25 to 65, and is infrequent at the two extremes of life. In young people under 30 years of age the disease is almost uniformly progressive toward a fatal issue in from a few months to four or five years. If the disease do not appear until between 40 and 50 years of age, it is often more amenable to treatment; after 50 it may usually be held in control by proper management. The disease is more severe in spare than in stout subjects, at all ages. In the young death is most frequent from diabetic coma, while in those advanced in life the end is often reached through cardiac degeneration, gangrene, or exhaustion.

The Urine.—The physical characters of the urine are characteristically altered in typical saccharine diabetes. The urine is light in color and of a greenish, rather than yellowish, tint. It remains perfectly transparent and froths much if poured from one vessel into another. The specific gravity is markedly increased, ranging from 1.030 to 1.045, or even higher. The reaction is sharply acid, and it long remains so upon standing. The quantity of urine is greatly increased, the increase being usually in direct ratio to the quantity of contained sugar. From 6 to 12 pints of urine are often voided in twenty-four hours, but in severe cases 25 to 40 pints are sometimes excreted.

The most characteristic feature of the urine is the presence of sugar, which forms the index to the disease. The quantity of sugar varies from 1 to 8 per cent., averaging perhaps 4 or 5 per cent. One and a half to two pounds of sugar per day constitute the highest range of sugar excreted in the more extreme cases, while half a pound is not uncommonly excreted in ordinary cases. The quantity of urea is markedly increased. A

21

marked decrease in the quantity of urea may be considered as an unfavorable indication. The uric acid is usually deficient, often reaching but half the normal amount or even less. Notwithstanding the above fact, uric-acid crystals are frequently deposited from freshly-voided diabetic urine; the deficiency of coloring matters and disproportion of salines permitting it to fall out of solution. The sulphates of the urine are not materially altered in quantity, probably because of the large amount of animal food usually eaten in these cases. The gross chlorides, like the sulphates, remain essentially unaltered in quantity, though often varying considerably from day to day. The phosphates vary greatly according to the quantity and quality of food taken, but, on the whole, the tendency is toward increase.

The urine often contains, in the advanced stages, acetone or an acetone-yielding substance. Diacetone is occasionally present in the urine, but only in serious cases, and it is usually the index of approaching *diabetic coma*. Albumin is often present in small quantity in chronic cases. It may be due to co-existing nephritis, but more often to disturbance of the renal circulation, impaired nutrition of the renal epithelium, or degeneration of the parenchyma of the kidneys.

Prominent Clinical Features.—The most prominent symptoms are: thirst, polyuria, lowered temperature, hunger, weakness, emaciation, and nervous disorders. The thirst is constant, and seemingly unquenchable in character. Although the amount of water consumed is sometimes enormous, the mouth and throat remain dry and parched. The appetite is always increased, at first sometimes inordinate, and but little appeased by food. As a result the stomach sooner or later becomes disordered under the strain of constant overloading, so that in late stages of the disease the appetite fails, and dyspepsia follows. Constipation and attacks of diarrhœa are common. The mouth, tongue, and fauces become intensely red and congested; the gums become tender and shrunken so that the teeth sometimes loosen. The temperature is lowered,—96° to 97° F. being common, but even a temperature of 93° F. has been observed. Chilly sensations are frequent; so that these patients instinctively seek the fire and require extra clothing. Colds are excited upon slight exposures.

Periods of somnolence are common, and various nervous manifestations appear, as neuralgia, cutaneous hyperæsthesia, sensations of abnormal heat of skin, or sudden spells of perspiration.
These patients become irritable, fretful, uneasy, vacillating, and
the mind deteriorates somewhat. The sexual power declines
or is completely lost. The skin is dry, harsh, unperspirable,
wrinkled, and loose, causing an early aged appearance. Emaciation progresses sometimes with rapidity ; the muscles feel weak
and tired, so that movements become laborious and exhausting,—
these patients do not care to exercise. The pronounced and persistent polyuria produces frequent micturition, which harasses
the patient both day and night. Tuberculosis sometimes sets in
in the late stages of the disease, often, however, preceded by
bronchitis or localized pneumonia. Gangrene of the extremities
is common in aged subjects. Cardiac enlargement, high-tension
pulse, and degenerative changes of the heart often supervene in
long-standing cases.

Finally, gastric pain, dyspnœa, and more or less drowsiness
announce the approach of *diabetic coma*, which quickly terminates life.

URINARY FEVER.

Various names have been applied to this disease, such as
*urethral fever, catheter fever, urinary fever, shock, urinary
poisoning, uræmic poisoning, urinary infection,*—names which
suggest the various and conflicting views held both of the
etiology and pathology of this condition, which, indeed, still
remain unharmonized.

The term " urinary fever " was first employed by Guyon to
denote the febrile disturbance and accompanying phenomena set
up by instrumentation of the urethra or bladder, or by operations upon the urinary organs, or by impressions upon the
urethra or bladder by other means. The morbid phenomena
evoked by instrumentation of the urethra and bladder may
become so wide-spread as to include septic inflammations of the
renal pelvis, the kidney itself, and even pyæmia ; or it may bring
about acute uræmia, with its attendant consequences, often ter-
minating in death. In most of these conditions some previous
disease existed either in the kidneys, bladder, or urethra, and

the instrumentation merely served to convert a chronic disease
into an acute and often highly-dangerous condition. Much of
the confusion in the past, and, to some extent, still existing, in
reference to the pathology of urinary fever, has arisen from the
mistake of describing the various inflammatory and septic processes set up in abnormal urinary organs by instrumentation, and
attempting to harmonize these with the temporary fever induced
by instrumentation of the urethra and bladder. We may, for
instance, have all the conditions present which tend toward the
development of septic nephritis, such as obstructive cystitis or
ascending pyelitis. The use of the catheter under such circumstances, especially in elderly men, is very apt to at once evoke
acute (septic) interstitial nephritis, resulting in death. While
the exciting cause in such case was instrumentation, pathological
conditions pre-existed, and the instrumentation merely served to
convert a chronic into an acute septic disease.

By urinary fever, as here considered, is meant the elevation
of temperature and accompanying symptoms evoked by the
passage of a sound or catheter, by operations or other impressions made upon the lower urinary tract, *the kidneys and urinary
organs being free from disease.*

The passage of a catheter into a healthy urethra, when the
bladder and kidneys are perfectly healthy, may evoke symptoms
and results of all grades, from a mere transient faintness, recovered from in a few minutes, to violent chill, elevated temperature (103° to 105° F.), suppression of urine, convulsions, and
even death in from six to forty-eight hours. Morris has especially pointed out that the nervous connections with the genito-
urinary tract are so peculiarly constituted that, if a local irritation be at all pronounced, conditions are favorable for the most
wide-spread nervous storm to prevail over the entire sympathetic
and cerebro-spinal systems, involving the cardiac, pulmonary,
and renal circulations to the extent of inducing syncope, impaired respiration, acute renal congestion, convulsions, and even
death.

But even the slighter forms of local irritation (measured by the
degree of instrumentation), as the gentle passage of a sound, are
as likely to evoke an attack of urinary fever as operations upon

the urinary organs of very considerable extent, such as lithotomy or lithotrity.

The Urine.—The quantity of urine is more or less diminished in urinary fever. The diminution is usually very decided and even complete suppression may occur, lasting for one to three days. Very decided diminution is the rule; complete suppression rather the exception. In cases of recurrent urinary fever unattended by suppression the volume of urine is much diminished during the febrile period, while during the intervals the volume increases considerably. The color of the urine is increased and often presents a bloody tint. The urine is smoky in appearance, the transparency being more or less diminished or absent. The reaction of the urine is acid and the specific gravity reduced. The solids are diminished, notably the urea. Blood is nearly always present in marked cases, varying from microscopical quantities to frank hæmaturia. Albumin is constantly to be found in the urine; the range, however, is usually moderate,—one to two grammes per litre,—although exceptionally two or three times that amount is present, and this is always of grave significance. Casts may or may not be present. They are always associated with high grades of albuminuria, and, like the latter, are of serious significance.

Prominent Clinical Features.—After passing a catheter, or some operative manipulation of the lower urinary tract, in from a few minutes to six or eight hours the patient is suddenly seized by a chill of various degrees of severity, from merely chilly sensations to pronounced and violent rigor, accompanied by chattering of the teeth and vibrations of the limbs or whole body. This is followed by pain in the back and limbs. The temperature rapidly rises from 2 to 7 degrees; headache and injection of the conjunctiva are present, and nausea, vomiting, and even delirium are common. Dyspnœa and cardiac irregularity are occasionally to be noted. The pulse becomes rapid, hard, and tense,—vibrating. After a time a pronounced perspiration succeeds, and with this the temperature lowers and more or less relief is experienced. The pulse grows less frequent and less tense; the temperature diminishes, but thirst continues unabated. After six to twelve hours the fever subsides, leaving the patient weak; but convalescence, as a rule, is established in a day or two. In some cases the patient has a recurrence of the paroxysm on the following day, or in two or three days, and these may be repeated a number of times. In the absence of definite lesions the attacks soon subside and the patient regains his normal condition.

Differential Features.—From pyelonephritis and suppurative nephritis urinary fever is distinguished by the sudden onset of the latter and the brief duration of the fever; by the history of the case, such as previously healthy kidneys and healthy state of the bladder and lower urinary tract.

From uræmia more difficulty is encountered in making a distinction, since suppression may occur for several days and death result, at least in part, from uræmia. In urinary fever sufficiently severe to cause death, it does so more rapidly than does uræmia. The absence of coma and convulsions, the retention of consciousness, etc., exclude uræmia. Septicæmia is distinguished by its slower onset, low typhoid character, and continuous progressive course without intermission.

HYDRONEPHROSIS.

The above term was first employed by Rayer to denote the overdistension of the kidney with urine. It is, in fact, a result of mechanical obstruction to the outflow of the urine, the obstruction being located in the ureter, bladder, or urethra. This disease should be carefully distinguished from pyonephrosis,—a condition also of distension of the kidney with urine plus purulent matter. It should also be distinguished from large cysts of the kidney the contents of which are fluid, but not urinous.

Hydronephrosis in its pathological significance has been best expressed by Terrier and Baudouin as " *an aseptic dilatation of the pelvis by urine, the flow of which is obstructed by some mechanical obstacle.*" About 35 to 40 per cent. of the cases are congenital, the remainder being acquired. The congenital causes comprise twists of the ureter upon its axis, undue obliquity of the ureteral opening into the bladder, reduplication, valve-like folds of the ureteral mucous membrane, and imperforate ureter. The acquired causes include cancer of the pelvic organs, notably of the ovaries; hydatids and other growths within or impinging upon the ureters; calculus in the ureter; traumatisms, including renal dislocations, twists, etc.; abdominal tumors, vesical growths involving the ureteral openings, and obstructive diseases of the prostate.

The Urine.—The quantity of urine varies according to the degree of obstruction, and whether the disease be confined to one or both kidneys. In the milder forms of obstruction the quantity of urine varies greatly, there being periods of diminution followed by periods of increased flow. On the whole, the volume of urine is diminished. The urine is of low specific gravity, and reduced in its solid constituents,—conditions which, as Dickinson has pointed out, always exist with urine secreted against pressure. The urine sometimes contains blood, which is discharged with great pain (renal colic), especially if clots be present. Slight albuminuria is usually present, though this is not invariably so. The urea is markedly reduced, both relatively and absolutely; the phosphates are greatly reduced in most cases, while the chlorides and sulphates suffer the least diminution. Sedimentation of the urine shows excess of epithelium, in which the spindle-shaped and angular cells predominate. Renal casts, as a rule, are absent, but a few scattering pus-corpuscles and blood-discs are usually present.

Prominent Clinical Features.—Dull, aching pain is usually present in the renal region, with some increased frequency of micturition. Tumor is present in most cases, gradually encroaching on the median line and downward toward the iliac fossa. About one-fourth of the cases of single hydronephrosis extend beyond the median line, and in a considerable number of these tumor occupies a very considerable area of the abdominal cavity. Sudden diminution in size of tumor, coincident with excretion of unusual quantity of non-purulent urine, may be considered diagnostic.

Vomiting sometimes occurs during periods of retention, and sometimes a urinous odor may be observed in the perspiration at such times. Constipation is a frequent result of pressure upon the colon; more rarely diarrhœa may be present from the same cause. So long as the hydronephrosis be single and the remaining kidney be sound, there is absence of uræmic symptoms. If, however, the remaining kidney be diseased or the hydronephrosis be bilateral, suppression of urine and uræmia are liable to result at any time and prove fatal.

Differential Features. — Hydronephrosis is to be distinguished from other abdominal tumors by the presence of urea and uric acid in the fluid withdrawn by aspiration, and by the abrupt diminution in the size of the tumor coincident with copious discharge of urine. Hydronephrosis may be confounded with renal cancer or cystic degeneration of the kidney. In hydronephrosis the tumor is evenly and distinctly fluctuant, no dullness on percussion being observable throughout its extent. The tumor, furthermore, does not conform to the shape of the kidney; it is usually unattended by dropsy, hæmaturia, or cachexia. In cystic disease the tumor is bilateral, non-fluctuant, preserves the form of the kidney, is painless, sometimes attended by dropsy, nearly always associated with hæmaturia, and the tumor does not rapidly change in size.

In cancer the tumor is unilateral, non-fluctuant, irregular in form, rapid in growth, attended by severe pain ; copious, recurrent, and persistent hæmaturia occurs, and in late stages pronounced cachexia is present.

PYONEPHROSIS.

Pyonephrosis is a dilatation of the renal pelvis and calices of the kidney with purulent urine, or, in other words, it is hydronephrosis with suppurative inflammation added. In marked cases the suppurative process extends beyond the calices and results in compression, atrophy, and destruction of the parenchyma of the kidney. The causes are the same as those of hydronephrosis plus suppurative inflammation.

The Urine.—Pus is always present if the obstruction be incomplete. If complete at times, and not at others, pus will appear intermittently in the urine if the disease be unilateral. The quantity of urine voided will depend upon the degree of pressure exerted. If the ureter be blocked, as often occurs for some periods of time, the urine will be greatly diminished in volume during the period of obstruction. If only partly occluded, the quantity of urine, as well as that of pus, will vary even during twenty-four hours. If the obstruction be temporarily relieved, large quantities of urine are voided which contains blood and pus, while during the period of occlusion the urine is clear and normal in appearance, unless the disease be bilateral.

In the early stages of this lesion the urine contains blood (sometimes only in microscopical quantity), more or less mucus, and epithelium and pus. The urine is usually acid in reaction, of low specific gravity, and contains albumin, as a rule corresponding with the quantity of blood and pus in the urine. As the disease advances, pyuria becomes more pronounced. The urine is still acid, unless in very advanced cases, when sacculation of the kidney occurs, in which case it may become ammoniacal. In all stages the urine is of lowered specific gravity, the solids more or less decreased, and micturition is somewhat more frequent than normal.

Prominent Clinical Features.—The prominent symptoms of pyonephrosis comprise pyuria with constitutional symptoms, such as rigors, evening temperature, emaciation, anæmia, prostration, and, in advanced cases, hectic. If tumor form, it may be elastic and fluctuant or hard, and extend both forward and downward. Pain is present, varying with the size of the tumor and degree of fluctuation. It often appears in paroxysms of intensity,— *renal colic*. Pressure over the anterior of the tumor greatly increases the pain, or develops it, if not before present. On the other hand, lateral pressure may relieve the pain when present. The bowels are usually disturbed, constipation or diarrhœa being frequent. When the ureter becomes suddenly and completely blocked, sharp constitutional symptoms often follow, such as chill followed by rise of temperature, which may reach 103° to 105° F.; profuse perspirations, rapid pulse, quickened respirations, and sharp pain in the affected side. These symptoms usually continue for some time, and are suddenly relieved by a copious flow of urine, which had previously been greatly reduced in quantity.

ACUTE INTERSTITIAL NEPHRITIS.

This disease has been described under several names, viz., "*suppurative nephritis,*" "*acute interstitial nephritis,*" "*pyelonephritis,*" and "*surgical kidney.*" It is, in fact, an acute interstitial nephritis with numerous points of suppuration in the kidney, varying in size from mere dots to large abscesses, which may occupy almost the entire organ. It is seldom observed as a primary disease; by far the greater number of cases are secondary and consequent to urethral stricture, prostatic enlargement, large vesical calculi, atony of the bladder, infectious emboli, or traumatisms. It sometimes complicates typhus, typhoid fever, diphtheria, carbuncles, pyæmia, cholera, and like infectious diseases. Obstructive diseases of the urinary conducting channels, with decomposing urine from retention, are strong predisposing as well as exciting causes. Under such conditions careless instrumentation of the urethra and bladder are exceedingly prone to induce acute interstitial nephritis; hence the misapplied term "*surgical kidney.*"

The Urine.—This is always cloudy, of pale, dirty-yellowish color, and of peculiarly foul odor. The specific gravity is reduced,—1.016 to 1.006,—and the quantity of urine is reduced. The reaction may be acid, neutral, or alkaline. If acid, the disease is likely to be limited to the kidney; but if alkaline, pyelitis and perhaps cystitis also exist. Whatever be the chemical reaction of the urine when voided, upon standing it rapidly undergoes ammoniacal decomposition. The normal constituents of the urine are reduced in quantity, notably the urea. Albumin is always present in the urine, though in variable quantity, the amount always exceeding that due to the contained pus and blood.

The urinary sediment is always abundant, and consists, for the most part, of pus, blood, bacteria, organic *débris*, epithelium, and usually casts. The presence of pus is an essential feature of the disease; it is always present and may be very copious in quantity.

The microscope reveals the presence of micro-organisms in abundance, and sometimes finely-formed casts of bacteria are to be seen. A sudden and marked increase in the quantity of pus in the urine, especially if containing recognizable remnants of glomeruli or urinary tubules, denotes the formation of renal abscess.

Prominent Clinical Features.—The commencement is marked by pronounced rigor or a succession of chills followed by rise of temperature, which in the evening may reach 103° to 105° F., while the morning temperature may be below 100°, F. Weakness, drowsiness, flatulence, and sense of abdominal fullness follow. Rapid emaciation, pinched features, and dull, leaden, or sallow appearance succeed. In the evening the skin becomes hot and great thirst is present. Profuse perspiration is common; the tongue may remain comparatively clean, but more often it becomes dry and coated with a brownish-white fur. Nausea is frequent, and vomiting occasionally follows. Renal pain and tenderness are usually absent, but considerable pain is present in the limbs and along the spine. In unfavorable cases the symptoms continue until exhaustion succeeds or renal suppuration becomes established, either of which usually terminates life within a few days.

Chronic Pyelitis.

The more acute forms of pyelitis are included in pyonephrosis and acute interstitial nephritis, just considered. The general causes include the acute infectious diseases, such as typhoid, diphtheria, pyæmia, cholera, puerperal septicæmia, and such diseases as carbuncle.

In addition to these we have chronic pyelitis of a more circumscribed character, which may result from an acute attack, or it may be induced by less active causes, being essentially chronic in nature from the beginning. Chronic pyelitis is rarely a primary disease, but is usually associated with other renal lesions or vesical diseases, especially septic inflammations of the lower urinary tract.

Of the local causes gravel constitutes the most frequent source, which may act either by direct irritation or through obstruction. Tuberculosis is a frequent local cause, notably the form of slowly-developing cheesy nodules which set up inflammatory changes both in the pelvic mucous membrane and neighboring tissues.

The various obstructive diseases of the lower urinary channels are frequent causes of inflammation of the mucous membrane of the renal pelvis. These include urethral stricture, enlarged prostate, large vesical calculi, atony of the bladder, etc. The retained urine in such cases is extremely prone to be contaminated with pyogenic germs introduced from without. These organisms, upon introduction, rapidly multiply and spread, causing changes both in the urine and mucous structures, including that of the renal pelvis. These micro-organisms are, for the most part, micrococci, but rod-like forms or bacilli may also be present, both in the urine and mucosa of the renal pelvis.

Pyelitis should not be confounded with suppurating nephritis. This is a common error, due, perhaps, to the fact that they are often, though by no means always, associated; but uncomplicated pyelitis is unattended by the usual general symptoms of septic absorption which are present in suppurating inflammation of the kidney.

The Urine.—The renal irritation consequent to pyelitis induces polyuria; so that the volume of urine is augmented in this lesion. The color of the urine is pale straw-colored, described as "*whey turbid*," and sometimes tends toward a greenish tint. The specific gravity is reduced and the chemical reaction is usually faintly acid. Pus is present in varying quantity, but always the most prominent feature of the sediment (Fig. 43). The solids are relatively reduced owing to the polyuria, but the absolute solids suffer no material changes in amount. Albuminuria is a constant feature of pyelitis, the quantity of albumin exceeding that depending upon the contained pus. The sediment is flocculent, of considerable volume, of greenish-yellow tint, not so viscid or sticky as in cystitis, and, as already stated, is chiefly composed of pus. The pus-corpuscles often exhibit tooth-like margins, notably so in chronic cases. They often become pressed together in the papillary ducts and form masses of round, oval, or long plugs, which are considered characteristic of pyelitis. Epithelium is present only in small quantities. This lesion cannot be diagnosticated by the presence of spindle-shaped epithelium in the urine alone, as has been frequently asserted. In truth, much of the epithelium lost by the mucosa of the renal pelvis breaks down into pus-cells and granular matter. Blood-corpuscles are rarely present, save when the pyelitis is due to calculus, tuberculosis, growths, or entozoa. If, as is usually the case, the pyelitis be unilateral, the urine sometimes presents one or two anomalous features in reference to its chemical reaction. Thus, the urine may be putrescent and offensive from decomposition, and yet remain acid. Again, it is not uncommon to meet with triple-phosphate crystals in the urine which gives

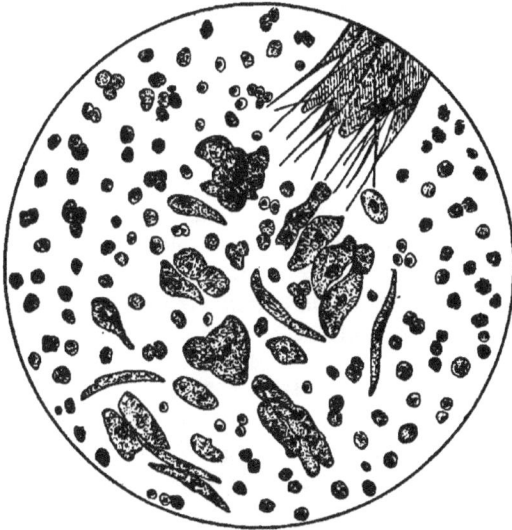

FIG. 43.—URINARY SEDIMENT IN PYELITIS. (After Peyer.)

a distinctly acid reaction. These features have been explained by the fact that the urine from the sound kidney overcomes the alkalinity of the urine from the diseased kidney as they mingle together in the bladder, and it is thus voided before it has had sufficient time to become ammoniacal. The triple-phosphate crystals met with in the urine under such circumstances often show evidences of their existence in acid urine in their superficial solution, their sharp angles being rounded. The urine in pyelitis is peculiarly offensive,—not merely ammoniacal, but rather also suggestive of sulphuretted hydrogen,—while in cystitis the odor is almost purely ammoniacal. Once experienced, the odor of the urine in pyelitis may afterward be recognized as characteristic when present. This property of the urine, however, varies from time to time, and may even be temporarily absent if the ureter of the affected side become blocked.

Prominent Clinical Features.—These comprise aching, dragging pains, central in the lumbar region, but often radiating along the course of the ureter toward the bladder. The pain is not usually severe, is always increased by deep pressure by the finger-tips, and it may be absent at limited, but considerable, periods of time, and it is increased by exercise. Renal colic is apt to form a feature of the history of this lesion, but especially so when the cause is connected with stone, hydatids, tuberculosis, or cancer. Micturition is more frequent than normal, but it is rarely attended by pain, being rather of reflex character. Occasionally, when the ureter becomes blocked and retention of urine occurs in the affected organ, elevation of temperature occurs with the usual febrile symptoms. The fever is nocturnal, usually of mild grade, and often accompanied by cutaneous eruptions not unlike rötheln.

MOVABLE KIDNEY.

The subject of misplacements of the kidney embraces three conditions of the organs : 1. *Simple misplacements of the kidney.* 2. *Movable kidney.* 3. *Floating kidney.*

Simple misplacement of the kidney is met with either as a congenital or acquired condition. Of the congenital order, the horseshoe kidney is most common ; the next most frequent congenital malposition is above the sacro-iliac synchondrosis. The suprarenal bodies are often found congenitally misplaced, especially when the kidneys are higher than normal. Congenital misplacement is most often unilateral, and most often involves the left kidney. Simple acquired misplacement may be brought about by pressure of adjacent organs, such as enlargement of the spleen or liver, or from pressure of large morbid growths. Tight lacing has been alleged as a frequent cause of downward displacement of the kidney, but this is now considered as an

exceptional cause. Both acquired and congenitally misplaced
kidney may be mistaken for new growths. As a rule, misplaced
kidney, if unattended by mobility, gives rise to no special symptoms, and is only observed at the autopsy. It therefore deserves
no extended study by the clinician.

Movable kidney differs from *floating kidney* in that the former
lies loosely between the peritoneum in front and the muscular
wall of the abdomen behind, while the floating kidney is inclosed in peritoneum, which forms a "*mesonephron.*" The floating kidney, therefore, floats within the abdominal cavity, as do
the intestines. Movable kidney, as a rule, is acquired, while
floating kidney is always congenital. Movable kidney is most
common in females. The right kidney is about twelve times
more frequently movable than the left, which is probably due to
the weight and unyielding nature of the liver, the greater length
of the right renal vessels, and the less firm attachments to the
colon on the right, together with the greater right tension and
strain in right-handed people. Movable kidney is bilateral
about once in every twenty cases. It is most frequent from
30 to 40 years of age, a large proportion of the cases occurring
in women who have rapidly borne children.

Absorption of the fatty elements of the renal capsule is now
regarded as the most common predisposing cause of movable
kidney, and this is borne out by the remarkable frequency of its
occurrence in spare subjects. Lastly, movable kidney is often
the result of dislocation from falls or blows.

The Urine.—It has been denied by many authors that movable
kidney produces any effect on the character of the urine. Those
who have seen much of such cases, however, will scarcely have
failed to observe frequent urinary disturbances, which form an
important, almost essential, part of the pathology of this condition. The urine is often scanty, high colored, and usually
deposits more or less sediment; at other times it is more copious
than normal. Hæmaturia is not unfrequent in movable kidney.
The twists of the ureter attendant upon renal mobility necessarily entail more or less interference with the circulation. The
duration and degree of this interference determine the degree of
suppression, the discharge of blood, presence of albumin, epi-

thelial denudation, and pyelitis. Suppression of urine usually occurs only at short periods, because the loosening of the renal capsule from the posterior abdominal wall permits the renal pelvis and part of the ureter to also separate from the abdominal wall and follow the renal movements. Thus, the obstacle to the escape of urine is more easily overcome by pressure of retained urine within the renal pelvis.

Diminished excretion (25 to 30 ounces) and short periods of suppression, followed by short periods of copious flow of pale-colored urine of low specific gravity, are common features of this condition, though these are not so marked as in pyonephrosis. During periods of suppression mild uræmic symptoms have been noted; these are rarely pronounced, because the flow is soon restored. The urine very often contains a small amount of albumin, owing to the circulatory disturbances in the kidney.

Prominent Clinical Features.—The chief clinical symptoms of movable kidney are: sensations of dragging weight in the side; occasional sense of movement; severe paroxysms of pain, like renal colic; pain of a dull, aching character; neuralgic pains in trunk-nerves of the affected side. Nausea and vomiting are often present, and may be severe. Intestinal disturbances are common, such as diarrhœa or constipation. Loss of appetite, weakness, debility, flatulence, emaciation, and depression are frequent features observable. Nearly all the symptoms are increased by movement or standing, while they are relieved by rest and quietude. Menstruation and pregnancy usually aggravate the symptoms. Manipulation of tumor causes peculiar sinking or fainting sensations, often attended by nausea. The tumor possesses more or less mobility.

Differential Features.—Limited contractions of the recti, transversalis, or abdominal oblique muscles sometimes give impressions of smooth, oval tumors, which may even disappear on pressure, not unlike movable kidney. Anæsthesia and strong percussion over tumor will usually distinguish the true condition.

A tumor of the liver, or distended gall-bladder, or a liver deformed from tight lacing, or an hypertrophied liver may give rise to similar percussion-sounds, as well as similar gastric disorders, and vague abdominal sensations and pain not unlike movable right kidney. Distinction is most difficult, but may be accomplished by palpation, urinary changes or their absence, etiological elements, and by results of treatment. Percussion cannot be depended on in these cases alone; if the tumor can be completely separated from the liver by palpation, it goes far toward establishing movable right kidney.

Movable spleen is distinguished from movable kidney by the former lying immediately beneath the abdominal wall throughout all movements, while a kidney when pushed upward retreats from the abdominal wall beneath the intestines. The peroussion-note over the spleen is, therefore, always dull, while over the kidney it is dull tympanitic or tympanitic.

The distinction between uterine or ovarian tumor and movable kidney is determined chiefly by the *direction* of mobility. If mobility can be made to extend to lumbar region without pain, tumor of the uterus or ovaries may be excluded. On the other hand, movable kidney must not be excluded by mobility toward the pelvis, as such tumor may exceptionally be renal. If the renal artery can be grasped and pulsations felt, the diagnosis becomes clear. Capability of replacement of kidney with disappearance of tumor renders renal source of tumor certain.

CYSTITIS.

Much confusion has existed as to the nature of cystitis in consequence of the diversified classifications of this affection. Writers speak of *purulent cystitis, hæmorrhagic cystitis, cystitis of the neck, cystitis of the body of the bladder*, and *catarrhal cystitis* of several grades. *Acute cystitis* and *chronic cystitis* are employed as denoting intensity or duration of inflammatory process, rather than the character of the lesions present, and as such the terms are convenient for descriptive purposes.

In the light of modern investigations and present pathological knowledge cystitis is coming to be better understood and its signification has assumed more accurate and definite meaning. In all its varying phases cystitis must now be considered as a local bladder infection by bacterial germs. Normal urine, as already pointed out, is an aseptic fluid characterized by the

22

absence of septic germs, while cystitis is always associated with the presence of septic micro-organisms in greater or less abundance.

The same causes which bring about local infection elsewhere may be looked for to explain the presence of bladder infection. Comprehensively considered, they include the following conditions: Weakened vitality of the tissues; favorable soil for development and the presence of micro-organisms possessing pyogenic powers. If we consider for a moment the many conditions to which the urinary bladder is exposed, which tend to weaken its resisting power to the action of germs; the many sources of exposure to local infection, and the favorable medium the urine constitutes as a culture medium for many pyogenic germs, it no longer seems strange that cystitis is one of the most frequent diseases of the urinary tract. Since the discovery of the micrococcus ureæ by Pasteur, thirty-five years ago, to the present time, a large number of micro-organisms have been discovered in pathological urine. Although many of these possess no pathogenic powers, a number of them are, without doubt, closely associated with cystitis in the relationship of cause and effect.

The older theory, that cystitis is set up by ammonia evolved in the decomposition of the urine by the micrococcus ureæ, is no longer accepted. It has been conclusively demonstrated by Guiard and Guyon that ammoniacal fermentation of the urine in the bladder takes place as a *result* of cystitis, and in the absence of the latter the former does not occur. The question, then, of the local etiology of cystitis must be narrowed down to local infection of the bladder by pathogenic germs, and the results of their action constitute cystitis.

Of the many etiological factors which favor bladder-infection may be mentioned the following: Retention of urine, trauma, deep gonorrhœa, stricture of the urethra, prostatitis, calculus, morbid growths, exposure to cold, sexual excesses, gout, paraplegia, etc.

The Urine.—Pus is present in the urine as the most essential feature of cystitis, its absence being proof that the bladder is uninflamed. The quantity of pus present depends upon the degree and extent of the cystitis and the length of time cystitis has been established. Early in the lesion the quantity is small, sometimes observable only under the microscope. If the urine be acid the cystitis is recent, and the pus readily settles. If the urine be alkaline—ammoniacal—the pus settles in a viscid, sticky mass, the cystitis is chronic, often termed "*chronic catarrhal cystitis.*" Blood is nearly always present in acute cystitis, and sometimes also in the chronic lesion. It is increased in quantity

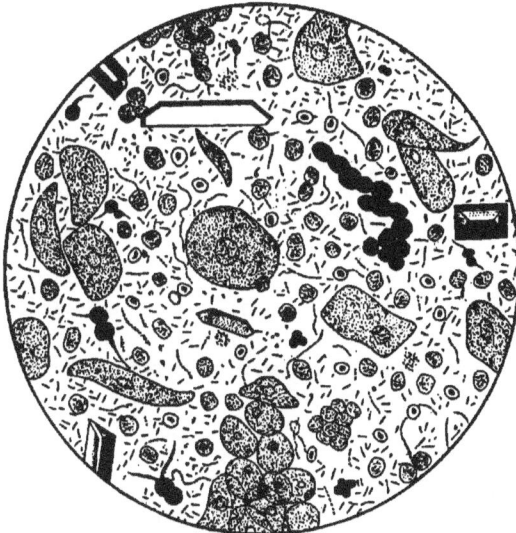

FIG. 44.—URINARY SEDIMENT IN CYSTITIS. (After Peyer.)

by movements or standing. In acute cases it is due to congestion; the quantity is comparatively small, and for the most part it appears at the close of micturition. In tumors and varicose conditions of the vessels, blood is mixed more uniformly with the urine, and the quantity may in such cases be large.

The quantity of albumin in the urine is inconstant in this lesion. In the early course of the disease it is usually due to pus or blood, and consequently the amount is small. In chronic cases the quantity is more considerable, since denudation of the vesical mucous surface permits of more ready escape of blood·

scrum. The reaction of the urine is usually acid in the early course of the disease as well as in mild cases; while chronic, long-continued cases are mostly characterized by alkaline urine from volatile alkali. Accompanying the ammoniacal changes, more or less copious deposits of triple-phosphate crystals are to be found in the urine. In addition to this, amorphous phosphates are often present in considerable quantity, and less frequently, also, are urates.

Recent cystitis differs from the chronic form as follows: In the former the quantity of urine is usually normal; color is dark; the reaction is acid *when voided*, quickly becoming alkaline after it is voided; the specific gravity of the urine is usually normal; the appearance is turbid and smoky. Microscopical inspection shows the presence of vesical epithelium, pus, often blood, and bacteria; and on standing a few crystals of triple phosphates usually are observed.

In chronic cases the quantity of urine is about normal; the color is light, the reaction alkaline, the specific gravity somewhat lowered, and albumin is usually present. Microscopical inspection shows the presence of pus usually in abundance, sometimes blood, bacteria always, and vesical epithelium. The appearance of the urine is turbid, the odor is ammoniacal, and the sediment sticks to the glass or vessel upon standing. The micro-organisms most often met with in cystitis are as follow: The *staphylococcus pyogenes aureus, albus,* and *citreus;* the *streptococcus pyogenes;* the *urobacillus liquifaciens septicus;* and the *bacillus coli communis.* These are all pathogenic germs. Those most frequently met with are the *bacillus coli communis,* and next is the *staphylococcus pyogenes aureus.*

Prominent Clinical Features.—Frequent micturition, pain, and pyuria are the most constant symptoms of cystitis. Frequent micturition varies with the intensity of the disease and the sensitiveness of the bladder. In acute cystitis the desire to micturate is nearly continuous. The frequency is increased upon standing or walking, while rest tends to relieve it. The pain also varies according to the acuteness of the attack. The pain is often very intense at or just before the beginning of micturition, is more or less relieved during the flow of urine, while at the close of micturition, and immediately following, for a few seconds the pain becomes greatly increased. The latter is most notably the case when cystitis is associated with stone or prostatitis. ' In cases of highly ammoniacal urine considerable pain is often experienced during the act of micturition. There is more or less discomfort, sometimes amounting to pain, with sensation of weight above the pubis, in cystitis, irrespective of the act of micturition. The bladder is sensitive to rectal touch, as well as to instrumentation, and also to distension by injections of fluids, however bland they may be. Hæmorrhage, if present, indicates, as a rule, the more acute grades of cystitis, unless in large amounts, which are more common in chronic forms. Constitutional symptoms are absent unless the disease be associated with tuberculosis, malignant disease, etc.

VESICAL STONE.

The etiology and symptomatology of calculous disease have already been considered in detail. More than one-half the cases of vesical stone met with in hospital practice occur before the age of puberty; about 22 per cent. occur between the ages of 50 and 70 years; while not above 2 per cent. are met with above 70 years of age. The statistics of Sir Henry Thompson, as well as those gathered from large hospitals, substantially confirm the above proportions. In private practice, however, the above proportions are almost directly reversed. Thus, in Sir Henry Thompson's private practice the patients between 50 and 70 years of age comprised 65 per cent. of the whole; and those over 70 comprised 22 per cent. of the whole; while those under 16 comprised less than 1 per cent. From these facts Sir Henry Thompson draws the following deductions : " Insufficient food, clothing, and fresh air, the necessary accompaniments of poverty, appear to encourage calculous formations among children, but not among adults. Habits of self-indulgence, in relation chiefly to diet and indolence, encourage calculous formations in early adult males, but the children of such parents are not affected. Hard physical labor and a regimen which necessarily contains simple diet, largely cereal, with animal food in small proportion, even although often associated with intemperate habits and unhealthy dwellings, discourage calculous formations among all classes of the community alike." The foregoing only applies to primary formations.

The Urine.—Before the onset of cystitis the urine may remain but little changed from the normal. The quantity is normal; the specific gravity is not materially altered ; the urine remains clear, tending to increased color, perhaps, and the reaction may be either acid or alkaline. The history often reveals habitual deposits of red sand or white, gritty powder for months or years before special symptoms appear. In the early stages, when, to all appearances, the urine remains normal, the microscope usually reveals the presence of uric-acid crystals or those of calcium oxalate, together with scattering blood- and pus- corpuscles. Cystitis sooner or later appears, when the urine becomes cloudy from increased pus formation, together with the presence of epithelium, crystals of uric acid, calcium oxalate, triple phosphates, or amorphous urates or phosphates. If the urine remain acid the crystalline deposit will be uric acid or calcium oxalate, often both together. If the urine be alkaline the crystalline deposit will consist of triple phosphates, and the pus deposit will be more or less viscid and sticky. The degree of pyuria indicates the extent of denudation or ulceration of the mucous surface of the bladder. Blood is often present in variable amount, but rarely excessive. Albumin is always present in the urine if cystitis co-exist, and frequently in excess of the contained pus or blood. Micro-organisms are present also in all cases attended by cystitis, and for the most part of the same order.

Clinical Features.—Pain is present as one of the most typical features, tending to reflection along the urethra, in the testes, or down the thighs. Pain is sharply increased at the close of micturition and for a few seconds afterward, especially noticed in the glans penis. Spasm of the bladder is common at the close of micturition. The pain is always increased by the erect posture and by motion, including riding, walking, and even by turning in bed sometimes. There is more or less increased frequency of micturition, augmented by motion and relieved by continued rest and recumbency. Micturition is sometimes suddenly shut off in the middle of the act, in consequence of the stone rolling forward over the urethral inlet. On lying upon the back this symptom is relieved.

Hæmaturia is the rule in this disease, the blood usually being clear, bright in color, and usually consists of a few drops at the close of micturition, the urine at the beginning of micturition being mostly free from blood. Sometimes hæmorrhage is more considerable, and the blood becomes more generally diffused through the urine in such cases. Hæmaturia is increased by movement, as walking and riding, etc., and is relieved by rest and recumbency, and therefore it is always less in the morning and more in the latter part of the day or in the evening. To complete the diagnosis exploration is often necessary. This may be done (a) *per vaginam* or *rectum* by means of the finger; (b) by means of the vesical sound; (c) by means of the cystoscope; (d) by digital examination through the dilated urethra in women. The most satisfactory methods are those of the sound and the cystoscope.

VESICAL TUBERCULOSIS.

Primary tuberculosis of the bladder is comparatively rare, and when present it invariably begins in the trigone. More often it results from extension of tubercular disease from the prostate, testes, or recto-vesical fold of the peritoneum, less often by infection from the upper portions of the urinary tract. The bladder may become infected from above either by direct extension of the disease along the ureter or by inoculation through the urine from tuberculous kidney.

When infection occurs from surface inoculations through the urine from above the symptoms rapidly supervene and the disease is acutely progressive. On the other hand, when the disease is primary the symptoms are more slow in development and the course of the disease is more apt to be protracted.

The Urine.—Blood is usually present in the urine in this lesion in small amount, often transitory and recurrent, a few drops often following the close of micturition, as in stone. Sometimes, however, hæmaturia is pronounced, in consequence of ulceration about the tubercular deposits. The urine contains small quantities of pus from the beginning, which gradually increase until appreciable deposits thereof are regularly present.

The urine is murky, of light color, of normal specific gravity, and it is usually feebly acid or neutral until it becomes ammoniacal from the presence of cystitis. The urine is not especially offensive, although it may be ammoniacal, as just stated.

With the occurrence of cystitis all the features of the urine characteristic of that disease supervene, and that in the most pronounced form. These include pyuria, hæmaturia, deposits of triple-phosphate crystals, amorphous phosphates, epithelium in states of progressive necrosis, and micro-organisms in abundance.

Prominent Clinical Features.—These patients are young, usually from 15 to 30 years of age, with family histories of tuberculous disease in some form. The early symptoms consist of increasing frequency of micturition for a few months, especially during the day. Blood appears in the urine, and with this the patient begins to rise at night to micturate. The pain in micturition is referred chiefly to the mid-penis. Sharp, vesical tenesmus becomes developed as the disease proceeds, and in such cases the pain is increased at the finish of micturition. The constitutional symptoms include debility, emaciation, weakness, elevation of temperature in the evening, and night-sweats. With these symptoms are often associated tuberculosis of the lungs, joints, glands, or elsewhere.

Differential Features.—Tubercular cystitis in many of its manifestations resembles stone in the bladder. Tuberculosis of the bladder, however, is more frequent in youth; the family history is often tuberculous; the irritability of the bladder is often marked at night, greatly disturbing the patient's rest; hæmaturia is often sudden and without apparent cause, being less dependent upon exercise; there is greater relief from pain at the close of micturition; pain is in mid-penis, rather than in glans penis; there is persistent post-scrotal perineal pain ; the urine is light in color, murky, inodorous, and purulent from the beginning, and periods of quiescence of the symptoms do not correspond to quiescence and rest of the patient. In late stages evening temperature is present, marked constitutional symptoms arise, implication of epididymis is common, and knobby or shotty feel of prostate *per rectum* points to the tubercular character of the disease.

CANCER OF THE BLADDER.

The great majority of malignant growths of the bladder are of the carcinomatous order. Carcinoma of the bladder usually begins at the base of the organ, or, more accurately speaking, at the lower third of the organ. It manifests a peculiar tendency to long remain a local disease, with little tendency of extension to neighboring structures. This has been explained by the scarcity—according to some, the absence—of lymphatics in the bladder-walls.

The Urine.—The urine contains blood in the vast majority of these cases. Seldom being entirely absent from the urine, it is subject to periods of marked augmentation without any apparent exciting cause. Attacks of marked hæmaturia are quite as likely, therefore, to occur at night as during the hours of exercise through the day. The quantity of blood is often greater than in almost any other disease of the urinary organs; so that the bladder is liable to become partly filled with large blood-clots. As a rule, the quantity of blood in the urine irregularly, but continuedly, increases as the disease advances. The color of the blood is usually bright red, especially when it is abundant, but if in small quantity it may be dull red or brownish in appearance. The blood is not intimately mingled with the urine as it is in renal hæmaturia.

The urine often contains shreds or detached bits of the morbid growth, which may even be noticeable to the patient. To the naked eye these appear as " washed-out " bits of tissue, which in reality they are, and they vary in size from minute specks to pieces as large as a pea or small bean. Their presence merely indicates the presence of some growth without indicating its character. No trustworthy inferences are to be drawn as to the nature of these shreds from microscopical examination, since both malignant and benign growths of the bladder may present papillary surfaces of practically the same microscopical appearance. Since, however, such shreds are most frequently the product of papillomata or malignant growths, their presence may be considered as simply presumptive of one or the other of these forms of growth.

The urine nearly always contains an abundant sediment in

malignant disease of the bladder, of which epithelium constitutes a prominent feature. The epithelium is of small or moderate size, but with very large and plainly-visible nuclei. The quantity of urine remains normal, as does the specific gravity, save in late stages, when the latter becomes somewhat reduced. The urine is turbid, often brownish red, and the reaction is acid until cystitis becomes established ; with the appearance of the latter there is pyuria, ammonuria, bacteriuria, and the usual accompanying symptoms.

Prominent Clinical Features.—Pain is a prominent and early symptom, usually preceding the hæmaturia. The character of the pain is often sharp, radiating to the thighs, above the symphysis pubis, or in the post-scrotal region. The pain is not especially increased by movements. Fenwick has recently called especial attention to the fact, as he claims that, if the disease do not invade the trigone, pain may be practically absent until a late period of the disease. This may account for the occasional absence of pain almost throughout the disease. Frequency of micturition is more or less pronounced from the beginning, and this is attended by pain, which is most notable just before the beginning of the flow. Rectal examination develops local tenderness and patches of induration.

BENIGN VESICAL GROWTHS.

The chief varieties of benign growths met with in the bladder are *papilloma, myxoma,* and *myoma* Of these, *papillomata* are by far the most frequent. These consist of proliferations of the natural structure of the vesical mucosa, forming papillæ or protrusions covered with cylindrical epithelium. Sometimes these papillæ are long and slender and float in the urine in numerous filaments from a common base or stalk. Examined in a dry state they collapse and form a soft, " *strawberry-like* " mass. The base or pedicle always contains more or less fibrous tissue, and usually some non-striped muscular fibres. Sometimes the growth expands to form a polypoid-like mass of more decided firmness, or it may have a wider attachment to the bladder. Again, it sometimes appears expanded into several bunches not unlike cauliflower. When the fibrous elements are numerous the structure is more dense, and this form has been termed "*fibrous papilloma*." The latter are, for the most part, tumors of some considerable solidity, and often have comparatively limited papillary margins.

The *myxoma*, or simple mucous polyp, is rarely met with in the bladder; thus far, only in childhood. Sometimes it is congenital. The growth is that of fibroid undergoing mucoid transformation. This growth is single, pedunculated in form, and resembles the ordinary nasal polypus.

Myoma is only occasionally met with in the bladder. It is usually of moderate size, with wide base, round or oval in form, rather firm in consistence, and chiefly made up of muscular fibre, evidently the outgrowth of the muscular coat of the bladder. The last two described growths are so uncommon in the bladder that for practical purposes papillomata only require special consideration.

The Urine.—The most important feature of the urine in papilloma of the bladder is the presence of blood,—hæmaturia. This occurs early, and at first in paroxysms, followed by more or less lengthy intervals of absence. Commencing insidiously, often appearing in the form of small dark clots, " *like flies*," appearing and disappearing alternately for a few days at each time; or again, the urine, instead, may appear slightly blood-stained, or a little clear blood may appear at the close of an otherwise micturition of clear urine; or, again, the whole urine voided may appear bloody,—dark " *coffee-colored*." The hæmaturia usually presents these various types in succession, without other prominent features, for long periods of time,—often for years. Sooner or later, however, attacks of more or less profuse hæmorrhage occur and recur as the growth increases. The blood is not of pronounced arterial color, but more inclined to darker shades. Aside from the presence of blood in the urine, the latter is not otherwise especially altered in its physical characters; the specific gravity and volume remain about normal; the solids are relatively and absolutely unchanged; and the urine remains acid, though often feebly so, until late stages, attended by cystitis, when the reaction may become alkaline.

The urine contains considerable sediment of a finely-flocculent

nature, brownish red in color for the most part, and contains
blood-corpuscles, more or less pus, epithelium, and often ragged
shreds of tissue. The deposit of epithelium from the vesical
mucosa is a prominent feature of the sediment, often exceeding
that in any other condition. With regard to the shreds often
voided with the urine in these cases, it was formerly held that
microscopical examination of these readily established the nature
of the growth. Unfortunately, however, more extended experience has shown that, while the epithelial character of these
pieces is easily enough recognized, yet, in consequence of the fact
that any bladder-growth, benign or malignant, may possess a
peripheral epithelial fringe, no positive deductions can be drawn
from their recognition. We are only able to say, presumptively
from these, that papilloma is most probable, because it is most
frequently attended by the appearance in the urine of those
structures recognized as epithelial formations.

Prominent Clinical Features.—The most prominent symptoms
are long duration of hæmaturia without marked pain or other
associated symptoms. In other words, hæmaturia precedes, for
longer or shorter periods, vesical symptoms, such as pain and frequent micturition Some slight increased frequency of micturition
develops after a time, with tendency to increase with the progress
of the disease. Pain is scarcely ever a pronounced feature of
the disease, usually only so when obstruction occurs from clot
or growth situated near the margin of the urethra, when it may
stop the flow of urine. In cases of this order tenesmus, frequent
micturition, and pain become prominent, and may even precede
the hæmaturia. The general condition of the patient remains
normal, save, perhaps, anæmia, which is apt to result to a greater
or less extent, according to the amount of the hæmaturia.

SECTION XII.

THE URINE IN OTHER DISEASES.

SIMPLE PYREXIA.

CERTAIN changes in the character of the urine in pyrexia are sufficiently constant to merit the term "*pyrexial urine*" often employed. Generally speaking, these comprise diminution in quantity or, more accurately speaking, deficiency of water; increased color, due to increase of pigment, both relative and absolute; increased acidity, and high specific gravity. In addition to these physical changes in the urine, provided no accidental circumstances interfere, more important changes take place in the composition of the urine. Among the most prominent of these changes is a decided increase, both relatively and absolutely, of the nitrogenous elements, viz., urea and uric acid. Since, for the most part, but little food is taken in pyrexial states, and the urea and uric acid rise considerably above the healthy range, we must conclude that the pyrexial state entails a more or less pronounced waste of tissue. While the amount of uric acid excreted varies in different forms of pyrexia as well as in different cases of the same fever, it is undoubtedly more uniformly increased than is any other urinary constituent. This increase of uric acid is proportionately independent of urea, which seems somewhat paradoxical considering that they are both derived from the same tissue-base. In addition to the increase of uric acid and urea, hippuric acid is often present in the urine in large quantity as a consequence of the pyrexial state, indicating, in all probability, a disturbed function of the liver.

The increased excretion of pigment is chiefly due to the destructive metamorphosis of the red blood-corpuscles, the latter being a well-known result of nearly all forms of fever. As might be expected, considering the excess of uric acid and urea, the sulphates are also markedly increased in pyrexia, constituting

(341)

another evidence of the destructive metamorphosis of nitrogenous
(sulphur-holding) tissues.

The chlorides and phosphates, as a rule, are reduced, both
relatively and absolutely, in pyrexia. With regard to the former
the diminution is probably due, at least in part, to retention; but
as regards the phosphates (save in cases of acute inflammations
of the nervous or muscular tissues) there is undoubtedly deficient excretion.

As a whole, then, pyrexia entails an excessive excretion of
solids by the urine. This excessive excretion begins with the
onset of fever, and usually maintains a parallel ratio with the
temperature; in short, the presence of the latter invariably
implies the former. The urine often contains a small percentage
of albumin in simple pyrexial conditions, and the same may be
said of acetone.

The urinary sediment in pyrexia is often considerable in
amount, and consists of uric-acid crystals, urates, sometimes
scattering hyaline casts, a few leucocytes, and epithelium. Aside
from the modifications of the above features of the urine by special
forms of fever, typical pyrexial urine is subject to certain variations through certain special circumstances, as follow: When,
during the course of a fever, an organ becomes the seat of disease,
where tissue-changes furnish special urinary products, the products may appear in the urine in excess. · As examples, note the
increase of phosphates in the urine in inflammatory diseases of
the nervous system, and the increase of bile-acids in hepatitis.
The overaction of other eliminating organs may disturb the
type of febrile urine. Thus, the diarrhœa of typhoid fever
modifies the character of the urine, as likewise does the sweating stage of intermittent fever. These are a few of the influences that may produce special variations of the febrile type of
urine to a limited degree; but, on the whole, the general pyrexial characters stand out with sufficient prominence for general
recognition.

It may be stated, as a general rule, that the amount of tissue-
metamorphosis, as indicated by the excretory waste in the urine
in fevers, constitutes a good indication of the severity of the
disease,—often, indeed, better than the thermometer or pulse.

Upon subsidence of pyrexia the urine assumes directly the opposite characters of the pyrexial state. The chlorides become increased, while the urea, uric acid, phosphates, and pigments are below the normal range. The volume of urine is increased. In short, everything points to delayed metamorphosis, conservation of tissue, and repair.

ACUTE INFECTIOUS DISEASES.

Typhoid Fever.—The quantity of urine is diminished about 50 per cent. during the febrile stage of typhoid, and the volume gradually rises to normal in the third, fourth, or fifth week. The specific gravity of the urine ranges from 1.025 to 1.030, occasionally reaching as high as 1.040. The quantity of urea is increased 25 per cent. or over; as high a range as 78 grammes per day has been recorded. In cases of marked splenic enlargement, copious hæmorrhages, and complicating nephritis the quantity of urea becomes lessened instead of increased. The chlorides suffer marked reduction, especially during the first week, but they gradually increase as convalescence approaches, while with urea this order is reversed. The uric acid is uniformly increased, the increase being relatively greater than that of urea. The gross amount of uric acid increases usually until about the end of the second week, when it reaches about double the normal range. Deposits of urates are common. The phosphates are somewhat diminished, and the pigments are increased relatively and absolutely. The reaction of the urine is sharply acid when voided, but often rapidly turns alkaline upon standing, owing to ammoniacal transformation of the large output of urea.

Albumin is present in the urine in a large proportion of the cases of typhoid fever, probably in the majority. Few observers place the proportion less than 20 per cent. of the whole, while some claim to have found it in nearly all cases. The author's observations lead him to conclude that albuminuria is present in from 70 to 80 per cent. of all cases of typhoid. The quantity of albumin in the urine is usually small; but occasionally it is decided in quantity, and is attended by nephritis. In such cases the prognosis of the disease is always to be considered grave. In most cases the albuminuria is temporary, setting in soon after pyrexia becomes established and subsiding soon after the latter disappears. In a very considerable proportion of the cases, however,—larger than has been generally supposed,—albuminuria fails to subside with pyrexia, and becomes permanent. Leucin and tyrosin are usually to be found in the urine in typhoid, although in no very decided quantities.

Renal casts, epithelium, and blood are frequently seen in the urine. Indeed, with regard to casts, it is probable that they are nearly always to be found during the first and second weeks, if carefully sought.. The typhoid bacillus is present in the urine in a large proportion of typhoid-fever patients,—in fact, it has been demonstrated thus far in about 20 to 25 per cent. of the cases examined. Moreover, the presence of this bacillus often continues in the urine for two and even three weeks after the temperature returns to normal and the patient is convalescent, which shows the dangers that may arise from infection if care be not exercised in disposing of the urine in these cases.

The following claims are made by Ehrlich regarding the diazo reaction in typhoid fever: "The reaction is found in typhoid fever after the fourth or fifth day, and if absent the diagnosis is doubtful. If the reaction be slight, and only found for a short time, the case will usually be very mild. In simple febrile intestinal catarrh, chlorosis, hydræmia, diabetes, and in diseases of the brain, spinal cord, liver, and kidneys, the reaction is never obtained. The reaction is obtained occasionally in phthisis pulmonalis, rarely in measles, pyæmia, scarlet fever, and erysipelas."

"The diazo reaction in typhoid has no dependence upon the height of temperature, nor is it influenced by the medication. The morning and evening urine give the same intensity of reaction. If the reaction cease in the second or third week, the rule is that the fever will early decline and the further course of the disease be mild. On the other hand, long-continued reaction indicates severity and long continuance of the disease. If relapse occur, the reaction returns if it has previously disappeared."

The importance of the subject justifies a repetition of the essentials of the test, especially since it is strongly probable that much of the diversity of views as to the value of the reaction in typhoid fever has arisen in consequence of faulty methods in performing the test.

The principle of the test depends upon the fact that diazosulphobenzol unites with certain aromatic substances met with in the urine in typhoid, which form analines. The diazosulphobenzol being unstable, Ehrlich obtains it fresh for testing by keeping sulpho-anilic acid in solution with hydrochloric acid. To this solution sodium nitrite is added, which liberates HNO_3 and forms diazosulphobenzol. A full description of the test and the proper method of its manipulation are described on pages 134 and 135.

Scarlatina.—The urine in scarlatina assumes the usual febrile characters, more or less marked in proportion to the degree of pyrexia present. During the first week the volume is reduced, the urea and uric acid are increased, and sediments of urates are precipitated. The chlorides are often decidedly reduced. From the sixth to the eighth day, if the disease proceed favorably, the urine becomes abundant, pale in color, and the general characters approach the normal standard.

But the urine in scarlatina should always be the subject of special observation, in consequence of the fact that it is often especially affected in this disease. This is due to the frequent—almost invariable—implication of the kidneys some time during the course of the disease. Recent observations on an extensive scale indicate that nephritis exists in scarlatina, either in acute form or in mild, evanescent attacks, almost as constantly as the rash or the angina. The number of acute cases of nephritis attended by dropsy and high grades of albuminuria in scarlatina is estimated as about one in every six cases met with,[1] though the proportion varies greatly in each special epidemic of the disease.

As a rule, albumin appears in the urine about the fifth to the eighth day, usually subsiding within about nine days if the disease proceed favorably. The quantity of albumin in the urine is subject to the widest variation, sometimes amounting to mere

---

[1] See section on "Scarlatinal Nephritis" in Bright's Disease and Allied Kidney Affections, by the author. Lea Bros. & Co., Philadelphia, 1886.

traces, while, again, the urine becomes almost solid with albuminous coagulum when boiled or treated with albumin precipitants. A few cases are on record in which both nephritis and dropsy were present in scarlatina, while albumin was absent from the urine.

Casts are almost invariably to be found in the urine if carefully sought for. In cases proceeding favorably the casts are chiefly hyaline, often preceded by the so-called cylindroids of Thomas. If nephritis become established, the sediment contains the usual elements characteristic of that lesion, more or less pronounced in proportion to its grade. Thus, we may meet with epithelial and bloody casts, as well as granular, in chronic cases. In addition to these, free blood-corpuscles and often large showers of round epithelium are observable. As occasional features, the urine in scarlatina has been found to contain sugar in small amount, hæmoglobin in severe—usually malignant— cases, and sometimes peptone or, more accurately speaking, deutero-proteose, which is usually taken for peptone.

Albuminuria sometimes subsides during the early course of scarlatina, to re-appear later with nephritis. If acute nephritis of pronounced type occur in scarlatina, it usually appears about the eighteenth to the twenty-fourth day; more rarely it arises in the latter part of the second week; more rarely still during the fifth week.

Cholera.—During the algid stage of cholera the urine is more or less completely suppressed. This is due, in the main, to collapse,—weakened circulation,—perhaps to some extent, also, to thickened blood and exudation into the renal tubules. After the cold stage is passed the volume of the urine slowly increases, or it may remain suppressed; in the latter case the patient dies comatose. If the secretion of urine become re-established, it usually does so during the third day from the attack,—forty-eighth to the seventy-second hour. The quantity of the first urine is small, gradually increasing if the patient recover; and on the fifth or sixth day the volume reaches the normal range, or often considerably above. The specific gravity of the urine at first, upon re-appearing, is below normal,—often 1.006 or 1.008. It may fall still lower at first, but gradually rises to normal as convalescence is established.

The quantity of urea in the urine is greatly reduced in cholera,—in fact, during the first day it may amount to but 2 or 3 grammes, and Bigbie even found it absent. It usually increases somewhat on the second day, and in favorable cases it increases on the third to the sixth day to considerably above the normal range. The prognosis may be considered favorable in proportion to the amount of urea excreted in cases which have passed the algid stage. Uric acid is usually present, though in reduced quantity. It often falls out of solution in a colorless state, owing to the absence of pigment in the urine. The phosphates are greatly reduced in the first urine passed; but, like urea, they become markedly increased with the re-establishment of the urine, and for several days even considerably exceed the normal amount. The chlorides at first are absent, or nearly so; after the fourth or fifth day they gradually return, and increase to, but rarely very much above, normal. The increase of chlorides is considered as even more favorable a prognostic sign than that of urea. As the urine becomes re-established its acidity becomes greatly increased. The normal urinary pigments are nearly absent for the first two days, gradually returning by the sixth day or so to the normal range. The first urine voided in all cases contains so-called indican—indoxyl-potassium sulphate—in marked quantities. Indeed, before the isolation of the cholera bacillus the presence of this substance in the urine was by many considered the best diagnostic indication of cholera where the symptoms were so mild as to cause doubts as to the nature of the disease. Our knowledge of so-called indicanuria, however, has so far increased that we know it is not uncommon in intestinal catarrhs of various kinds, as well as in certain general diseases attended by pronounced albuminous transformation.

The first urine passed in cholera almost invariably contains albumin. So constantly is this the case that, if the urine were always free from albumin in other forms of diarrhœa, cholera might almost be diagnosticated by the accompanying albuminuria. As a rule, the albuminuria of cholera is of comparatively brief duration, subsiding within a week after its appearance. There is no doubt, however, that the albuminuria sometimes pursues an acute course, accompanied by nephritis, which latter is the direct cause of death by uræmic coma. As a general rule, the quantity of albumin in the urine is proportional to the degree and duration of the algid stage.

Renal casts are invariably to be found in the urine of cholera patients, as well as large deposits of epithelium. With these, blood-corpuscles and uric-acid crystals are usually found. Later on, calcium-oxalate crystals are deposited, either alone or associated with amorphous urates.

Diphtheria.—The urine in diphtheria is decidedly reduced in volume, of high specific gravity, of sharply-acid reaction, and it deposits a copious sediment of uric acid, amorphous urates, oxalates, and sometimes phosphates. The urine contains albumin in over 50 per cent. of the cases of diphtheria, varying in quantity from mere traces to the most pronounced types of albuminuria. The albumin usually appears early in this disease. If of mild grade it sometimes subsides temporarily, and may occur even repeatedly. There is no doubt that the kidneys become implicated in many of these cases, but the tendency to spontaneous recovery from nephritis seems greater than in scarlatina or pneumonia. The quantity of urea is largely increased, and this continues throughout, unless the function of the kidneys becomes much crippled in consequence of associated nephritis. At the height of the disease the quantity of urea is often double the normal range. Renal casts are frequently present in the urinary sediment; less often, blood. Pus sometimes appears in considerable quantity as a consequence of pyelitis, which is often evoked by this disease.

Variola.—In small-pox the urine assumes the usual typical features common to pyrexia, which continue until about the twelfth day. The urea is moderately increased, and reaches its highest average with the highest temperature. Even in the absence of marked temperature and upon a spare diet the urea is still above the normal range. The uric acid is excreted pretty uniformly in excess throughout the course of the disease, and deposits of urates are constant. The chlorides are somewhat, though not decidedly, reduced. The sulphates suffer slight reduction, though to no marked degree, probably only consequent to the lessened quantity of food ingested. The pigment is very considerably increased.

Albumin appears in the urine in about 30 per cent. of the cases of small-pox. It usually appears at the height of pyrexia, though in severe cases it often appears at the onset of the disease. As a rule, albuminuria is only temporary in variola, the tendency to leave behind serious nephritis being comparatively slight, although such results sometimes occur. The presence of bile-pigment in the urine is a frequent and noteworthy fact first observed by Schonlein in these cases. In malignant forms of the disease the urine contains hæmoglobin. In less severe cases hæmaturia is not uncommon, and often of pronounced degree. Casts, epithelium, and other evidences of nephritis are to be found in the urinary sediment in cases associated with albuminuria.

Yellow Fever.—The quantity of urine is markedly diminished from the onset of this disease. In many cases it becomes nearly or even quite suppressed, and in such cases all the manifestations of uræmia follow. The reaction of the urine is usually acid throughout the first stage, and becomes alkaline during convalescence. The color of the urine varies: it may be bright yellow, dirty orange-colored, greenish brown, olive black, or sometimes red from the presence of blood. The urine becomes albuminous almost without exception, and very often highly so. A very close relationship undoubtedly exists between the more serious symptoms of the disease and the associated renal lesions, for in such cases albuminuria is *always* present in the urine, together with tube-casts, epithelium, blood, and the morphological elements common to acute nephritis. In addition to this, all the usual phenomena of uræmia follow in most of such cases, but especially coma. The urea is greatly diminished; sometimes it is totally absent. The uric acid is also greatly diminished, and is even said to be absent in some cases.

Typhus Fever.—The volume of urine is gradually, though not profoundly, diminished up to the third week of this disease. The color is increased up to the crisis, after which it becomes normal, and during convalescence it is lighter than in health. The reaction of the urine is sharply acid. The urine may remain practically free from deposit, although about the crisis a deposit of urates is frequent. The quantity of uric acid is increased, though less marked, and less uniformly so than urea. The most notable feature of the urine, however, in typhus is the marked reduction in the quantity of chlorides, which in severe cases practically amounts to complete absence or retention. This is not due to diarrhœa, lack of food, or other readily explainable cause, but seems a constant feature of the disease itself.

The urine in typhus is frequently albuminous, perhaps more frequently so than in typhoid, although there is some difference of opinion upon this point. Most are agreed, however, that in the more severe forms of the disease albuminuria is the rule. The quantity of albumin in the urine varies much, but is frequently in large amount in serious cases; appearing usually toward the crisis of the disease, most often on or about the sixteenth day.

DISEASES OF THE LIVER.

Cirrhosis.—The quantity of urine is constantly diminished in cirrhosis of the liver, so constantly indeed that a copious flow of urine may be considered strongly presumptive evidence of the absence of this disease. Diuretics act with difficulty or fail to be effective until the congestion of the liver be modified by mercurials or purgatives, and it is in such cases that calomel acts as a diuretic. The color of the urine is markedly increased, being dark, red, brown, and even blackish. The acidity of the urine is increased, and, unlike in pyrexial states, the acidity rapidly intensifies after the urine is voided. The dark coloration of the urine is due to the presence of bile-pigments in the urine. So constant, indeed, is the association of bile-pigment in the urine in this disease that it may serve to distinguish ascites of hepatic origin from that of peritoneal effusion, especially in non-febrile forms of the latter. The excessive bile-pigmentation of the urine in cirrhosis of the liver receives its most plausible explanation in the facts of long detention of the blood in the liver, where the red corpuscles are subjected to prolonged metamorphosis.

The solids of the urine vary widely in quantity in cirrhosis of the liver, chiefly in consequence of the great variability of the appetite and digestion, the stomach being subject to more or less disturbance in these cases through implication with the cirrhotic process. With a normal appetite and digestion the quantity of urea is subject to little change. On the other hand, the quantity of uric acid is pretty uniformly above normal, and often very considerably so. The chlorides are somewhat deficient; most so in cases attended by ascites, the ascitic fluid becoming heavily charged with sodium chloride. Deposits of amorphous urates and calcium-oxalate crystals are common. Albuminuria is rare, save in cases depending upon valvular lesions of the heart. The urinary sediment, as a rule, does not contain renal casts and other evidences of nephritis, save in very exceptional cases, mostly due to cardiac disease.

Jaundice.—As a rule, the volume of urine suffers some reduction in jaundice, although it is often quite up to the normal range. The quantity of urea, as a rule, is diminished; the uric acid is increased; the sulphates are not especially altered in quantity; hippuric acid is usually absent; the urine is always highly acid, the acidity rapidly intensifying after the urine is voided. The color of the urine depends upon the presence of bile-pigments, and varies from saffron-yellow to dark greenish brown or sometimes porter color. The bile-pigments are sometimes present in the urine in very large quantity, as well as the bile-acids. Benzoic acid fails to be eliminated by the urine as hippuric acid, contrary to conditions of health. In marked cases sugar sometimes appears in the urine, and this may be regarded as an unfavorable indication.

The chief value of examinations of the urine in jaundice is to establish the diagnosis at an early period. Our present tests for bile-pigments enable us to ascertain when the disease is coming on, as these products appear in the urine very early. The presence or absence of bile-elements, as well as leucin and tyrosin, in doubtful cases assists in distinguishing between obstructive and non-obstructive jaundice.

Acute Yellow Atrophy.—This somewhat rare and rapidly-fatal disease is attended by marked icterus and by extensive destruction of the hepatic cells. The urine is strongly acid in this lesion, and contains both bile-pigments and the bile-acids. The quantity of the urine is much diminished, though not suppressed, and the color is usually very dark brown. The urine usually contains leucin or tyrosin or both, the former nearly always, and sometimes in very large quantity. The quantity of urea is greatly lessened and sometimes nearly absent; uric acid and phosphates are usually reduced in quantity. The urine sometimes contains albumin and casts, although with no great constancy, and when casts are present they often appear yellow from staining with bile-pigment.

ARTICULAR DISEASES.

Acute Rheumatism.—In rheumatic fever the general features of the urine are typically those of pyrexia. The quantity of urine is diminished; the specific gravity is increased, as is also the color; the reaction is sharply acid, and on cooling the urine deposits sediments of deeply-colored urates. The quantity of solids is increased, but chiefly that of urea. The increase of urea usually reaches from 150 to 250 grains above the normal daily range. The quantity of uric acid is somewhat increased, though to a much less extent than urea. A copious precipitate of urates usually occurs at or shortly following the crisis of the disease, and as convalescence becomes established this tendency subsides. The chlorides are subject to a considerable reduction in quantity, though not to so extreme a degree as in pneumonia. The return of chlorides is comparatively early, usually as soon as the temperature declines to 100° F. and the joint-swellings begin to subside. The sulphates are increased in quantity to a marked degree, usually reaching double the normal range. In acute rheumatism and pneumonia the blood contains a marked excess of fibrin, and these two diseases furnish the most marked examples of increase of sulphates in the urine of all acute fevers. The quantity of phosphates in the urine is not materially altered in acute rheumatism.

Albumin often appears in the urine in small amount, but it is usually transient. Exceptionally it occurs in marked quantity, attended by nephritis. In such cases, morphological elements are found in the urinary sediment characteristic of nephritis. The kidneys, however, are less frequently, as well as less profoundly, affected than in pneumonia.

Acute Gout.—Before the attack of gout, or, in other words, between the paroxysms, the urine is more or less deficient in solids, especially uric acid, urea, extractives, and phosphates. In the case of uric acid the reduction is greatest immediately before the attack, when it may, indeed, be totally absent. The same thing occasionally occurs in chronic gout during the formation of tophaceous deposits.

During the attack the volume of urine diminishes more or less markedly. Exceptions to this rule occur in cases characterized by chronicity. The frequent co-existence of cardiac hypertrophy and early renal cirrhosis in chronic gouty subjects furnishes the key to the solution of the occasional polyuric form of gout, as well as those exceptional cases attended by deficiency of urea in the urine. As a rule, the excretion of urea is not materially altered during the paroxysm of gout. Garrod first pointed out the now well-known fact that the urine is greatly deficient in uric acid in gout; and this applies both to the attack and the intervals between the paroxysms. Sir William Roberts, pursuing the chemistry of the subject further, has shown[1] that the uric acid is retained and precipitated in the tissues in a state of combination as biurate. The urinary pigment, on the whole, is somewhat deficient in gout. Exceptions to this occur in cases attended by pyrexia. The phosphates of the urine are markedly deficient, notably those of sodium, which doubtless goes to make up the tophaceous deposits in the joints. The sulphates are not essentially altered in quantity.

The urine frequently contains albumin, nearly always in minute quantity, and sometimes it is a permanent condition. Casts are often to be seen in the urine, nearly always of the small, narrow, hyaline order. These features are usually the result of accompanying cirrhosis of the kidney, which is very frequent in gout. Crystals of calcium oxalate are frequent features of the urinary sediment. In chronic gout this same deficiency of uric acid occurs throughout, though more intermittently so.

---

[1] Croonian Lectures for 1892.

DISEASES OF THE NERVOUS SYSTEM.

Epilepsy.—In the intervals between the epileptic seizures observations thus far have failed to establish any changes of a constant character in the urine, although the urea, uric acid, chlorides, and phosphates have been claimed to be diminished. Immediately succeeding the attacks the volume of the urine is often markedly increased, of pale color, of low specific gravity, feebly acid in reaction, and sometimes contains albumin, more rarely sugar. During and immediately succeeding the attacks the urea and phosphates are increased and deposits of urates and uric acid are common.

Hysteria.—Perhaps no physical phenomenon is more widely known than the marked increase and pale, aqueous appearance of the urine during or immediately succeeding an attack of hysteria. The urine resembles precisely that voided after copious libations of water. The color is pale, "watery," the specific gravity is greatly lessened, the acidity is diminished, and the solids are relatively reduced. The quantity of urine voided by hysterical patients is sometimes very large; two or even three pints is not at all uncommon at one passage. It is, perhaps, not generally or widely known that in hysterical states the urine is sometimes totally suppressed. Laycock was the first to point out this fact, and more recently Charcot has called special attention to this matter, and has recorded a case in which no urine whatever was secreted for eleven days. The utmost watchfulness of the patient was ordered, so that deception was not possible. The patient suffered much of the time from vomiting, and the ejections contained urea. No other serious symptoms ensued, and the urine was ultimately re-established spontaneously. The author has observed, in cases of hysteria due to nervous exhaustion, that the volume of urine for twenty-four hours often becomes reduced to 15 ounces or even less, which he attributes to the deficient vascular tension and insufficient supply of nervous force to the kidneys.

Meningitis.—The urine in meningitis is more or less highly concentrated. The specific gravity is accordingly high, but the reaction is weakly acid, sometimes alkaline. The phosphates are greatly in excess, and are readily precipitated in large amount upon boiling the urine. The chlorides are normal or slightly increased. The quantity of urea is uniformly above normal, the increase usually amounting to 25 per cent. or more. The urine usually contains a small quantity of albumin as a transient condition.

DISEASES OF THE RESPIRATORY ORGANS.

Pulmonary Tuberculosis.—The urine is subject to considerable variation in its characters in tuberculosis of the lungs, in consequence of the numerous incidental complicating conditions present,—such as diarrhœa, pyrexial periods, copious diaphoresis or expectoration,—as well as the great variability in the quantity of food taken. As a rule, the volume of urine is somewhat augmented. This is largely due to the increased quantity of water consumed, thirst being more or less constant in consequence of the pyrexia. During attacks of diarrhœa the volume of urine decreases temporarily, often to 30 ounces and sometimes even to 15 ounces. The same decrease precedes a fatal termination of the disease. If the appetite remain good, and the disease be proceeding without marked disturbing features, the urea remains about normal in quantity, perhaps slightly below. During marked hectic the quantity of urea diminishes decidedly, especially just before the rigors; after the rigors it rises rapidly and reaches its highest excretion about an hour before the sweating stage begins. In cases of intestinal irritation accompanied by vomiting, diarrhœa, anorexia, etc., the quantity of urea is subject to sudden and decided decrease. Uric acid suffers little, if any, reduction; in fact, it usually rises above the normal range. The sulphates are little, if any, affected,—perhaps slightly reduced. The excretion of chlorides varies very much, the variation depending chiefly upon the quantity of food taken, the degree of pyrexia, and the degree of elimination by the skin and bowels. The quantity of pigment varies in phthisical urine owing to the disturbing influences of hectic, diarrhœa, and pyrexia. A pink sediment is often seen in phthisical urine, especially during hectic stages, and the precipitated urates often become of a pink or carmine hue. In cases rapidly progressing toward a fatal issue the diazo reaction of Ehrlich is sometimes obtained in the urine. If this reaction be present for any length of time the prognosis may be regarded as unfavorable.

Albumin is often present in the urine of phthisis, the quantity being usually small, except in cases complicated by amyloid kidneys. The milder grades of albuminuria are chiefly due to impaired nutrition of the renal epithelium, though occasionally, perhaps, the result of pyrexia. Renal casts, epithelium, and blood are found in the urine exceptionally, the casts chiefly in cases complicated by renal lesions, of which amyloid disease is the most frequent, and this is usually associated with cases of marked chronicity.

Pneumonia.—The general pyrexial characters of the urine are well marked in pneumonia. The quantity of urine is diminished one-third to one-half. The quantity of urea is increased, as is also uric acid; the greatest increase occurs on the so-called critical days. It is at such times that enormous deposits of amorphous urates are so often observable in pneumonia,—more marked than in other febrile diseases of equal degree of pyrexia. The specific gravity of the urine is increased,—1.025 to 1.035. Pigmentation of the urine is markedly increased, which intensifies the color of the precipitated urates to deep brown, red, or even carmine. The increase of pigment often reaches two or three times the normal range.

The chlorides are invariably greatly diminished or absent during the early stages and commencing hepatization. This deficiency of chlorides continues until convalescence is well established, sometimes for several days after, when they re-appear in great excess, indicating their retention during the active stages of the disease. The absence of chlorides indicates a period of danger in pneumonia. The sulphates are uniformly increased in pneumonia from one-fourth to one-third above the normal range; only exceptionally do cases occur without being attended by this increase. The quantity of phosphates in the urine suffers more or less reduction.

Of the morbid products met with in the urine in pneumonia, albumin is by far the most constant, the average being about 45 per cent. of all cases. Albumin usually appears in the urine at the height of the disease, especially during the stage of consolidation. Its appearance in notable quantity must be looked upon as an unfavorable indication; the death ratio reaches from 45 to 50 per cent. in such cases, while in non-albuminuric pneumonia the death ratio is only about 15 per cent. The fact that albuminuria is so frequent in pneumonia—perhaps as frequent as in any other febrile disease—strongly favors the now generally-accepted view of the infectious nature of the disease, for albuminuria is now known to be one of the most frequent features of infectious fevers. Furthermore, the albuminuria of pneumonia is evidently independent of the intensity or extent of the local pulmonary lesions and of the disturbed function of the lungs, since frequently cases of the most extensive consolidation and urgent dyspnœa are unattended by albuminuria; while, on the other hand, cases of comparatively mild grade are often accompanied by both albuminuria and nephritis. The kidneys are often seriously damaged in pneumonia; the nephritis some-times remains comparatively latent throughout convalescence, as in scarlatina, to be discovered weeks or months after the pneumonia has subsided, attention being called to the condition through uræmia, dropsy, or some of the usual symptoms of nephritis. From the above considerations, as might be expected, the urine of pneumonia often contains blood, epithelium, casts, and the usual products associated with nephritis. The urine also contains an excess of mucus, which renders the urine unstable; hence, the urine in pneumonia possesses a decided tendency toward alkaline decomposition, and its reaction quickly becomes alkaline upon standing.

Exceptionally pneumonia is attended not only by diminished volume of urine, but also by decrease of the solids, including urea and uric acid. Such cases are almost invariably characterized by delayed convalescence, diarrhœal attacks, and slow recovery. The proportion of solids in the urine is, in fact, one of the most trustworthy guides for prognosis in pneumonia, and cases attended by diminished excretion never proceed so favorably as those in which the solids are excessive.

During convalescence the volume of the urine increases; the quantity of uric acid, urea, and sulphates, previously excessive, now gradually diminish to or below the normal range; while the chlorides, previously held back, now increase considerably above normal, often, in fact, to an extraordinary degree.

Acute Pleurisy.—The urine in pleurisy presents the usual febrile type, though not so pronouncedly as in pneumonia. The volume is reduced; the organic solids are increased, notably during pyrexia. Both urea and uric acid are somewhat excessive. The chlorides, sulphates, and phosphates are but little changed; sometimes there is a reduction in the quantity of chlorides, but not to any marked degree. Albuminuria is uncommon in pleurisy; sometimes it appears in mild form, but it is usually of temporary duration. The urine often contains a very considerable amount of peptone, especially during the stage of resolution, when absorption of large serous effusion is in progress. Pleurisy presents an example of the fact that intense pain often causes but little effect over tissue metamorphosis, since few pyrexial states present so little alterations in urine indicative of tissue waste as does simple pleurisy.

Acute Bronchitis.—The urine in bronchitis varies much in character in different cases. This, indeed, is to be expected, since the grades of the disease are of all degrees, from slight catarrh of the large tubes to disease including nearly all the small tubes of both lungs, with collapse of the air-cells, to the extent of entailing sharp dyspnœa. In the latter case the urine approaches the same characters as those in pneumonia.

In cases of severe diffuse capillary bronchitis accompanied by dyspnœa, there is usually a marked diminution of nearly all the solid constituents of the urine, more so than perhaps in any other disease save cholera. Parkes records the case of a young man who, during two days, voided but 294 cubic centimetres of urine per day, containing only 244 grains of total solids, 176 grains of which were urea, 10.8 grains of sulphates, and no chlorides. There was no albuminuria or symptoms of uræmia, and the urine became gradually established as the patient recovered. Moos has recorded a similar case in a girl who, during the height of the disease, excreted only 9 grammes of urea (139 grains) and 2 grammes of sodium chloride. The conditions favoring the retention of chlorides are those interfering with proper aëration of the blood, and there is reason to believe that when these conditions are extreme the same influence extends to the entire solids of the urine. The urine is occasionally albuminous in bronchitis, but this condition is nearly always transient; implication of the kidneys being rare.

DISEASES OF THE DIGESTIVE SYSTEM.

Under the term "dyspepsia" a number of symptoms are commonly grouped which are often of widely different causation, and many of which are only recently becoming understood, notably some of the forms associated with auto-intoxication. In chronic disorders of the stomach and intestines the urine is often increased in quantity, of pale color, of lowered specific gravity, with a tendency toward alkaline reaction from fixed alkali. In consequence of the alkaline tendency of the urine, there is a proneness to precipitation of the earthy phosphates, and such urine is often turbid from this cause when voided. As a secondary consequence, vesical irritation is a common accompaniment which not infrequently leads to the formation of gravel. The quantity of urea depends upon the digestive power of the stomach. Usually the quantity of urea in the urine is below normal, often markedly so; the quantity of chlorides correspond closely with the amount of food taken.

In disorders of the intestinal tract attended by increased albuminous decomposition the urine is apt to contain large amounts of so-called indican, and this is especially favored by obstinate constipation or obstructive diseases.

In organic diseases of the stomach and intestines, especially if associated with ulcerative changes, the urine often contains notable quantities of peptone. A mild type of albuminuria is sometimes associated with disorders of the stomach, and it is, indeed, remarkable how frequently the so-called functional albuminuria is found associated with disorders of the stomach. In such cases, though the albuminuria is persistent, often extending over periods of years, it is rare to find casts in the urine or other evidences of nephritis. Small quantities of sugar are sometimes found in the urine in dyspeptic conditions. Of the urinary sediments found in digestive disorders, calcium-oxalate crystals and amorphous phosphates are the most common.

APPENDIX A.

EXAMINATION OF URINE FOR LIFE-INSURANCE.

THE examination of the urine for life-insurance has for its object the determination of the presence or absence of diseases of the urinary organs which tend to abridge the normal expectancy of life. As a special field of urinary diagnosis, this has grown to very extensive proportions in nearly all civilized countries, and, since the interests involved are so important and wide-spread, the subject is well deserving of special consideration.

The questions involved in the examination of the urine for purposes of life-insurance often call for a high order of skill and judgment for their accurate determination, and experience has demonstrated that the adoption of systematic methods of conducting the examination not only simplifies the subject, but also renders the conclusions reached more trustworthy. It has, therefore, become the custom with many insurance associations to furnish certain rules as a guide in conducting examinations of the urine, which are intended to cover the more important points of information desired. Notwithstanding these precautions, life-insurance associations still find a very large percentage of their unprofitable risks arise through diseases of the kidneys which have escaped detection. It is with a view of contributing to the avoidance of such losses, on the one hand, and, on the other, of securing to applicants the privileges of life-insurance to which they may be fairly entitled, that the following suggestions are presented as a guide for medical examiners.

A due regard for the interests of his company, as well as for his own reputation, should prompt the medical examiner to personally ascertain that the urine about to be examined has been voided by the applicant. Substitutions of healthy urine by un-

healthy applicants are matters of undoubted facts in the history
of insurance circles, and, though fortunately of infrequent occurrence, the examiner should be on his guard against such possible
source of imposition. The applicant should be directed to retain
his urine for, say, two or three hours before presenting himself
for examination ; and upon his arrival he should be given a clean
glass vessel and requested to void therein his urine, which is
always preferably done in the presence of the medical examiner ;
but if this be impracticable, the temperature of the urine should
be immediately noted in order to guard against possible imposition. This method also secures a perfectly fresh sample of urine
—always a matter of prime importance—and, moreover, voided
at a time most desirable for the purpose, viz., after food and
exercise ; for it should not be forgotten that the urine voided
on rising in the morning (which it has become somewhat
the custom to furnish for these purposes) is the least likely to
contain either albumin or sugar, when these are present in
minute quantities.

If the examination of the urine be unsatisfactory, or if there
be reason to believe that the sample examined be not a fair average, it will be advisable to have the whole twenty-four hours'
product of the kidneys *collected*, *mixed*, and *measured*, and a
sample of this mixture examined, and compared with a freshly-
voided sample of the urine.

PHYSICAL EXAMINATION OF THE URINE.

Having secured a perfectly-fresh sample of the applicant's
urine, it should be allowed to stand until it cools down to a temperature of about 75° or 78° F., and careful observation made of
its appearance and physical characters. If the color be very
light (watery), it suggests diminished specific gravity, possibly
hydruria,—diabetes insipidus. Should the color be of decided
greenish tint, the possible presence of sugar is suggested :
should the color be unduly increased,—reddish,—excess of
urates or the presence of blood is inferred; in the first case
suggesting rheumatic or gouty conditions ; in the latter, calculi or some organic renal or vesical lesion. The transparency
or opacity of the urine should be carefully noted : if cloudy, add

a few drops of acetic or other acid to a sample in a test-tube, and if the urine become perfectly cleared thereby the earthy phosphates in suspension was the cause of the opacity, their suspension being due to diminished acidity of the urine, and it suggests some such condition as fasting, dyspepsia, or general debility. If the opacity of the urine fail to yield to the action of an acid, gently warm the upper layers of the urine by holding the test-tube over a spirit-flame, and if it now clear the opacity was due to suspension of the amorphous urates, the significance of which has already been stated. If, however, the urine still remain cloudy after treating it with both heat and acid, the opacity is due to the presence of pus, bacteria or cellular elements, and the necessity of microscopical examination is suggested for diagnostic purposes.

The specific gravity of the urine should next be observed by the aid of a good instrument,—preferably a Squibb's specific gravity instrument (urinometer),—and the range should be near 1.020. If the specific gravity, however, be above 1.025, it suggests the possible presence of sugar; if below 1.018, it is suspicious of the presence of albumin, and should lead to further investigations in these special directions.

The chemical reaction of the urine should next be ascertained by means of litmus-paper. If found to be very sharply acid, as indicated by intense reddening of blue litmus, the possibility of the presence of sugar is suggested, since the urine is more sharply acid when it contains sugar than in most other conditions. Should the urine be frankly alkaline in reaction, as indicated by decided blue coloration of red litmus-paper, the paper should be dried in order to ascertain if the alkalinity of the urine be due to fixed or volatile alkali. If the red color return upon drying, after having turned blue in contact with the freshly-voided urine, ammonia is present or volatile alkali, and this suggests the presence of some chronic inflammatory condition of the lower urinary tract, most likely the bladder. If, on the other hand, the blue color remain permanent upon drying the paper, the urine is alkaline from fixed alkali, and indicates unusual alkalescence of the blood, which may suggest fasting, dyspepsia, or vegetarian habits of eating.

A careful survey of the physical characters of the urine, as just indicated, will often lead further investigation in certain directions, and also lend confirmation to the points brought out by further investigations of the urine. But even though all the physical characters of the urine be found perfectly normal, the urine cannot be positively asserted to be healthy until search has been made at least for sugar and albumin, since either or both of these may exceptionally be present in small amounts without their presence being indicated by a physical inspection of the urine.

CHEMICAL EXAMINATION.

The most simple chemical examination of the urine for purposes of life-insurance should, at least, include a search for sugar and albumin and a quantitative estimation of the urea.

Albumin.—If the urine contain albumin in large or even medium quantity, it will be made apparent by almost any of the ordinary tests for albumin in use. When, however, the urine contains but minute quantities of albumin, the case is altogether different, and the greatest care is absolutely necessary not only to be able, in all cases, to positively identify it, but also to interpret its true significance. To the search for small quantities of albumin in the urine, therefore, the following considerations are intended chiefly to apply :—

The urine in all cases should first be filtered before submitting it to albumin-reagents. After filtration of the urine one or both of the following methods may be followed :—

1. Fill an ordinary test-tube half-full of the urine, and to this add ferrocyanide-of-potassium solution (1 to 20) to the depth of about an inch; after mingling the urine and the reagent thoroughly by inverting the tube a few times, add a few drops of acetic acid and again invert the test-tube a few times until the urine and reagents are well mingled. Finally, stand the tube in a good light and note any changes appearing. If albumin be present, in a half-minute or so a diffuse, milk-like turbidity will gradually appear throughout the test, more or less pronounced according to the quantity of albumin present. If the reaction seems doubtful, it will be found useful to stand another

tube filled with the filtered urine beside the one containing the test for purposes of comparison.

2. Fill a test-tube about half or two-thirds full of the filtered urine, and to this add about one-sixth its volume of saturated solution of chemically-pure sodium chloride. Next, mingle the urine and sodium-chloride solution by inverting the tube a number of times, and then add a few drops of acetic acid. Lastly, heat the upper third of the test over a spirit-flame until it gently boils, and then stand the test in a good light for observation. If albumin be present, a white cloud, more or less dense, will appear in the upper, boiled portion of the test, while the lower, unboiled portion will remain clear and unclouded.

Both of these tests possess the great advantage over most other albumin reagents of giving no reaction with nucleo-albumin or mucin. In point of delicacy they are probably as sensitive as at present attainable, coupled with trustworthiness.

Heller's nitric-acid method has been much employed heretofore in insurance circles; indeed, few, if any, albumin tests have become so popular and so generally adopted. Its simplicity leaves little to be desired. Then, too, many observers are wedded to the contact method of testing, and, certainly, when the eye is trained to this method it requires practice to acquire a new method. However, accuracy and trustworthiness are, after all, the prime *desiderata* in insurance as in all other urinary work, and after many years' careful and patient investigation the author does not hesitate to give the preference to the two tests above described over all other methods for detecting *small quantities of albumin* in the urine.

The nitric-acid method of Heller is a ready and excellent test in the presence of considerable amounts of albumin in the urine. When mere traces of albumin are present, however, it requires from twenty minutes to half an hour to bring it to light with certainty. In addition to this the reactions of this test with mucin, oleoresins, urates, etc., require corrections and precautions which demand skillful manipulation and interpretation to reach trustworthy results; and, therefore, the interests both of the company and the applicant are best served by the use of the more trustworthy methods. The quantity of albumin in the urine is most rapidly determined by the centrifugal method described on page 83.[1]

Significance of Albuminuria.—Having positively identified the presence of albumin in the urine, in all cases its true significance should be next traced.

As a rule, minute quantities of albumin in the urine are less likely to be the result of serious disease of the kidneys in young people than when met with in people beyond middle age. A very considerable percentage of such cases in the young belong to that class which has been termed *physiological* or *functional albuminuria*, terms which are meant to indicate that the kidneys are not structurally damaged. In such cases, in addition to the fact that the subjects are mostly young, the urine presents certain fairly uniform features, viz.: the specific gravity of the urine is usually increased to 1.025 or above; albumin is often absent from the urine on rising in the morning, but appears plainly after exercise, food, or mental excitement; the urine is free from renal casts and significant morphological elements; and the quantity of urea in the urine remains normal. While many such cases continue for years without any marked changes either in the urine or in the general health of these subjects, yet a certain proportion of them ultimately develop into serious nephritis. The fact that we possess no positive data by means of which we can with certainty distinguish the special cases of this class which will ultimately terminate unfavorably from those which will pursue a favorable course, and, furthermore, since nephritis is more liable to arise from slighter causes in these cases than ordinarily, it cannot be affirmed that these cases are safe risks. It has been suggested by some authorities that these cases might be accepted for a limited endowment insurance of, say, five or ten years, but

---

[1] It has been shown (pages 83 and 84) that the unit of measurement of Esbach's albuminometer-tubes, as ordinarily supplied by the dealers, is too large (1 gramme) for accurate measurement of small amounts of albumin in the urine. Messrs. Eimer & Amend, of 205 and 211 Third Avenue, New York, now manufacture and supply these albuminometers, the first three grammes of the reading being graduated in tenths of a gramme, which obviates the above-named objection. Moreover, these tubes are specially adapted for use with the author's electric centrifuge, so that those who prefer Esbach's quantitative method to the author's may carry out the test in two or three minutes.

the safer course for the company would be delay for sufficient time to determine the ultimate course of the case. These questions, however, will be determined by the medical director at the home office, to whom all the facts and particulars of such cases should be carefully reported.

Traces of albumin are sometimes observable in the urine of perfectly healthy people, the reaction being due to dissolved morphological elements, and such is most frequently observed in women the subjects of leucorrhœa, or in men with slight bladder irritation. If such urine be permitted to stand in a conical glass for a few hours, or if it be submitted to the centrifugal apparatus, a deposit, mostly of epithelium and mucous corpuscles, becomes plainly visible to the naked eye. The nucleoalbumin of these structures becoming dissolved out, often renders the urine sufficiently albuminous to cause slight reaction in testing if the quantity of epithelium, etc., be abundant. It may be necessary, therefore, in the cases of women, to direct a vaginal douche to be used previous to voiding the urine for examination, while in other cases it may be necessary to quell the vesical irritation by medical treatment before passing opinion upon the true state of the urine.

Minute quantities of albumin are often observed in the urine of men at and beyond middle age, who not only appear perfectly healthy, but who have, as a matter of fact, enjoyed the most typical robust health all their lives. *Among this class will be found the largest number of those cases which have always proved so unprofitable to life-insurance associations, through concealed or overlooked disease of the kidneys.*

Two features in this class of cases stand out so prominently that they are well calculated to mislead the medical examiner, viz., (a) the robust general condition of seemingly perfect health of the applicant and (b) the minute (often doubtful without careful testing) traces of albumin in the urine often accompanying these cases. These facts teach us two highly-important lessons which the medical examiner would do well to remember: 1. That a healthy appearance or healthy personal record of the subject carries less weight, in reaching conclusions as to certain pathological conditions of the kidneys, than in any other disease. 2. That the presence of traces of albumin in the urine, however minute, are often the index of irretrievably-damaged kidneys.

Chronic interstitial nephritis, or so-called chronic Bright's disease, to which the foregoing facts especially apply, is nearly always the outgrowth and sequel to robust life, the kidneys being the first organs to fail under the stress of long-continued functional activity in eliminating the waste-products, which are always excessive in people of large appetites and ample nourishment. These cases present the following typical features: The subjects are, as a rule, over 40 years of age; usual history of robust health; appetite always good, often heavy; and the food has consisted largely of meat and highly-nitrogenous products. These people usually rise regularly at night—once, twice, or oftener—to void their urine, which, to all appearance, is normal and free from sediment. The urine, however, is usually of lowered specific gravity,—1.018 to 1.014,—more or less deficient in urea, and contains a small amount—often mere traces—of albumin. Microscopical examination merely shows the presence of a few small, perfectly hyaline casts. The pulse is always full, hard, and resisting to the finger,—almost characteristically so; the second sound of the heart is abnormally loud, and in many cases enlargement of the heart is plainly to be observed. If the above-named points be kept in view, the medical examiner will have no difficulty in recognizing this dangerous class of risks.

Sugar.—In searching for sugar in the urine, it is of prime importance to have on hand some trustworthy and stable test that may be depended upon when required. Fehling's solution has been much depended upon, but its well-known instability greatly detracts from its usefulness for the purposes under consideration. Fehling's solution will not keep, for reasons explained (page 111), and its preparation requires some time and pains. A better test, more simple in preparation and sufficiently stable that it may be kept on hand for months without impairment of its qualities for testing, is that devised by Professor Haines (page 103). This test, if properly manipulated, will yield as trustworthy results as it is possible for any copper test to give.

In searching for sugar with Haines's test, 1 drachm of the test-solution, in an ordinary test-tube, should be raised to the boiling-point over a spirit-lamp, and the suspected urine should be added, drop by drop, until 8 or 10 drops are added, *but not more.* The test should now be boiled for about half a minute, and, if sugar be present, a copious yellow or yellowish-red precipitate will suddenly appear throughout the whole mixture. If no such reaction take place, sugar is absent. The chief feature in the manipulation to be kept in mind is not to add more than the stated limit (8 or 10 drops) of urine. If this be disregarded and the urine be added (as in Fehling's test) to a volume equal to the test-solution, or thereabout, any urine may cause reaction when sugar is absent. This is due to the fact that normal urine contains certain substances (chiefly uric acid, creatinin, etc.) which possess feebly reducing powers over copper tests; and, therefore, if the urine be added in excess the test is liable to respond to these agents.

In exceptional cases the urine may contain an excess of uric acid or other reducing substances just named, or such foreign elements as tannin, carbolic acid, or vegetable alkaloids, which may cause slight reaction with this or any copper test. For the most part, this reaction is usually an imperfect one, the test-solution turning green rather than yellow or red, though exceptionally a frank yellow precipitate may be formed. Such cases, however, in reality are very rare, if the test be manipulated as directed. Should any doubts, however, arise as to the presence of sugar, after thorough cleansing of the test-tube and carefully repeating the test in all particulars, as directed, an appeal may be made to the phenyl-hydrazin test, as described (page 105), which may be considered conclusive. Having identified the presence of sugar in the urine, its exact quantity may be readily and rapidly determined by the author's method, already described in the text (page 108).

Significance of Sugar in the Urine.—The presence of sugar in the urine, in a general sense, is nowise less serious in its signification than is that of albumin. As in the case of albumin, the tendency has been to look upon small quantities of sugar in the urine as of no grave import; but recently this view of the subject is giving way to a belief that, as a rule, the presence of sugar in the urine, regardless of the quantity, means serious defect, either in the brain or the liver, or in both, and this view will prove the safer one to follow. It is true that in a few conditions, notably in some forms of indigestion, as well as over-ingestion of highly-saccharine or amylaceous foods, small quantities of sugar may appear in the urine in the absence of a diabetic state. On the other hand, genuine diabetes mellitus is often preceded for a time by precisely these symptoms, viz.: the appearance of small quantities of sugar in the urine, often, indeed, intermittent, but always aggravated by indulgence in saccharine or starchy foods. The author is unable to identify the special cases of the former class which subsequently do or do not terminate in diabetes, but he is able to affirm, from observation, that many of these cases of so-called " *digestive glycosuria* " end in fatal diabetes, more especially in young subjects. It has been truthfully said, by an author of wide experience on this subject, that " a man with sugar in his urine is like a house that is undermined ; he will surely fall, but no one can predict the precise time that the disaster will occur." It will be safer to accept this assertion as a guide in such cases.

Urea.—An examination of the urine can scarcely be considered complete which does not include an estimate of the quantity of its contained urea. Representing, as it does, by far the greater bulk of the organic output of the kidneys, the quantity of urea in the urine becomes a valuable index of the functional capacity of the kidneys, and, therefore, serious forms of disease of these organs are usually quickly and markedly reflected in the diminished excretion of urea.

The estimation of urea in the urine is now a matter of such simplicity and rapidity that it ranks among the more simple manipulative methods, such as testing for albumin and sugar, and, in reality, requires but little more time. The method best suited for the purposes under consideration is that known as the hypobromite test, which is most readily performed with Dr. Doremus's ureometer. A fresh solution of sodium hypobromite is necessary for testing, as the solution does not keep well. This is best obtained as follows : Have on hand a quarter- or half-pound bottle of bromine, and, after once opened with care, set it aside for use. Next prepare a solution of sodium hydroxid (caustic soda) by dissolving 3 ounces of caustic soda (in sticks) in 8 ounces of distilled water, which may also be kept on hand for use. In preparing the solution of sodium hypobromite for testing, pour 10 cubic centimetres of the caustic soda solution into a graduated glass, and with the pipette (furnished with the ureometer) take up 1 cubic centimetre of bromine[1] and mix thoroughly with the caustic soda solution; next add an equal volume of water, and, after thoroughly mixing until the solution becomes *transparent*, fill the bulb of the ureometer with the solution, and incline the instrument until the hypobromite solution fills the long arm. Next thoroughly cleanse the pipette, and take up 1 cubic centimetre of the urine and slowly discharge it into the hypobromite solution, in such position that the disengaged nitrogen-gas will all ascend the long arm of the instrument, where it is measured. As soon as the urea is all decomposed, as indicated by no further ascending of bubbles, read off the quantity as indicated by the amount of nitrogen-gas marked on the long arm of the instrument. Each number represents the fractions of a gramme of urea per cubic centimetre of urine, of which 0.02 is the normal proportion. If preferred, these ureometers are furnished with a scale indicating the number of grains of urea per ounce of urine instead of grammes per cubic centimetre.[2]

The normal quantity of urea is about 512 grains in twenty-four hours for a man of 145 pounds weight, upon a mixed diet and moderate amount of exercise. This gives an approximate proportion of 10 grains of urea per ounce of urine, the whole quantity of the urine being 50 ounces. It will be necessary to make an allowance of about 25 or 30 per cent. from this standard in order to cover variations caused by differences in weight, age, diet, and exercise. If, therefore, the gross quantity of urea sink below 350 grains (7 grains per ounce), there is reason to appre-

---

[1] Avoid inhaling the strong vapor of the bromine, which is very irritating to the air-passages.

[2] These instruments are furnished by Elmer & Amend, 205 and 211 Third Avenue, New York, at moderate cost.

hend the presence of some organic disease of the kidneys.
Should the deficiency be still more marked, the quantity of urea
diminishing to 250 or 200 grains (5 grains per ounce), it furnishes strong evidence of diseased kidneys. On the other hand,
a normal amount of urea in the urine, coupled with the absence
of albumin therefrom, strongly indicates that the kidneys are
healthy. Certain it is that no very advanced renal disease can
be present under such circumstances.

MICROSCOPICAL EXAMINATION.

A microscopical examination of the urine is often required
of the medical examiner in cases of applications for heavy
amounts of insurance. In addition to this, in cases in which a
chemical examination of the urine has not been conclusive, a
microscopical investigation should be made. It is not proposed
to here enter into the technique of microscopy, which every one
has access to in the numerous works especially devoted to the
subject. A few general suggestions in reference to the special
subject herein considered may, however, be of use.

The urine for microscopical examination should in all cases be
freshly voided, and it is better to have it somewhat concentrated
by directing the applicant to abstain from the use of fluids for a
few hours previous to voiding the urine for examination. Where
practicable, the sediment is preferably obtained by the centrifugal method. Should the centrifugal apparatus not be available,
proceed by adding 10 grains of resorcin, chloral hydrate, or
salicylic acid to the urine, in a conical glass, to preserve it from
change, and, after covering the glass, stand it aside for from
twenty-four to forty-eight hours, until the sediment subsides.
Then take up from 4 to 6 drops of the sediment, by means of a
nipple pipette, from the bottom of the deposit, and place them
in a shallow cell; cover the cell with a cover-glass, take up the
overflow of urine with the torn edge of a piece of blotting-
paper, place the slide under the microscope, and examine deliberately with a one-fourth-inch objective, avoiding too brilliant
illumination of the field, since too much light tends to render
hyaline casts invisible, as they are feebly refractive. Careful
search should be made over at least two slides prepared as above

directed, noting the presence and relative number of pus- or blood- corpuscles, but more especially the presence, number, and characters of any renal casts. If no pathological products be found after examination of two slides, the evidence may be considered conclusive in the negative.

PYURIA.

The medical examiner will encounter but little difficulty in detecting the presence of pus in the urine of the applicant if it be present in any notable quantity. Such urine is always more or less cloudy when voided, and the turbidity does not clear up by the addition of an acid or the application of heat; on the contrary, the cloudiness is rather increased by these agents. A sediment, more or less pronounced, quickly settles to the bottom of the vessel, and if the urine be decanted from the sediment and liquor potassæ added to the latter, it thickens into a jelly-like consistence, and becomes ropy and sticky, as will be observed in pouring it from the vessel. Should any doubts arise, however, as to the nature of this deposit, the microscope will readily reveal the presence of characteristic corpuscles, described in the text (page 182).

The significance of pyuria may be very grave; it can scarcely ever be considered trivial, inasmuch as it points to bacterial infection of the urinary tract. In people beyond middle age it is nearly always a serious matter, since in such people the resisting powers of the system against pyogenic germs are much diminished, and consequently the infection under such circumstances is more likely to extend than to subside. Moreover, the causes which most often provoke pyuria in elderly people are of a permanent nature.

On the other hand, pyuria in younger subjects is often the index of tubercular pyelitis, pyonephrosis, etc., which are not only, as a rule, incurable diseases, but often rapidly fatal ones. Pyuria in its least serious signification is the associate of mild forms of cystitis in otherwise healthy subjects, and as such it is usually soon recovered from, in young or even middle-aged subjects, under properly-directed treatment. On the whole, pyuria should be looked upon as more or less serious,

and the only safe course to pursue in these cases is to recommend
delay of the application for a few weeks until it be determined
if the condition be a permanent one.

HÆMATURIA.

Application will scarcely be made for life-insurance if blood
be present in the urine in sufficient amount to attract the attention of the applicant. Should such be the case, however, the
examiner will not be likely to overlook cases of pronounced
hæmaturia, as they will be apparent upon the most superficial
inspection of the urine. But the urine may become so highly
colored from concentration as to conceal minute quantities of
blood; or, again, blood may be present in quantities so minute
that it merely lends to the urine a deep shade of normal coloration. Close inspection will usually distinguish the abnormal
tint due to blood, and upon standing the distinction will usually
be more marked in the sediment. In cases of doubt, however,
the microscope will readily detect the presence of the blood, in
quantities however minute it may be present.

The examiner should always inquire for a history of attacks
of hæmaturia, with the circumstances associated therewith. The
appearance of blood in the urine, occurring at irregular intervals,
often indicates the presence of gravel, or the beginning of such
still more serious diseases as renal tuberculosis and cancer of
the kidney. If hæmaturia has appeared, inquiries should be
made as to its character, the frequency of the attacks, as well as
its extent. Thus, if the blood appeared intimately mingled with
the urine, of dark color, and without clots, it may be concluded
that its source was renal, and it would be presumptive evidence
of the presence of more or less serious disease. On the other
hand, if the blood appeared somewhat separate from the urine,
of a brighter and more arterial tint, and associated with small
clots, it may be concluded that it originated from the lower
urinary tract, most likely the bladder, and further inquiry is
likely to elicit symptoms of stone, villous growths, early tuberculosis, or even malignant disease. Unless in the early stages of
the conditions just named, the accompanying symptoms will be so
pronounced that they could scarcely be overlooked. There is a

period, however, in the early stages of most of these diseases, in which mild hæmaturia may be almost the only symptom observable; hence the importance of ascertaining its occurrence. It will be a safe rule for the medical examiner to consider a subject in danger of serious disease of the kidneys or bladder who presents a history of hæmaturia, unless five years have passed since the last appearance of such attacks, the urine remaining in the interval in all respects normal.

CALCULI.

An examination of the urinary organs cannot be considered complete without inquiring for any attacks of renal colic or the passage of calculi by the applicant. A number of such attacks may have occurred, and even small calculi may exist and be in process of growth in the kidneys without any evidence thereof being present in the urine.

The passage of renal calculus, however small, is usually attended by symptoms of so marked and characteristic an order that it will be distinctly remembered by the applicant upon interrogation. The sudden onset of pain in the side without any apparent cause; the intense suffering for some hours, in which the pain extends along the course of the ureter, often retracting the testicle on the affected side; frequently accompanied by nausea, vomiting, cold perspiration, and even some degree of collapse, and followed by sudden and complete relief. These form a chain of phenomena so pronounced and distinctive that they are neither likely to be overlooked by the applicant nor misinterpreted by the medical examiner.

A man may have one or more such attacks, and the symptoms may permanently disappear without calculus lodging and developing into calculous disease. Such cases are termed spontaneous recovery; but it is not safe to class cases in this category until at least five years have passed since the last attack, the urine remaining constantly free from pus and blood. Under the latter circumstances it may be concluded that the conditions which favored the formation of calculi have passed away. If, on the other hand, attacks have been frequent, and especially if there have been recent attacks, it may be concluded that the conditions favorable for the development of calculous disease in the urinary tract still continue, and such cases cannot be safely recommended for life-insurance.

From the sum of these considerations the following rules may be formulated, as a concise guide in conducting examinations of the urine and reaching conclusive data as to the condition of the kidneys and urinary organs in life-insurance:—

RULES.

1. Secure a freshly-voided sample of urine for examination, preferably after food and exercise, and be certain that the urine examined has been voided by the applicant.

If the examination result in any doubts as to the true conditions present, procure a sample of a mixture of the whole twenty-four hours' product of the kidneys, and also another freshly-voided sample of the urine; examine both of these separately and compare carefully the results.

2. Carefully observe first the physical characters of the urine, more especially: (a) *The appearance.* If cloudy, ascertain the cause, whether due to phosphates, urates, pus, blood, etc. The color, whether light (watery) like hydruria, or greenish (diabetic-like), or normal straw-yellow. (b) *The specific gravity.* If 1.025 or above, search for sugar. If below 1.020, search for albumin, and also ascertain the total amount of urea. (c) *The chemical reaction.* If very sharply acid (possibly diabetic, rheumatic, or gouty conditions are present). If alkaline from fixed alkali (probably debility, dyspepsia or fasting, or vegetable diet). If ammoniacal (cystitis is suggested).

3. Examine next for albumin with the ferrocyanic test as directed. Do not ignore the presence of albumin, *however minute the quantity* may be. If in any doubt as to the presence of minute traces, take two perfectly-clean test-tubes half full of the suspected urine. To one apply the ferrocyanic test, but to the other add no reagent whatever. Stand the two side by side in a good light for ten minutes. If they remain alike, albumin is absent; if faint opacity occur in the one with the test, albumin is present.

4. Examine next for sugar with Haines's test as directed.

Boil 1 drachm of this test; add 8 or 10 drops of the urine,—*no more;* boil again half a minute. If sugar be present, a yellow or yellowish-red precipitate will appear. If no such precipitate appear, sugar is absent. If a gray or whitish precipitate occur, it is caused by earthy phosphates, and not sugar. If any doubts arise as to the presence of sugar by the copper test, appeal to the phenyl-hydrazin test for confirmation.

5. An estimation should next be made of the quantity of urea in the urine with the Doremus ureometer, but especially if the specific gravity of the urine be materially reduced. Should the proportion of urea be 25 per cent. below normal, or lower, estimate the whole excretion of urea for twenty-four hours, and if below 300 grains report the fact to the home office.

6. In cases in which a microscopical examination of the urine is requested by the company, carefully report the following features of the urinary sediment: (*a*) The presence, quantity, and features of pus- or blood- corpuscles, *i.e.*, whether well preserved or partly broken down. (*b*) The presence, number, and characters of any renal casts, especially noting their size, whether clear or granulated, and if any epithelium or blood-corpuscles are seen attached to the casts.

7. Inquire for a history of attacks of renal colic (passage of calculi) or hæmaturia. Report the number and special features of such attacks, especially the severity and length of time they may have continued, and the *date of the last attack.*

APPENDIX B.

REAGENTS AND APPARATUS FOR QUALITATIVE AND DETERMINATE URANALYSIS.

ALL liquid-reagent bottles should be made of the purest glass and fitted with carefully-ground glass stoppers. The four-ounce bottles furnished by Whitall, Tatum & Co., of Philadelphia, are the best for the purpose. The glass from which these bottles are made is free from lead or other impurities, and, as a double check in laboratory work, each bottle has the name of the contained reagent upon it in raised-glass letters with ground tops, and the chemical symbol of the reagent below and separate from the lettering. An additional advantage of such bottles is the great facility with which they can be cleaned and kept in order. For efficient laboratory work the following list of reagents and apparatus should be kept in stock; but for the general practitioner, who only does the main essentials of urinary testing, the list from 1 to 14 and from 34 to 42 and 66 to 75, inclusive, will answer his purposes very well:—

LIQUID REAGENTS.

1. Nitric Acid, C. P. (HNO_3).
2. Hydrochloric Acid, C. P. (HCl).
3. Acetic Acid ($C_2H_4O_2$).
4. Potassium Hydroxid (KOH).
5. Sodium Hydroxid [(NaOH), 40-per-cent. solution].
6. Bromine ($\frac{1}{2}$-pound bottle).
7. Sat. Sol. Sodium Chloride.
8. Alcohol (95 per cent.).
9. Glycerin (C. P. ; free from lead).
10. Aqua Destill.
11. Sol. Potassium Ferrocyanide (1 in 20).
12. Sol. Barium Chloride (4 oz. barium-chloride crystals, 16 oz. distilled water, 1 oz. hydrochloric acid).

(378)

13. Magnesium Mixture (magnesium sulphate, ammonium chloride, āā $\mathfrak{Z}$j ; distilled water, $\mathfrak{Z}$viij ; ammonia-water, $\mathfrak{Z}$j).
14. Sol. Silver Nitrate [(standard aqueous solution, 1 in 8) $\mathfrak{Z}$j to $\mathfrak{Z}$j].
15. Strong Ammonia (U. S. P. ; sp. gr., 0.90).
16. Nitrous Acid (HNO_2).
17. Sulphuric Acid (H_2SO_4).
18. Sol. Potass. Ferrocyanide [1 in 10 (for quantitative determination of albumin, Purdy's method)].
19. Millon's Reagent.
20. Sol. Ferric Chloride (Fe_2Cl_6).
21. Sol. Calcium Chloride ($CaCl_2$).
22. Sol. Potassio-mercuric Iodide [(Tanret's test) Potass. iodidi, 3.32 grammes ; hyd. bichloridi, 1.35 grammes ; distilled water, to 100 cubic centimetres. (Dissolve the two salts separately, mix, and make up to 100 cubic centimetres with distilled water.)].
23. Sol. Lead Acetate [$Pb(C_2H_3O_2)_2$] ; 1 part lead acetate to 4 parts distilled water.
24. Sol. Basic Lead Acetate [$Pb(C_2H_3O_2)_2.2PbO$] ; 1 part basic lead acetate to 4 parts distilled water.
25. Hydrogen Dioxide (Oakland Chemical Company).
26. Sat. Sol. Ammonium Chloride.
27. Sat. Sol. Ammonium Sulphate.
28. Sat. Sol. Barium Nitrate.
29. Cupric-Sulphate Sol.
30. Sodium Nitroprusside (2-per-cent. solution).
31. Tinct. Guaiaci.
32. Turpentine.
33. Uranium Nitrate (5-per-cent. solution).

Solid Reagents.

34. Potassium Ferrocyanide (C. P.).
35. Cupric Sulphate (C. P.).
36. Sodium Hydroxid (purified by alcohol).
37. Potassium Hydroxid (purified by alcohol).
38. Phenyl-hydrazin Hydrochlorate.
39. Sodium Acetate.
40. Sodium Chloride (C. P.).
41. Ferric Chloride.
42. Picric Acid.
43. Sulphanilic Acid.
44. Barium Chloride (crystals).
45. Sodium Carbonate (C. P.).
46. Magnesium Sulphate (C. P.).
47. Sodium Nitrite.

48. Lead Acetate.
49. Basic Lead Acetate.
50. Ammonium Chloride.
51. Ammonium Sulphate.
52. Citric Acid.
53. Mercuric Chloride.
54. Potassium Iodide.
55. Barium Nitrate (crystals).
56. Uranium Nitrate.
57. Ammonium Nitrate.
58. Barium Hydrate.
59. Calcium Carbonate.
60. Oxalic Acid.
61. Potassium Bichromate.
62. Potassium Chlorate.
63. Potassium Permanganate.
64. Resorcin.
65. Tannin.

APPARATUS.

66. Test-tubes. Several sizes. Some with bases so that they will stand on a table or shelf. Some should be graduated for the purpose of approximate bulk determinations.
67. Spirit-lamp and Bunsen burner.
68. Urinometer (preferably Squibb's).
69. Test-tube rack and brush.
70. Graduate glasses,—one for 100 cubic centimetres' measurement and one for 500 cubic centimetres.
71. Doremus's ureometer.
72. Nipple pipettes.
73. Litmus-paper (blue and red) and Swedish filtering-paper (two sizes).
74. Glass funnels (two sizes).
75. An accurate thermometer.
76. Set of porcelain capsules.
77. Set of beaker glasses.
78. Long funnel for filtering through animal charcoal.
79. Glass rods.
80. Retort-stand with water-bath.
81. Platinum spoon and foil.
82. Blow-pipe.
83. Burettes graduated in fractions of a cubic centimetre ; also in minims.
84. Volume pipettes (set from 5 to 50 cubic centimetres).
85. A litre flask.
86. An accurate scale, turning at $\frac{1}{50}$ of a grain.

87. A good centrifuge capable of 2000 revolutions per minute, with 12 to 14 inches from tip to tip of tubes when in motion, and armed with both sediment and percentage tubes of capacity of at least 15 cubic centimetres each.

88. Microscope with ¼-inch and ½-inch objectives, glass slides, covers, shallow cells, etc.

For the purpose of recording the results of urinary analysis some systematic form of blank should be employed, both to expedite and systematize work, as well as to preserve the results in the form of a permanent record. These blanks may be kept in separate sheets or bound in book-form, or, better still, both. In the latter case the original analysis may be recorded in the volume as a permanent record, and a copy may be furnished the patient on one of the separate sheets or slips. For practical clinical purposes the record of analysis should have prominently in view the leading or more important features of the urine, normal and abnormal, to which *suggestive headings* should be added for those features less commonly met with, and those of minor importance · or significance. Any attempt at comprehensive analysis of the urine must include quantitative as well as qualitative data. The blank form on the succeeding page (No. 1, Regular Form) will answer most of the usual requirements for both qualitative and quantitative data. It aims at a reasonable measure of completeness without unnecessary elaboration. In case only qualitative data are worked out, the more simple form on the succeeding page will be found useful (No. 2, Special Form).

ANALYSIS OF URINE.
(NO. 1. REGULAR FORM.)

CASE..

....

CHEMICAL EXAMINATION.

SPECIMEN.. Total 24 hours..............c. c. ....ounces.

TRANSPARENCY..

COLOR...(...........Vogel's scale).

SPECIFIC GRAVITY.. at 15° C. (Westphal balance).

CHEMICAL REACTION..

ALBUMIN $\left\{\begin{array}{l}\text{(Heat test, Purdy's method)}...................... \\ \text{(Ferrocyanide test)}...................................... \\ \text{(Tanret's test)}..\end{array}\right\}$

NUCLEO-ALBUMIN (Mucin)..

SUGAR $\left\{\begin{array}{l}\text{(Copper test)}.. \\ \text{(Phenyl-hydrazin test)}..............................\end{array}\right\}$

ACETONE.. BILE...

DIACETIC ACID.................................. HÆMOGLOBIN.................................

INDICAN.. DIAZO REACTION...............................

....

....

QUANTITATIVE ESTIMATIONS.

	Volumetric Percentage.	Gravimetric Percentage.	Grains per Fluidounce.	Amount in 24 Hours.	
				Grammes.	Grains.
UREA (Knop-Hœfner method)					
ACIDITY (Expressed as oxalic acid) .					
SUGAR· · · · · · · · · · · · · · · · · ·					
ALBUMIN · · · · · · · · · · · · · · ·					
CHLORIDES· · · · · · · · · · · · ·					
PHOSPHATES· · · · · · · · · · · · ·					
SULPHATES· · · · · · · · · · · · ·					
SEDIMENT· · · · · · · · · · · · · ·					

(382)

DATE ...

MICROSCOPIC EXAMINATION.

NUMBER OF SLIDES EXAMINED ...

ORGANIZED SEDIMENT.

CASTS
- Epithelial
- Narrow Hyaline
- Medium Hyaline
- Broad Hyaline

- Finely Granular
- Dark Granular
- Bloody
- Amyloid

FALSE CASTS .... (Cylindroids of Thomas.)

EPITHELIA
- Small Round
- Spindle Form
- Pavement Form

PUS .... SPERMATOZOA ....

BLOOD-CORPUSCLES .... OTHER PRODUCTS ....

UNORGANIZED SEDIMENT.
CRYSTALLINE.

URIC ACID .... TRIPLE PHOSPHATE ....

CALCIUM OXALATE .... OTHER FORMS ....

AMORPHOUS.

URATES ....

PHOSPHATES ....

OTHER FORMS ....

MICRO-ORGANISMS.

....

....

REMARKS.

---

(383)

ANALYSIS OF URINE.

(No. 2. Special Form.)

CASE...

....

CHEMICAL EXAMINATION.

SPECIMEN

TRANSPARENCY

COLOR   (............Vogel's scale.)

CHEMICAL REACTION

SPECIFIC GRAVITY   at 15° C. (Westphal balance).

UREA Percentage   Grains per fluidounce.

ALBUMIN
- Heat test (Purdy's method)................................
- Ferrocyanide test...
- Tanret's test...

NUCLEO-ALBUMIN (Mucin)

SUGAR
- Copper test..
- Phenyl-hydrazin test...

ACETONE

DIACETIC ACID

INDICAN

BILE

HÆMOGLOBIN

DIAZO TEST (Ehrlich's)

OTHER PRODUCTS

(384)

MICROSCOPIC EXAMINATION.

NUMBER OF SLIDES EXAMINED.............

ORGANIZED SEDIMENT.

CASTS	Epithelial	Finely Granular
	Narrow Hyaline	Dark Granular
	Medium Hyaline	Bloody
	Broad Hyaline	Amyloid

FALSE CASTS ....(Cylindroids of Thomas.)

EPITHELIA { Small Round
Spindle Form
Pavement Form

PUS .... SPERMATOZOA ....

BLOOD-CORPUSCLES.............................. OTHER PRODUCTS ....

UNORGANIZED SEDIMENT.
CRYSTALLINE.

URIC ACID... TRIPLE PHOSPHATE ....

CALCIUM OXALATE................................ OTHER FORMS...

AMORPHOUS.

URATES...

PHOSPHATES ....

OTHER FORMS ....

MICRO-ORGANISMS.

....

....

REMARKS.

TERMS USED.
EXPRESSING

REACTION.		QUANTITY.		NUMBER.	
Negative.	Decided.	Absent.	Moderate amount.	None.	Moderate number.
Very faint.	Strong.	Traces.	Large amount.	Very few.	Numerous.
Faint.	Very strong.	Small amount.	Excessive amount.	Few.	Very numerous.

(385)

INDEX.

www.ingramcontent.com/pod-product-compliance
ightning Source LLC
mbersburg PA
W021346210326
9CB00011B/774